Satellite Derived Global Ocean Product Validation/Evaluation

Satellite Derived Global Ocean Product Validation/Evaluation

Editors

SeungHyun Son
Trevor Platt
Shubha Sathyendranath

MDPI • Basel • Beijing • Wuhan • Barcelona • Belgrade • Manchester • Tokyo • Cluj • Tianjin

Editors

SeungHyun Son
Colorado State University
USA

Trevor Platt
Plymouth Marine Laboratory
UK

Shubha Sathyendranath
Plymouth Marine Laboratory
UK

Editorial Office
MDPI
St. Alban-Anlage 66
4052 Basel, Switzerland

This is a reprint of articles from the Special Issue published online in the open access journal *Remote Sensing* (ISSN 2072-4292) (available at: www.mdpi.com/journal/remotesensing/special_issues/Ocean_Product).

For citation purposes, cite each article independently as indicated on the article page online and as indicated below:

LastName, A.A.; LastName, B.B.; LastName, C.C. Article Title. *Journal Name* **Year**, *Article Number, Page Range.*

ISBN 978-3-03943-645-3 (Hbk)
ISBN 978-3-03943-646-0 (PDF)

Contents

About the Editors

SeungHyun Son (Research Scientist) received BS and MS degrees in oceanography from Busan National University, in Busan, South Korea in 1996 and 1998, respectively. He received a Ph.D. degree in oceanography from the University of New Hampshire in 2004. After postdoctoral appointments at the Bedford Institute of Oceanography in Canada and the University of Maine, he joined NOAA/NESDIS/STAR in 2007. His current research is focused on validation/evaluation of the satellite ocean color products, such as VIIRS on SNPP and NOAA-20, and algorithm development and application for ocean color remote-sensing. His broader interests span monitoring water quality properties and the marine ecosystem using various satellite remote sensing and sea-measured data, human/climate-induced changes in the coastal ecosystems, understanding the spatial/temporal variation in phytoplankton productivity at various scales, and understanding interactions between the physical structure of the marine ecosystem and phytoplankton.

Trevor Platt (PML Professorial Fellow) is a Professorial Fellow at Plymouth Marine Laboratory in UK. His research interests include: the physiological ecology of marine phytoplankton; thermodynamics of open-ocean ecosystems; the influence of physical structure of the marine environment on populations living within it; submarine optics; size structure of marine communities; remote sensing of ocean colour; the ocean carbon cycle and climate change, and the ecological approach to fisheries' management.

Shubha Sathyendranath (Merit Remote Sensing Scientist) is a scientist in the Remote Sensing Group at Plymouth Marine Laboratory in UK. Dr. Satheydnranath's research interests include ocean colour modelling, spectral characteristics of light penetration underwater, bio-optical properties of phytoplankton, modelling primary production, bio-geochemical cycles in the sea, climate change, biological–physical interactions in the marine system, ecological provinces in the sea, ecological indicators and phytoplankton functional types.

Preface to "Satellite Derived Global Ocean Product Validation/Evaluation"

Ocean satellite instruments provide short-term to long-term (hourly to decadal) observations of physical and biogeochemical phenomena and properties in the coastal and global open ocean at high spatial resolution. Ocean-observing satellite sensors have been launched recently by international space agencies such as the US National Aeronautics and Space Administration (NASA), the US National Oceanic and Atmospheric Administration (NOAA), the European Space Agency (ESA) and Japanese Space Agency (JAXA), and operationally measure the various physical, biological, and biogeochemical variables in the ocean.

Validation/evaluation efforts and uncertainty assessments are crucial to providing more accurate satellite-derived ocean products. Validation of the satellite products requires a combination of ground field measurements, instrumented surface sites, inter-satellite comparisons, and research and modeling efforts with robust methodologies. This book—*Validation of Satellite Ocean Products*—presents the validation/evaluation results of the various satellite-derived ocean products including sea surface temperature, sea surface salinity, wave height, altimetry, ocean surface wind, sea ice, ice surface temperature, chlorophyll-a, particulate organic carbon, particulate backscattering, and marine fishery resources in the global and various regional seas, as well as some applications with the ocean products. This book is aimed at various audiences, including graduate students, university professors, junior to senior scientists and decision makers.

SeungHyun Son, Trevor Platt, Shubha Sathyendranath
Editors

In Memoriam—Professor Trevor Platt

The very renowned scientist in oceanography, Professor Trevor Platt, FRS, in the Plymouth Marine Laboratory in the UK, passed away on April 5, 2020. Professor Platt was born in 1942, in Salford, England, and finished an undergraduate degree at the University of Nottingham in the UK (1963). Then, he moved to Canada to study for his MA at the University of Toronto (1965) and his Ph.D. at Dalhousie University (1970). He was Head of Biological Oceanography from 1972 to 2001, acting Director in 1976, and senior scientist from 2001 to 2008 at the Bedford Institute of Oceanography, Nova Scotia, Canada. In 2008, he moved back to the UK and joined the Plymouth Marine Laboratory as a Professorial Fellow as well as executive director of the Partnership for Observation of the Global Ocean (POGO).

He provided remarkable contributions to the fundamental and applied research in the broad fields of oceanography, phytoplankton ecology, marine ecosystem dynamics, marine optics, ocean colour remote sensing, ocean carbon cycle and climate change, with numerous scientific publications (>320) during his scientific career.

Professor Platt received numerous honors, medals and awards in recognition of his contributions to oceanography, including 'APICS-Fraser Gold medal (1982)', 'Rosenstiel Gold medal (1984)', 'G. Evenly Hutchinson medal (1988)', 'Fellowship of the Royal Society of Canada, Academy of Science (1990)', 'A. G. Huntsman medal (1992)', 'Fellowship of the Royal Society of London (1998)', 'Plymouth Marine medal (1999)', 'Timothy Parsons medal (2006)', 'Prix de Distinction (2007), Department of Fisheries and Oceans, Canada', ' Prix d'Excellence, Department of Fisheries and Oceans (2008)', and 'Jawaharlal Nehru Science Fellow of the Government of India (2014)'.

Professor Platt also showed his strong leadership in numerous international scientific organizations, and contributed to: NATO Advance Study Institute, Scientific Committee on Oceanic Research (SCOR), International Joint Global Ocean Flux Study (JGOFS), American Society of Limnology and Oceanography, International Council for the Exploration of the Sea (ICES), International Geosphere-Biosphere Program (IGBP), International Ocean Colour Coordinating Group (IOCCG), and the Partnership for Observation of the Global Oceans (POGO).

In particular, as a founder and the first chairman of IOCCG, he played significant roles in raising the recognition of the importance of ocean colour remote-sensing. He had a special interest in educating and encouraging young scientists and scientists from developing countries. He organized international training courses to promote capacity-building for young scientists through IOCCG and POGO from 1997. He is survived by his loving wife and colleague of over 30 years, Dr. Shubha Sathyendranath.

Prof. Trevor Platt was one of the guest editors of this Special Issue, and we dedicate it to him.

 *remote sensing*

Article

Ice Surface Temperature Retrieval from a Single Satellite Imager Band

Yinghui Liu [1],*, Richard Dworak [1] and Jeffrey Key [2]

1 Cooperative Institute for Meteorological Satellite Studies, University of Wisconsin-Madison,
1225 West Dayton St., Madison, WI 53706, USA; rdworak@ssec.wisc.edu
2 NOAA/NESDIS, 1225 West Dayton St., Madison, WI 53706, USA; jeff.key@noaa.gov
* Correspondence: yinghuil@ssec.wisc.edu; Tel.: +1-608-890-1893

Received: 7 October 2018; Accepted: 28 November 2018; Published: 29 November 2018

Abstract: Current methods for estimating the surface temperature of sea and lake ice—the ice surface temperature (IST)—utilize two satellite imager thermal bands (11 and 12 μm) at moderate spatial resolution. These "split-window" or dual-band methods have been shown to have low biases and uncertainties. A single-band algorithm would be useful for satellite imagers that have only the 11 μm band at high resolution, such as the Visible Infrared Imaging Radiometer Suite (VIIRS), or that do not have a fully functional 12 μm band, such as the Thermal Infrared Sensor onboard the Landsat 8. This study presents a method for single-band IST retrievals, and validation of the retrievals using IST measurements from an airborne infrared radiation pyrometer during the NASA IceBridge campaign in the Arctic. Results show that IST with a single thermal band from the VIIRS has comparable performance to IST with the VIIRS dual-band (split-window) method, with a bias of 0.22 K and root-mean-square error of 1.03 K.

Keywords: sea ice; ice surface temperature; Suomi NPP; JPSS; remote sensing

---

1. Introduction

The surface temperature of sea ice, referred to as ice surface temperature (IST), is one of the most fundamental variables for assessing changes in Arctic climate [1]. Ice surface temperature integrates changes in the surface energy budget, which controls sea ice growth/melt and the exchange of heat and moisture between the surface and the atmosphere. The Arctic is warming at a higher rate than the midlatitudes [2], a pattern known as Arctic amplification. Accurate and consistent measurement of the ice surface temperature across the Arctic is key to understanding Arctic climate change. While measurements at the surface are critical to monitoring climate, they are sparse in polar regions. In contrast, satellite-derived ice surface temperature provides broad spatial coverage, frequent temporal sampling, and relatively high accuracy.

Algorithms have been developed to retrieve IST under clear conditions with longwave infrared bands from satellite imagers. A dual-band regression algorithm utilizes the brightness temperature difference between the "split window" 11 and 12 μm thermal bands to account for the atmospheric absorption [3,4]. This approach is most common, and has been applied to retrieve IST using data from the Advanced Very High Resolution Radiometer (AVHRR) [3], the Moderate Resolution Imaging Spectroradiometer (MODIS) [5], and the Visible Infrared Imager Radiometer Suite (VIIRS) onboard the Suomi National Polar-orbiting Partnership (S-NPP) satellite and the Joint Polar Satellite System (JPSS) [6,7]. Validation studies have shown good performance, with a root-mean-square error (RMSE) of 1.2 K or less [5,6,8–12].

Single-band regression algorithms that use the 11 μm brightness temperature alone have also been developed to retrieve ice surface temperature, either for the lack of a 12 μm band in the earlier AVHRR sensors [13,14], or as a backup method if one of the VIIRS thermal window bands fails [7].

Limited validation of these single-band approaches indicates performance similar to the dual-band method [14].

Some practical applications would benefit from the ability to do single-band IST retrievals with high quality. The S-NPP and JPSS VIIRS instruments have 16 moderate resolution bands (M-bands) at 750 m and five "image" resolution bands (I-bands) at 375 m. The 16 M-bands include split-window bands, an 11 μm band (M15) with a wavelength range from 10.26 to 11.26 μm and a 12 μm band (M16) with a wavelength range from 11.54 to 12.49 μm. The five I-bands include a relatively broad 11 μm band (I5) with a wavelength range from 10.5 to 12.4 μm. The dual-band regression algorithms have been developed utilizing the M15 and M16 bands to retrieve IST at 750 m resolution [6,7]. Extensive validation studies have shown this product has an absolute bias less than 0.5 K and an uncertainty less than 1.0 K [6,10]. These IST products are key inputs for other VIIRS ice product algorithms, notably ice concentration and ice thickness [15,16].

While 750 m ice products are relatively high resolution by today's standards, even higher resolution products are desirable for operational ice applications where small-scale features need to be resolved. VIIRS ice concentration and ice thickness products rely on IST, so it is necessary to create a single-band IST product. Similarly, Landsat 8 has been used in case studies of sea ice for its very high spatial resolution, with 30 m resolution for visible bands and 100 m resolution for thermal bands [15]. An IST product from Landsat 8 using either a dual-band or single-band algorithm is desirable for studying sea ice in greater detail, and for deriving other products. Unfortunately, Landsat 8 band 11 (12 μm) has a large uncertainty, thus it is not recommended for quantitative analysis, including retrieval of surface temperature (https://landsat.usgs.gov/landsat-8-data-users-handbook-appendix-a). Therefore, a single-band regression approach for Landsat 8 IST is also needed.

In this paper we formulate and validate single-band IST products for S-NPP VIIRS and Landsat 8 using high-resolution field survey mission IST measurements. The IST products and their uncertainty information will be useful for monitoring small-scale variations in surface temperature and for generating other high spatial resolution ice products with IST as inputs. We first describe the single-band IST algorithm and the validation dataset, then show the results of the validation.

2. Materials and Methods

The dual-band IST algorithm has the following form:

$$T_s = a + b \times T_{11} + c \times (T_{11} - T_{12}) + d(T_{11} - T_{12}) \times (sec\theta - 1) \tag{1}$$

where T_s is the estimated IST (K), T_{11} and T_{12} are the brightness temperatures (K) at 11 μm and 12 μm bands, θ is the sensor scan angle, and a, b, c, and d are retrieval coefficients [5,8].

The single-band IST algorithm in previous studies has the following form:

$$T_s = a + b \times T_{11} \tag{2}$$

where T_s is the estimated IST (K), a and b are retrieval coefficients; and T_{11} is the brightness temperature in the 11 μm band (K) [13,14]. The atmospheric absorption and surface emissivity effects are accounted for implicitly by the coefficients a and b. The IST in (1) and (2) is the "skin" temperature, or the radiative temperature of the surface. The ice surface is generally covered with snow, though it may also be bare ice or melt ponds in the summer.

Coefficients of the algorithms can be derived in two ways. One approach is to collocate the satellite brightness temperature with in situ IST measurements, and to determine the retrieval coefficients through linear regression as suggested in [14]. However, collecting sufficient in situ IST measurements with extensive spatial and temporal representation over the sea ice is challenging. The other approach is the use of a forward model as in [3], where a radiative transfer model is used with observed temperature and humidity profiles and a sensor's spectral response functions to simulate the 11 and 12 μm brightness temperatures that would be observed under a broad range of conditions. We will apply the second approach in this study as described in the following. More details of this approach can be found in [3,4].

A total of 1338 atmospheric temperature and humidity profiles from 1976 to 1991 were sampled from three archives: the North Pole archive, the National Center for Atmospheric Research archive, and the Historical Arctic Rawinsonde archive [17]. These include 553 profiles over the Arctic Ocean and 785 profiles from coastal weather stations. They are more or less evenly distributed over 12 months. All profiles have at least 10 levels of pressure, air temperature, and dew point temperature. VIIRS thermal band response functions for all bands were obtained from the National Oceanic and Atmospheric Administration (NOAA) at https://ncc.nesdis.noaa.gov/VIIRS/VIIRSSpectralResponseFunctions.php. The Landsat 8 thermal band response functions were obtained from the National Aeronautics and Space Administration (NASA) at https://landsat.gsfc.nasa.gov/preliminary-spectral-response-of-the-thermal-infrared-sensor/. Here the ice surface is assumed to be snow covered, and the wavelength-dependent snow emissivities were parameterized for all sensor scan angles following the procedure in [18]. Thermal band radiances at 5 cm^{-1} intervals for scan angels from 0° to 60° every 10° were simulated using the LOWTRAN 7 radiative transfer model [19], with inputs of temperature and humidity profiles and snow emissivity. The temperature at the lowest layer is assigned as the ice surface temperature. The background tropospheric and stratospheric aerosols for subarctic winter and summer were used.

Simulated radiances were integrated for the thermal band based on its response function and then converted to brightness temperature. Linear regression was used to determine the coefficients *a*, *b*, *c*, and *d* in Equation (1) for Landsat 8, and *a* and *b* for S-NPP VIIRS I-band I5 and M-band M15, and Landsat 8 band 10 (11 µm) in Equation (2) for three 11 µm brightness temperature ranges: less than 240 K, 240–260 K, and 260–273 K. Values of the retrieval coefficients and their standard errors are given in Tables 1 and 2. The coefficients for the S-NPP VIIRS dual-band algorithm can be found at https://www.star.nesdis.noaa.gov/jpss/documents/ATBD/ATBD_EPS_Cryosphere_IST_IceCover_v1.0.pdf. It should be noted that the M15 band covers wavelength from 10.26 to 11.26 µm, while the I5 band has a broader wavelength range from 10.5 to 12.4 µm. Retrieval performance of the single-band algorithm for M15 and I5 may be different due to the differences in wavelength range, although the spatial resolution of retrieval products using I5 band is higher.

Cross-validation was used to test the performance of the retrieval algorithms. Of the 1338 profiles, 90% were randomly selected to derive the dual-band and single-band retrieval coefficients for Landsat 8 and VIIRS, and the remaining 10% were used to determine the bias and standard deviation (RMSE$_{nobias}$). This process was repeated 100 times. The statistics are shown in Table 3. For the single-band retrieval algorithm in Equation (2), the retrieval biases have a sensor scan angle dependence, e.g., biases range from 0.202 to −0.480 K for Landsat 8, from 0.269 to −0.699 K for VIIRS M15, and from 0.344 to −0.819 K for VIIRS I5 with sensor scan angles from 0 to 60 degree. Also, the differences of retrieved and true surface temperature appear to have linear relationships with column total water vapor (Figure 1). It should be noted that such an angle dependence and linear relationship between retrieval bias and total water vapor for Landsat 8 (Figure 1) and VIIRS (similar as that of Landsat 8, not shown) do not exist for the dual-band retrieval algorithm, as shown in Table 3 and Figure 1. These features suggest an addition of an extra term in the single-band retrieval equation could account for the sensor scan angle dependence. We propose an updated single-band IST retrieval equation:

$$T_s = a + b \times T_{11} + c \times sec\theta \tag{3}$$

where T_s is the estimated IST, *a*, *b*, and *c* are retrieval coefficients, T_{11} is brightness temperature at 11 µm band, and θ is the sensor scan angle. The retrieval coefficients are derived in the same way described above. The addition of the new term alleviates the sensor scan angle dependence of the retrieval bias and the linear relationship between retrieval bias and total water vapor (Table 4, Figure 2). The retrieval coefficients for Equation (3) are listed in Table 5, with their standard errors given in parenthesis.

Table 1. Landsat 8 dual-band ice surface temperature retrieval coefficients and the standard errors of the coefficients (in parenthesis) in the Arctic as in Equation (1).

Satellite	Temperature Range	a (K)	b	c	d
Landsat 8	<240 K	−0.40	1.00 (0.019)	1.59 (0.53)	−0.76 (0.46)
	240–260 K	−0.77	1.00 (0.0050)	1.51 (0.21)	−0.32 (0.15)
	260–273 K	−3.49	1.01 (0.0050)	1.46 (0.13)	0.06 (0.080)

Table 2. Suomi National Polar-orbiting Partnership (S-NPP) Visible Infrared Imaging Radiometer Suite (VIIRS) and Landsat 8 single-band ice surface temperature retrieval coefficients and the standard errors (in parenthesis) of the coefficients in the Arctic as in Equation (2).

Satellite	Temperature Range	a (K)	b
Landsat 8	<240 K	−5.39	1.023 (0.0047)
	240–260 K	−8.49	1.035 (0.0088)
	260–273 K	−12.47	1.051 (0.0046)
S-NPP VIIRS single-band I5	<240 K	−8.61	1.037 (0.0047)
	240–260 K	−15.40	1.063 (0.0088)
	260–273 K	−14.36	1.060 (0.0045)
S-NPP VIIRS single-band M15	<240 K	−7.25	1.031 (0.0047)
	240–260 K	−11.56	1.048 (0.0088)
	260–273 K	−11.78	1.049 (0.0045)

Table 3. Bias (top row in a cell) and standard deviation (bottom row in a cell) of the differences between dual-band and single-band ice surface temperature retrievals using Equations (1) and (2) and true ice surface temperatures for Landsat 8 and S-NPP VIIRS in the Arctic.

Equation	Sensor Scanning Angle (Degree)						
	0	10	20	30	40	50	60
Landsat 8 dual-band	−0.068	−0.06	−0.035	0.01	0.045	0.09	0.042
	0.049	0.049	0.047	0.048	0.065	0.074	0.174
Landsat 8 single-band	0.202	0.192	0.163	0.107	0.017	−0.132	−0.480
	0.194	0.193	0.190	0.186	0.189	0.226	0.437
S-NPP VIIRS dual-band	−0.017	−0.02	−0.017	−0.008	0.028	0.124	−0.088
	0.137	0.138	0.139	0.143	0.152	0.183	0.392
S-NPP VIIRS single-band M15	0.269	0.257	0.227	0.166	0.063	−0.118	−0.699
	0.252	0.250	0.244	0.234	0.227	0.255	0.596
S-NPP VIIRS single-band I5	0.344	0.327	0.278	0.187	0.040	−0.205	−0.819
	0.317	0.314	0.306	0.293	0.281	0.299	0.617

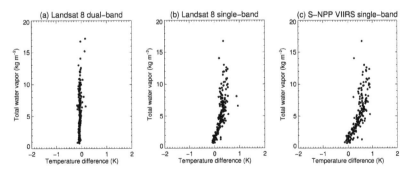

Figure 1. Relationship of the temperature difference between the retrieved and true ice surface temperature and total column water vapor for the (**a**) Landsat 8 dual-band algorithm using Equation (1), (**b**) Landsat 8 single-band using Equation (2), and (**c**) S-NPP VIIRS single-band for I5 band using Equation (2).

Table 4. Bias (top row in a cell) and standard deviation (bottom row in a cell) of differences between single-band retrievals using Equation (3) and true surface temperatures for Landsat 8 and S-NPP VIIRS in the Arctic.

Equation	Sensor Scanning Angle (Degree)						
	0	10	20	30	40	50	60
Landsat 8 single-band	−0.030	−0.028	−0.019	−0.004	0.021	0.067	0.054
	0.156	0.156	0.157	0.165	0.186	0.242	0.363
S-NPP VIIRS single-M-band	−0.070	−0.064	−0.041	−0.002	0.060	0.150	0.004
	0.178	0.177	0.174	0.178	0.206	0.279	0.460
S-NPP VIIRS single-I-band	−0.061	−0.057	−0.038	−0.003	0.06	0.149	0.040
	0.220	0.219	0.217	0.225	0.262	0.333	0.541

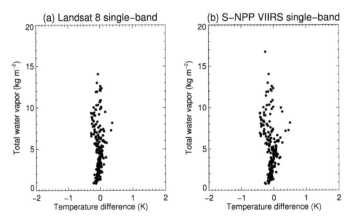

Figure 2. Relationship of the temperature difference between the retrieved and true ice surface temperature and total column water vapor for (**a**) Landsat 8 single-band algorithm, and (**b**) S-NPP VIIRS single-band for I5 band using Equation (3).

Table 5. S-NPP VIIRS and Landsat 8 single-band ice surface temperature retrieval coefficients and the standard errors (in parenthesis) of the coefficients in the Arctic as in Equation (3).

Satellite	Temperature Range	a (K)	b	c
Landsat 8	<240 K	4.92	1.020 (0.017)	0.147 (0.12)
	240–260 K	−7.93	1.031 (0.0035)	0.505 (0.056)
	260–273 K	−15.19	1.054 (0.0046)	1.438 (0.047)
S-NPP VIIRS single-band I5	<240 K	−7.29	1.029 (0.017)	0.316 (0.11)
	240–260 K	−12.65	1.048 (0.0035)	0.943 (0.055)
	260–273 K	−21.89	1.076 (0.0045)	2.550 (0.045)
S-NPP VIIRS single-band M15	<240 K	−6.51	1.027 (0.018)	0.149 (0.12)
	240–260 K	−10.37	1.040 (0.0035)	0.727 (0.056)
	260–273 K	−16.55	1.057 (0.0045)	2.055 (0.046)

The primary validation dataset in this study consists of IST measurements made with an airborne Heitronics KT-19.85 Series II infrared radiation pyrometer, hereinafter simply "KT-19", flown on a NASA P3 aircraft during the NASA Operation IceBridge campaigns (https://www.nasa.gov/mission_pages/icebridge/mission/) [20]. The surface temperature measured by the KT-19 is provided by the IceBridge science team in a dataset available from the National Snow and Ice Data Center (NSIDC; https://nsidc.org/data/iakst1b). The KT-19 measurements have a temperature resolution of ±0.1 °C and an accuracy of ±0.5 °C plus 0.7% of the difference between target temperature and

KT-19's internal temperature [21] (https://www.wintron.com/heitronics-kt19-infrared-thermometer). The internal temperature may vary as much as 30 °C during the course of a flight; no correction for these changes has been applied to the KT-19 surface temperature data. The P3 aircraft usually flies only a few hundred meters above ground level, so atmospheric absorption is minimal because of the low water vapor content over sea ice, though it may not be negligible. In deriving surface temperature from KT-19 measurements, the IceBridge science team used a surface emissivity of 0.97 starting in 2012. Before 2012, a surface emissivity of 1.0 was used. This change by the IceBridge team suggests that 0.97 provides the best estimate of surface temperature [21]. More details of KT-19 surface temperature measurements can be found in IceBridge Data Products Manual [21]. All these factors contribute to the overall uncertainty in the KT-19 measurements, which has not been assessed extensively. Previous validation studies show close agreement between KT-19 measurements and dual-band IST retrievals [6,10], and between dual-band IST retrievals and other validation data sets [5,8,9,11,12], which indicates the usefulness of KT-19 measurements in IST validation.

All available cases of KT-19 observations in the Arctic from 2013 to 2017 were used to validate the collocated S-NPP VIIRS and Landsat 8 retrieved IST [22]. Also, a rigorous visual check of both cloud mask products was done to ensure that only clear observations were used. The VIIRS and Landsat 8 cloud masks and a visual inspection of the data were applied to eliminate cloudy and foggy cases for both satellite datasets and the KT-19 [23]. All available KT-19 temperature measurements within 100 m (Landsat) or 375 m (VIIRS) of the center of the pixel were averaged. The maximum time difference for matchups between KT-19 and S-NPP VIIRS and Landsat 8 was 60 min. This time difference can contribute to the overall differences between the satellite-retrieved IST and the KT-19 temperature shown below due to surface temperature changes over 60 min.

3. Results

On 19 March 2014 from 2237–2238 UTC, a concurrent Landsat 8 overpass and an Ice-Bridge P-3 flight off the Northwest coast of Alaska near Point Lay were acquired (Figure 3). Only clear pixels over sea ice in Landsat 8 were used in analysis. The clear area was determined with the cloud mask, a high value of the Normalized Difference Snow Index (NDSI), and visual inspection. Comparison with KT-19 measurements showed a very small overall cold bias of -0.03 K, and an $RMSE_{nobias}$ of 0.76 K (Figure 4). Towards to the end of the KT-19 track, possible cloud contamination near cloud edges in the Landsat 8 data led to lower retrieved IST (Figures 3 and 4).

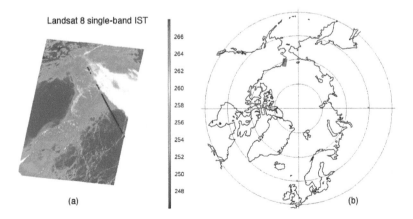

Landsat 8 single-band IST

(a)   (b)

Figure 3. Landsat retrieved ice surface temperature (IST) with a P-3 Ice Bridge flight overlaid (red line) from 22:37–22:38 UTC on 19 March 2014 (**a**), off the Northwest coast of Alaska (**b**). The white color in IST field represents cloud cover.

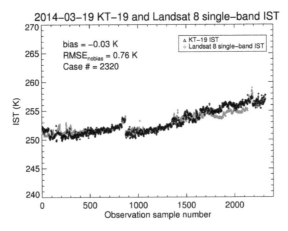

Figure 4. KT-19 ice surface temperature (IST) (black) and retrieved IST from Landsat 8 (green) along the P-3 Ice Bridge flight track from 22:37–22:38 UTC on 19 March 2014.

A mostly clear scene over the Beaufort Sea when an Ice Bridge flight coincided with a S-NPP VIIRS overpass on 30 Mar 2015 was also analyzed (Figure 5). The S-NPP VIIRS derived IST from the M-band dual-band approach [6], and from the single-band approach using the I5 band (11 μm) covered the same area. The dual-band IST was quality controlled to screen out clouds with the VIIRS M-band cloud mask (there is no I-band cloud mask) and visual inspection (Figure 5); I-band IST used the same cloud screening. Compared to the dual-band IST, the single-band IST at image resolution shows more detail of the sea ice leads (fractures) because of its higher spatial resolution. The time difference of KT-19 and VIIRS ISTs was 30 min or less for the comparison in Figures 6 and 7. The inter-comparison of VIIRS retrieved IST and KT-19 IST showed very small biases and $RMSE_{nobias}$ for both approaches, with a bias of 0.29 K and $RMSE_{nobias}$ of 0.58 K for the dual-band approach and a bias of -0.14 K and $RMSE_{nobias}$ of 0.62 K for the single-band approach.

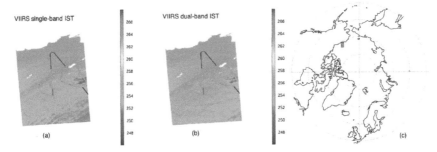

Figure 5. VIIRS IST with the single-band approach using the I-5 band (**a**) and with the M-band dual-band approach (**b**), with the IceBridge flight shown (red line) around 22:17 UTC on 30 March 2015 off the North coast of Alaska (**c**). The white color in IST field represents cloud cover in the M-band cloud mask.

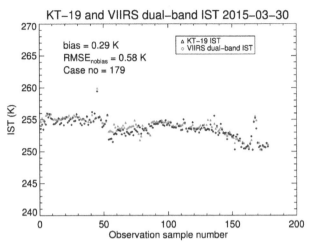

Figure 6. KT-19 ice surface temperature (IST) (black) and dual-band IST from S-NPP VIIRS (green) along the P-3 Ice Bridge flight track around 22:17 UTC on 30 March 2015.

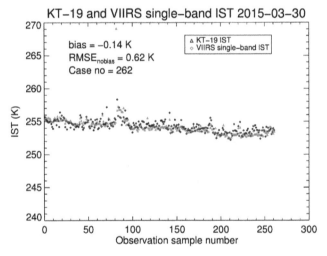

Figure 7. KT-19 ice surface temperature (IST) (black) and single-band IST from S-NPP VIIRS (green) along the P-3 Ice Bridge flight track around 22:17 UTC on 30 March 2015.

Besides the case studies, we also collected all 809 collocated cases of KT-19 and VIIRS ISTs from 2013 to 2017 covering a large area of the Arctic. For easier comparison, both dual-band and single-band VIIRS ISTs were calculated using the M-band brightness temperature. The comparison results are shown in Figure 8. A comparison of these 809 collocated KT-19 IST measurements and VIIRS derived ISTs gave an overall bias of 0.28 K and a $RMSE_{nobias}$ of 0.99 K for the dual-band algorithm, and an overall bias of 0.22 K and $RMSE_{nobias}$ of 1.03 K for the single-band algorithm. The positive biases for both approaches increased slightly with increasing 11 μm brightness temperature. To investigate the remaining impacts of the atmospheric effect in the retrieval bias we calculated the correlations between the retrieval bias, the difference of retrieved IST and KT-19 IST, and KT-19 IST, which is a proxy of total column water vapor. The correlations are −0.08 and −0.13 for dual-band and single-band algorithms, which indicates that the atmospheric effect can only explain 0.6% and 1.7% of the variance in the retrieval bias for dual-band and single-band algorithms. Both algorithms compensate for the atmospheric effects reasonably well. Calculations also show that both bias and RMSE$_{nobias}$ are not dependent upon the satellite sensor scan angle.

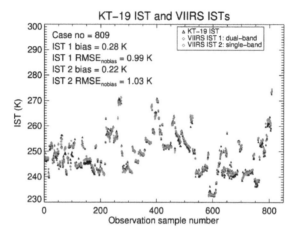

Figure 8. Comparison of IceBridge KT-19 ice surface temperature (IST) and VIIRS derived IST from various scenes using dual-band and single-band algorithms from the S-NPP moderate resolution bands.

4. Discussion

Comparison of satellite-derived IST with KT-19 measurements has its limitations. The spatial resolutions of VIIRS and Landsat 8 pixels at nadir are 375 m for the I-bands (750 m for M-bands) and 100 m, respectively, while the KT-19 is a 15 m point-track observation. The collocation of derived ISTs and averaged KT-19 measurements are not always optimal, as the observations of VIIRS, Landsat 8 and KT-19 are not always concurrent in time and cloud contamination may introduce biases and larger uncertainties in the satellite-retrieved ISTs.

Our use of LOWTRAN 7 is primarily a function of heritage, as our earlier work also used this model [3,4]. More advanced radiative transfer models are available, primarily providing higher spectral resolution, e.g., the community radiative transfer model (CRTM) and MODTRAN5 [24,25]. Previous tests with MODTRAN5, however, did not indicate that the resulting IST regression coefficients would result in significantly more accurate retrievals.

The dual-band retrieval approach is expected to perform better than the single-band approach because the split-window difference accounts for atmospheric absorption, especially under warmer and more moist conditions. The single-band retrieval approach is therefore expected to be more sensitive to the atmospheric effect and surface emissivity variations. However, results in this study show that both single-band and dual-band retrieval biases do not have an apparent dependence on total column water or sensor scan angle. This might be due to (1) both algorithms have a term related to the sensor scan angle specifically to account for the scan angle and column water vapor effects; and (2) the column water vapor in the Arctic is low, thus its effect on the retrieval is small. In deriving the retrieval coefficients, surface emissivity is prescribed as a function of satellite viewing angle. Snow emissivity changes with snow grain size and shape, and thus depends on surface snow type [26]. Both single-band and dual-band algorithms have been developed to retrieve land surface temperature with an explicit dependence on the surface emissivity [27]. With prior knowledge of surface type and thus surface emissivity, a similar approach can be applied to retrieve the ice surface temperature. Whether this algorithm can be used to retrieval surface temperature over other surface types needs further investigation [28].

It should be acknowledged that there has not been a robust assessment of the uncertainty of the KT-19 surface temperature measurements. The overall bias and RMSE$_{nobias}$ of the single-band and dual-band IST retrievals may need to be refined when additional information becomes available on the

KT-19 absolute bias and uncertainty. However, the lack of information on the absolute KT-19 surface temperature uncertainty does not change the conclusion of this study that the single-band IST retrieval approach can provide results comparable to traditional dual-band IST retrieval approach.

In this study the retrieval coefficients were derived using atmospheric profiles in the Arctic, and performance of the retrievals were evaluated with KT-19 measurements in the Arctic. The application of the same retrieval algorithm on sea ice in the Antarctic is valid because of the similar surface snow type and relatively low water vapor content in the Antarctic. For the same reason, the retrieval algorithm may be applicable to estimating lake ice surface temperature. However, the application of the methods described above over the Antarctic continent and over lake ice needs to be evaluated using in situ measurements and other remote sensed observations with comparable or higher spatial resolution (for example KT-19 measurements) in order to obtain a quantitative assessment of the retrieval algorithm performance. It is possible that new retrieval coefficients would be needed based on atmospheric profiles collected over high-altitude portions of Antarctica and over or near lakes.

5. Conclusions

Methods for estimating ice surface temperature from space using a split-window approach similar to those for sea surface temperature have been available since the early 1990's (Key and Haefliger 1992). However, a single-band approach is desirable for special cases, notably for the VIIRS instrument that has a higher resolution band at 11 μm but no corresponding 12 μm band, and for Landsat 8 where the 12 μm band has a large calibration uncertainty. The higher-resolution VIIRS I-band (375 m) is particularly attractive for operational ice applications where small-scale features are important.

In this study, we developed and validated VIIRS and Landsat 8 IST algorithms that utilize a single thermal band at 11 μm. For VIIRS IST, a comparison with KT-19 IST measurements in the Arctic from 2013–2017 showed an overall bias of 0.22 K and $\mathrm{RMSE_{nobias}}$ of 1.03 K, with only a slightly higher uncertainty than those of dual-band IST. Limited case studies of Landsat 8 IST showed similarly low bias and uncertainty values. The IST from the single-band approach can provide high-quality inputs to algorithms that estimate other sea ice characteristics, such as sea ice concentration and thickness.

Author Contributions: Conceptualization and methodology, Y.L. and J.K.; software, validation and analysis, Y.L. and R.D.; writing—original draft preparation, Y.L. and R.D.; writing—review and editing, Y.L. and J.K.

Funding: This research was funded by NOAA, grant number NA15NES4320001.

Acknowledgments: This work was supported by the JPSS Program Office and the GOES-R Program Office. We thank the anonymous reviewers for their valuable comments and suggestions. The views, opinions, and findings contained in this report are those of the author(s) and should not be construed as an official National Oceanic and Atmospheric Administration or U.S. Government position, policy, or decision.

Conflicts of Interest: The authors declare no conflict of interest.

References

1. Chapman, W.L.; Walsh, J.E. Simulations of arctic temperature and pressure by global coupled models. *J. Clim.* **2007**, *20*, 609–632. [CrossRef]
2. Manabe, S.; Spelman, M.J.; Stouffer, R.J. Transient responses of a coupled ocean-atmosphere model to gradual changes of atmospheric CO_2. Part II: Seasonal response. *J. Clim.* **1992**, *5*, 105–126. [CrossRef]
3. Key, J.; Haefliger, M. Arctic ice surface-temperature retrieval from avhrr thermal channels. *J. Geophys. Res. Atmos.* **1992**, *97*, 5885–5893. [CrossRef]
4. Key, J.R.; Collins, J.B.; Fowler, C.; Stone, R.S. High-latitude surface temperature estimates from thermal satellite data. *Remote Sens. Environ.* **1997**, *61*, 302–309. [CrossRef]
5. Hall, D.K.; Key, J.R.; Casey, K.A.; Riggs, G.A.; Cavalieri, D.J. Sea ice surface temperature product from modis. *IEEE Trans. Geosci. Remote Sens.* **2004**, *42*, 1076–1087. [CrossRef]

6. Key, J.R.; Mahoney, R.; Liu, Y.; Romanov, P.; Tschudi, M.; Appel, I.; Maslanik, J.; Baldwin, D.; Wang, X.; Meade, P. Snow and ice products from suomi npp viirs. *J. Geophys. Res. Atmos.* **2013**, *118*, 12816–12830. [CrossRef]
7. Baker, N. Joint Polar Satellite System (JPSS) VIIRS Sea Ice Characterization Algorithm Theoretical Basis Document (ATBD). 2011. Available online: http://npp.gsfc.nasa.gov/sciencedocs/2015-06/474-00047_VIIRS_Sea_Ice_ATBD_Rev-20110422.pdf (accessed on 1 June 2018).
8. Key, J.; Maslanik, J.A.; Papakyriakou, T.; Serreze, M.C.; Schweiger, A.J. On the validation of satellite-derived sea ice surface temperature. *Arctic* **1994**, *47*, 280–287. [CrossRef]
9. Scambos, T.A.; Haran, T.M.; Massom, R. Validation of avhrr and modis ice surface temperature products using in situ radiometers. *Ann. Glaciol.* **2006**, *44*, 345–351. [CrossRef]
10. Liu, Y.; Key, J.; Tschudi, M.; Dworak, R.; Mahoney, R.; Baldwin, D. Validation of the Suomi NPP VIIRS ice surface temperature environmental data record. *Remote Sens.* **2015**, *7*, 17258–17271. [CrossRef]
11. Hall, D.K.; Box, J.E.; Casey, K.A.; Hook, S.J.; Shuman, C.A.; Steffen, K. Comparison of satellite-derived and in-situ observations of ice and snow surface temperatures over greenland. *Remote Sens. Environ.* **2008**, *112*, 3739–3749. [CrossRef]
12. Shuman, C.A.; Hall, D.K.; DiGirolamo, N.E.; Mefford, T.K.; Schnaubelt, M.J. Comparison of near-surface air temperatures and modis ice-surface temperatures at summit, greenland (2008–13). *J. Appl. Meteorol. Climatol.* **2014**, *53*, 2171–2180. [CrossRef]
13. Comiso, J.C. Surface temperatures in the polar-regions from nimbus-7 temperature humidity infrared radiometer. *J. Geophys. Res. Oceans* **1994**, *99*, 5181–5200. [CrossRef]
14. Yu, Y.; Rothrock, D.A.; Lindsay, R.W. Accuracy of sea-ice temperature derived from the advanced very high-resolution radiometer. *J. Geophys. Res. Oceans* **1995**, *100*, 4525–4532. [CrossRef]
15. Liu, Y.; Key, J.; Mahoney, R. Sea and freshwater ice concentration from VIIRS on Suomi NPP and the future JPSS satellites. *Remote Sens.* **2016**, *8*, 523. [CrossRef]
16. Wang, X.; Key, J.; Liu, Y. A thermodynamic model for estimating sea and lake ice thickness with optical satellite data. *J. Geophys. Res. Oceans* **2010**, *115*, C12035. [CrossRef]
17. Serreze, M.C.; Schnell, R.C.; Kahl, J.D. Low-level temperature inversions of the eurasian arctic and comparisons with soviet drifting station data. *J. Clim.* **1992**, *5*, 615–629. [CrossRef]
18. Dozier, J.; Warren, S.G. Effect of viewing angle on the infrared brightness temperature of snow. *Water Resour. Res.* **1982**, *18*, 1424–1434. [CrossRef]
19. Kneizys, F.X.; Shettle, E.P.; Abreu, L.W.; Chetwynd, J.H.; Anderson, G.P. *Users Guide to Lowtran 7*; Air Force Geophysics Lab: Hanscom AFB, MA, USA, 1988.
20. Studinger, M.; Koenig, L.; Martin, S.; Sonntag, J. Operation icebridge: Using instrumented aircraft to bridge the observational gap between icesat and icesat-2. In Proceedings of the 2010 IEEE International Geoscience and Remote Sensing Symposium, Honolulu, HI, USA, 25–30 July 2010; pp. 1918–1919.
21. Kurtz, N.; Harbeck, J. *Operation Icebridge Sea Ice Freeboard, Snow Depth, and Thickness Data Products Manual, Version 2 Processing*; Technique Report; NSIDC: Boulder, CO, USA, 2015.
22. Bennett, R.; Studinger, M. *IceBridge KT19 IR Surface Temperature, Version 1. [2013–2017]*; NASA National Snow and Ice Data Center Distributed Active Archive Center: Boulder, CO, USA, 2012; Updated 2018.
23. Hutchison, K.D.; Roskovensky, J.K.; Jackson, J.M.; Heidinger, A.K.; Kopp, T.J.; Pavolonis, M.J.; Frey, R. Automated cloud detection and classification of data collected by the visible infrared imager radiometer suite (VIIRS). *Int. J. Remote Sens.* **2005**, *26*, 4681–4706. [CrossRef]
24. Berk, A.; Anderson, G.P.; Acharya, P.K.; Bernstein, L.S.; Muratov, L.; Lee, J.; Fox, M.J.; Adler-Golden, S.M.; Chetwynd, J.H.; Hoke, M.L. Modtran5: A Reformulated Atmospheric Band Model with Auxiliary Species and Practical Multiple Scattering Options. In Proceedings of the SPIE, Remote Sensing of Clouds and the Atmosphere IX, Maspalomas, Canary Islands, Spain, 13–16 September 2004; Volume 5571, pp. 341–348.
25. Han, Y.; Van Delst, P.; Weng, F.; Liu, Q.; Groff, D.; Yan, B.; Chen, Y.; Vogel, R. Current Status of the JCSDA Community Radiative Transfer Model (CRTM). In Proceedings of the 17th International TOVS Study Conference (ITSC-17), Monterey, CA, USA, 14–20 April 2010; pp. 1–20.
26. Hori, M.; Aoki, T.; Tanikawa, T.; Motoyoshi, H.; Hachikubo, A.; Sugiura, K.; Yasunari, T.J.; Eide, H.; Storvold, R.; Nakajima, Y. In-situ measured spectral directional emissivity of snow and ice in the 8–14 μm atmospheric window. *Remote Sens. Environ.* **2006**, *100*, 486–502. [CrossRef]

27. Jiménez-Muñoz, J.C.; Sobrino, J.A.; Skoković, D.; Mattar, C.; Cristóbal, J. Land surface temperature retrieval methods from landsat-8 thermal infrared sensor data. *IEEE Geosci. Remote Sens. Lett.* **2014**, *11*, 1840–1843. [CrossRef]
28. Aubry-Wake, C.; Baraer, M.; McKenzie, J.M.; Mark, B.G.; Wigmore, O.; Hellström, R.Å.; Lautz, L.; Somers, L. Measuring glacier surface temperatures with ground-based thermal infrared imaging. *Geophys. Res. Lett.* **2015**, *42*, 8489–8497. [CrossRef]

Article

The Detection and Characterization of Arctic Sea Ice Leads with Satellite Imagers

Jay P. Hoffman [1,*], Steven A. Ackerman [1], Yinghui Liu [1] and Jeffrey R. Key [2]

[1]   Cooperative Institute for Meteorological Satellite Studies (CIMSS), University of Wisconsin-Madison, Madison, WI 53706, USA; stevea@ssec.wisc.edu (S.A.A.); yinghuil@ssec.wisc.edu (Y.L.)

[2]   National Oceanic and Atmospheric Administration (NOAA), Madison, WI 53706, USA; jeff.key@noaa.gov

*   Correspondence: jay.hoffman@ssec.wisc.edu; Tel.: +01-608-890-1690

Received: 29 January 2019; Accepted: 27 February 2019; Published: 4 March 2019

Abstract: Sea ice leads (fractures) play a critical role in the exchange of mass and energy between the ocean and atmosphere in the polar regions. The thinning of Arctic sea ice over the last few decades will likely result in changes in lead distributions, so monitoring their characteristics is increasingly important. Here we present a methodology to detect and characterize sea ice leads using satellite imager thermal infrared window channels. A thermal contrast method is first used to identify possible sea ice lead pixels, then a number of geometric and image analysis tests are applied to build a subset of positively identified leads. Finally, characteristics such as width, length and orientation are derived. This methodology is applied to Moderate Resolution Imaging Spectroradiometer (MODIS) observations for the months of January through April over the period of 2003 to 2018. The algorithm results are compared to other satellite estimates of lead distribution. Lead coverage maps and statistics over the Arctic illustrate spatial and temporal lead patterns.

Keywords: leads; sea ice; MODIS

1. Introduction

Leads are elongated fractures in the sea ice cover. They form under stresses on the sea ice forced by wind and ocean currents [1]. The open water refreezes in the cold environment, so leads may contain unfrozen water or ice of varying thicknesses. Leads may be a few meters or a few kilometers in width and may be tens of kilometers in length. While leads occupy a relatively small area of the pack ice overall (e.g., less than 5% of the pack ice surface area), the open waters provide a significant source of heat and moisture to the atmosphere, particularly during winter [2,3]. In the summer, leads absorb more solar energy than the surrounding ice, warming the water and accelerating melt. In the winter, spring, and autumn, leads impact the local boundary layer structure [4] and cloud properties because of the large heat and moisture fluxes into the atmosphere. Leads affect atmosphere and ocean chemical exchanges, such as carbon dioxide, mercury and bromine (e.g., [5,6]). The spatial and temporal descriptions of sea ice lead characteristics can provide useful information in studying the chemical properties of the Artic region. From an operational perspective, knowledge of lead characteristics can aid in navigation, with direct benefits to security, subsistence hunting, and recreation. Given the rapid thinning and loss of Arctic sea ice over the last few decades [7–9], changes in leads can be expected. Lead formation is driven by stress imposed on the sea ice by wind and ocean currents, so they respond to changes in the wind and the resulting ice deformation [10]. From a climate perspective, identifying changes in lead characteristics (width, orientation, area coverage and spatial distribution) will advance our understanding of both thermodynamic and mechanical processes in the Arctic.

A number of studies used satellite data to detect sea ice leads, dating back at least to the early 1990s. Key et al. [11] developed a method to detect and characterize leads using thermal infrared and visible satellite imagery, primarily from Landsat, and explored the sensitivity of lead detection using the Advanced Very High Resolution Radiometer (AVHRR) thermal imagery under various atmospheric conditions. A follow-on study examined the effect of sensor pixel resolution on lead detection [12]. Lindsay and Rothrock [13] produced binary lead maps for specific Arctic regions from the AVHRR data in 1989. Miles and Barry [14] used AVHRR data to study leads in the western Arctic for the winters from 1979 to 1985. Drüe and Heinemann [15] applied the potential open water concept of Lindsay and Rothrock [13] to Moderate Resolution Imaging Spectroradiometer (MODIS) data to derive sea ice concentration based on infrared window brightness temperatures. Willmes and Heinemann [16,17] presented a technique for lead detection using the thermal channels of MODIS.

Satellite microwave observations, both passive and active, have also been employed in leads detection, the advantage being that most clouds are transparent in the microwave spectral range. Röhrs and Kaleschke [18] and Röhrs et al. [19] used the low-resolution Advanced Microwave Scanning Radiometer-Earth Observation System (AMSR-E) where the 18.7 and 89 GHz brightness temperatures were mapped to a 6.25 km grid and an emissivity ratio method was used to detect thin ice. A spatial high-pass filter was applied to retain linear thin-ice areas. It was determined that subpixel-resolution leads could be identified. This work was extended by Bröhan and Kaleschke [20] to develop a nine-year "climatology" of lead orientation and frequency. Synthetic Aperture Radar (SAR) provides the best spatial resolution of microwave sensors but is limited in coverage, both spatial and temporal. Nevertheless, SAR observations have been used to characterize leads and to validate lead information derived from other sensors [21]. Murashkin et al. [22] found that using both horizontally-polarized channels of Sentinel-1 SAR identified more leads than single co-polarized observations. Their lead classification algorithm uses a random forest classifier based on polarimetric features and textural features.

Satellite altimetry can also be used for the detection of leads. For example, Zakharova et al. [23] used the SARAL/AltiKa (satellite with Argos and ALitKa) altimeter to detect leads. They defined a threshold of the maximal power of waveform to discriminate the leads from 200 m to 2-4 km in width. Lead area fraction and widths have been examined using the Cryosat2 radar altimeter [24], where a supervised classification of CryoSat-2 data was performed by a comparison with visual scenes. As in [23], the maximum power of the waveform showed the best classification properties.

This paper presents a methodology for detecting and estimating characteristics of sea ice leads using infrared satellite data. It describes the algorithm, demonstrates applications, and provides examples of the analysis. Our approach improves upon the Key et al. [11,12] algorithm, and provides lead characteristics such as width, orientation, and area that are not produced by more recent thermal infrared algorithms. While the application of the methodology described here is demonstrated using MODIS on NASA's Terra and Aqua satellites, it is equally applicable to other satellite imagers.

2. Data and Methods

The methodology described below is applicable to data from any relatively high-resolution (spatial and temporal) satellite borne imager. In this study, the algorithm is applied to MODIS data. MODIS has 36 spectral bands covering solar to infrared wavelengths, with spatial resolutions of 250 m to 1 km. The robust spectral coverage improves cloud detection in the polar regions [25,26]. Sea ice lead detection and characterization uses the MODIS 1 km resolution 11 μm infrared window band (band 31). Because the pixel size increases away from nadir, the algorithm only uses observations with satellite scan angles within 30° of nadir with a pixel size of approximately 1.5 km. With the scan angle limit, the advantage is constrained pixel size, the tradeoff is that due to orbital geometry, the satellite imager has no coverage north of approximately 81° N (polar coverage is achieved only with a scan angles above 30°). As an example, an enhanced 11 μm brightness temperature image is shown in Figure 1 from 15 February 2018 at 0545 UTC. The warm (bright) leads are distinct against a relatively colder (darker) background of ice and clouds.

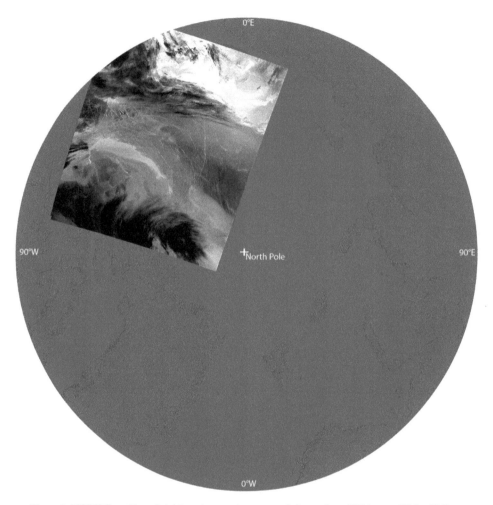

Figure 1. MODIS-Terra 11 μm brightness temperature greyscale image from 15 February 2018 at 0545 UTC. Notice leads appear as bright (warm) features relative to the darker (colder) ice and clouds. The granule projection onto a 1 km Equal-Area Scalable Earth Grid version 2 (EASE2-Grid, [27]). The region shown is north of 65° N, with the North Pole in the center of the map.

2.1. Algorithm Description

The lead detection and characterization algorithm is divided into three steps. Figure 2 is a flowchart that summarizes the steps of this algorithm. Table 1 describes the different mask categories in the lead product, references will be given to the codes when describing the algorithm tests.

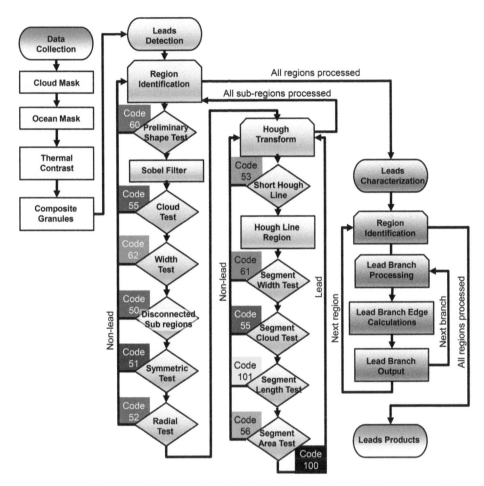

Figure 2. Lead detection and characterization algorithm flowchart. Each color-coded column corresponds to a step in the process (step 1 in green, step 2 in blue, and step 3 in grey). Mask code numbers are used in NetCDF output files and the mask code colors form the color table used in the output imagery.

Table 1. Summary of lead product codes and color table.

Color	Name	Code	Description
	Water	10	Water or ice, fails thermal contrast test; not a lead.
	Disconnected Sub Regions	50	Less than 1% of potential lead area: A continuous object after Sobel filter is applies contains more than 1 discontinuous region in the unfiltered mask, and more than 50% of the object area comes from sub-regions smaller than 5 km^2, and there is between 3–5 sub regions with more 5 km^2 or more.
	Symmetric	51	Less than 1% of potential lead area: A circumscribed rectangle over the object and is divided into 4 equal quadrants. If each quadrant contains $+/-5\%$ of 25% of the object area, the object is too symmetric.
	Radial	52	Approximately 1% of potential lead area: A circle is defined where the radius as half of the average of the span of the object in the x and y direction. If more than half of the object area is within 1.5 km of the edge of the test circle, the object fails to be a lead.
	Short Hough Line	53	Much less than 1% of potential lead area: Hough line contains 3 points or less.
	Cloud	55	Approximately 55% of potential lead area: Greater than 90% of object area occurred in 2 overpasses or less.
	Segment Area	56	Much less than 1% of potential lead area: Object area less than 4 km^2.
	Large region	60	Approximately 8% of potential lead area: Object area divided by diagonal length greater than 60 km. Similar to code 62 test, this test is performed on the object before the image filter is applied.
	Segment Width	61	Less than 1% of potential lead area: Object area divided by length of the diagonal line greater than 25 and object area divided by circumscribed rectangle over the object greater than 5.
	Width	62	Approximately 6% of potential lead area: Object area divided by length of the diagonal line greater than 60; same test as code 60, this test is performed on the object after the Sobel filter is applied.
	Lead	100	Approximately 28% of potential lead area: All tests pass.
	Segment Length	101	Much less than 1% of potential lead area: Great-circle length of lead squared divided by area less than 2.
	Land	200	Land, not tested for leads.
	No coverage	201	Outside of domain, south of 65°N.

2.1.1. Step 1: Aggregate Imager Data, Identify Potential Leads

The Terra and Aqua MODIS 11 μm brightness temperature observations north of 65°N in each individual satellite overpass are remapped to a daily 1 km Equal-Area Scalable Earth Grid version 2 (EASE2-Grid, [27]) grid using a nearest neighbor approach. The brightness temperatures are derived from the Level1B Calibrated Radiances within the Collection 6 MxD021KM files [28,29] ("x" in the file name is "O" for Terra and "Y" for Aqua). For reference, Figure 1 shows an example granule using the EASE2-Grid. Land/Sea Mask within the Collection 6 MxD03 1km Geolocation files [30,31] is applied to exclude the land pixels. Cloud screening uses the standard MODIS cloud mask (MxD35; [25,32,33]). For our application, clear is defined when the cloud mask category is "confident clear", with all other mask values as cloudy. To the nighttime overpasses (solar zenith angle greater than 85°), we apply a cloud removal filter developed by Fraser et al. [34] to remove false cloud detection over sea ice leads. Note that with no filtering the cloud mask flags likely leads as clouds, and the spatial filtering technique is used to reclassify these false clouds as clear, so that these features can be subjected to further lead detection testing.

Grid cells with cloud coverage less than a threshold of 25% in a 5 × 5 convolution kernel are reclassified as clear in Fraser et al. [34]. Using the example Terra overpass from 15 February 2018 at

0545 UTC, we test thresholds of 25% (green), 50% (blue) and 75% (red) of the spatial filter (Figure 3). Note that this image is not geolocated. The example is given to illustrate that filter performance is best for a threshold of 50% in a 5 × 5 pixel window (pixels in blue and green). When the threshold is low (pixels in green: 25% threshold), leads of larger size remain flagged as clouds. With a high threshold (red: 75%), degradation along cloud edges becomes prominent.

Figure 3. Sample portion of MODIS-Terra cloud mask from 15 February 2018, at 0545 UTC. Gray areas are clear, unmodified clouds are white, and the original cloud mask defines clouds as all non-gray areas. A spatial filter is applied to remove false cloud detection over sea ice leads from the mask. In green (blue, red), 25% (50%, 75%) or less of the local mean is cloudy. The sample is a small portion of the same granule as Figure 1, the image is in the native, unnavigated satellite projection.

After the remapping, the algorithm detects possible sea ice lead pixels. A thermal contrast method is a primary component of possible lead detection. A pixel is identified as a potential sea ice lead if it has an 11 μm brightness temperature that is both 1.5 K greater than the mean, and greater than the standard deviation of the brightness temperature of its 25 by 25 pixel surrounding area. This approach is similar to what Willmes and Heinemann [16,17] described, except that we use the 11 μm brightness temperature instead of the derived surface temperature product. In addition, we constrain the MODIS scan angle to 30° within nadir, due to the degradation of spatial resolution at larger sensor viewing angles. The algorithm also assumes that a lead pixel will have a brightness temperature less than 271 K—the nominal freezing point of salt water (271.15 K is the freezing point of salt with salinity of 35 ppt). Due to the presence of open water or very thin sea ice, lead pixels will be below freezing but still warmer than the surrounding sea ice (where sea ice concentrations are 100%). At this point, the algorithm has identified potential leads; an example is shown in Figure 4. The algorithm detects most of the leads (black in Figure 4) that appeared as warm (white) features in the brightness temperature image (Figure 1).

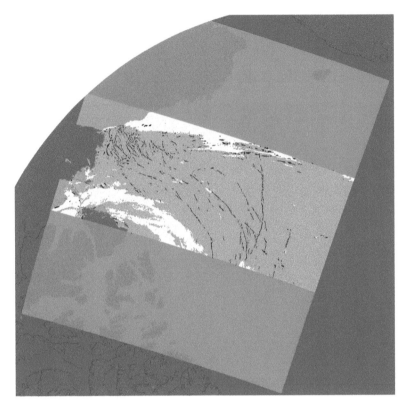

Figure 4. MODIS-Terra potential lead mask from the single overpass at 0545 UTC on 15 February 2018 in the EASE2 projection. Potential leads are in black, clouds are white, land is brown, water not meeting the thermal contrast characteristics is grey, water outside of the image granule is blue, and the scan angle block-out is illustrated in red.

The procedure described above is applied to all Terra and Aqua overpasses on a given day. Rather than generating a composite of all possible sea ice leads pixels from all individual overpasses, our technique produces a map of the number of overpasses in a day where the thermal contrast tests are satisfied. Figure 5 shows a cropped section of a daily composite. Most of the areas where the thermal contrast test is satisfied in only one overpass (red) appear to be false leads - likely caused by short duration features such as clouds (cloud mask omission errors) or cloud shadows. The majority of the features that appear to be leads correspond to a color (shades other than gray and red in Figure 5) that indicates that the detection of the thermal feature occurred in multiple overpasses within the same day. A composite of the number of cloud-free overpasses at each location is also recorded (Figure 6). This map illustrates areas where sampling is limited or missing due to the scan angle constraints (over the pole and lower latitudes) or due to cloud cover. Lead detection is more likely in regions with frequent clear observations and less likely in regions with fewer overpasses or cloud contamination. The counts of clear, cloudy, and potential lead observations are provided in the output file. A user can infer higher confidence in detections when a potential lead (high thermal contrast) is detected in several overpasses, and lower confidence in regions with persistent clouds (a potential lead detection may be a cloud mask omission error).

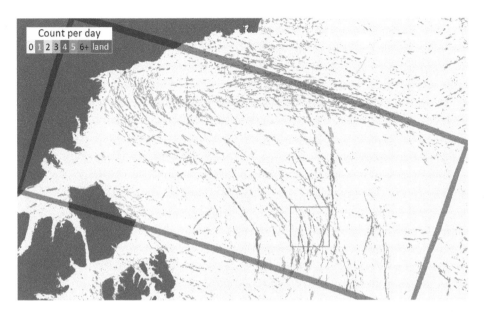

Figure 5. The number of overpasses when a potential lead is detected on 15 February 2018, a composite of all overpasses over the Beaufort Sea, remapped to the EASE2-Grid. The large boxed region corresponds with the location of the 0545 UTC granule shown in Figure 1. The small square is 200 km by 200 km and used for reference in Figure 7.

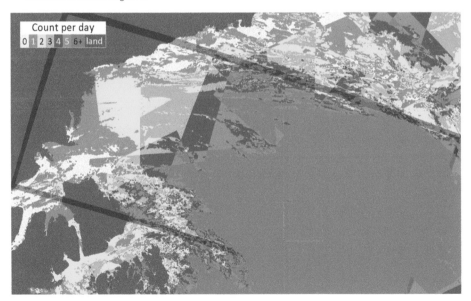

Figure 6. Daily composite of the number of cloud-free overpasses with a MODIS scan angle less than 30° on 15 February 2018 in the Beaufort Sea. The large boxed region corresponds with the location of the 0545 UTC granule shown in Figure 1. The small square is used for reference in Figure 7.

2.1.2. Step 2: Bulk Lead Detection

The daily composite map from Step 1—the composite of the number of times the thermal contrast suggests a potential lead may exist—will be referred to as the map of "potential leads", this is represented in panel (a) in Figure 7 where a 200 by 200 km subset of the 15 February 2018 example is shown. In Step 2, the algorithm performs a number of geometric and image analysis tests on the mask of potential leads and builds a subset of positively identified bulk leads. The iterative process is depicted in Figure 7. The image analysis begins by identifying connected points in the potential lead mask (Figure 7, panel b), where each continuous object has a unique color. A rough estimate of the object width is used to test if an object is a non-lead open water feature (e.g., a polynya). This width is defined as the object area divided by the square root of the sum of the square of the span of the object in the x and y directions (in the equal area grid), as the line from point 1 to 4 in panel c. The region is considered as open water and no further lead processing performed if the width estimate is greater than 60 km (code 60 in Table 1). The lead detection algorithm next rejects potential lead regions that contain only one or two pixels as being too small. Applying these preliminary tests reduces algorithm runtime time by removing some features before applying computational expensive image analysis techniques.

A Sobel image filter [35] is applied to the mask of the remaining potential leads, resulting in each continuous object being assigned a unique color (Figure 7, panel d). The Sobel image filter is an edge detection image-processing tool. For lead detection, the advantage of using the Sobel filter is that it can connect discontinuous features. The width of a lead may fluctuate above and below the nominal resolution of the MODIS imager along the path of the lead. The result would be a single object (lead) that appears as a series of discontinuous (sub-resolution) points in the native imagery. Applying the Sobel filter to the binary mask of the number of times a potential lead is detected, related groups of discontinuous objects are combined into a single continuous object that is subjected to further lead testing.

Cloud artifacts (cloud shadows, cloud mask omission errors, and cloud edges) can cause thermal contrast features that look similar to the thermal contrast from a lead. However, cloud artifacts tend to be short-lived either because the cloud moves, the satellite and/or solar view angle changes, or cloud mask omission errors might not persist in subsequent observations. Therefore, the algorithm contains a test that checks the number of overpass the thermal contrast conditions have been satisfied. For the area of each continuous potential lead feature, a test requires that more than 90% of the observed area within a lead must occur in two or more overpasses within a day. This test (Table 1, code 55) removes features that are most likely cloud contamination (cloud mask omission error) from becoming detected leads, but still allows detection of a lead that is changing (growing, shrinking, or moving) over the course of a day (providing that no more than 10% of the lead area is detected in only one overpass). This 90%/10% threshold was derived empirically by examining several test scenes over several days.

The potential lead undergoes additional geometric testing. First, the original mask (panel b of Figure 7) is compared against the Sobel filtered mask (panel d of Figure 7) to form panel (f) of Figure 7. The number of disconnected features in the original mask (number of colors in panel (f)) are counted that correspond to the single potential lead object region from the Sobel filter (coral color object in panel (d)). A non-lead flag (code 50) is assigned if an object has too many sub-regions or if too much of the object area is made up of small (less than 5 km^2) sub-regions. There is also a test for the object symmetry to test if the object area is too symmetrical (Table 1, code 51), it fails to be a lead. The symmetry test defines four equal-sized quadrants that encompass the test object, black rectangles in panel (c) of Figure 7 and an object fails to be a lead if each of the quadrants contain between 17%-33% of the object area. Similarly, the radial test defines a green circle in panel (c) of Figure 7 centered on the center of the rectangle (point 5) that encompasses the object undergoing analysis. The radius of the circle is half of the average of the length and width of the rectangle that encompasses the object. For the radial test, the number of object points that overlap with the edge of test circle versus the number of points where the object does not overlap with edge of the test circle are calculated. The test object fails to be a lead if more than 50% of the circumference of the test circle overlaps with the test object (Table 1, code 52). The geometric thresholds described in this paragraph were all established empirically by testing a range of values on the synthetic dataset and examining several test scenes over several days.

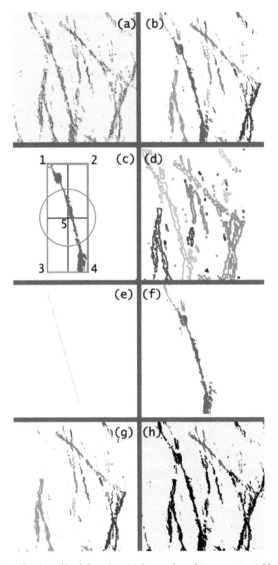

Figure 7. Example application of lead detection. (**a**) the number of times a potential lead detected from a small section (gray square) in Figure 5. (**b**) the region divided into connected objects, each continuous object having a different color. (**c**) the largest continuous object from panel (**b**) being subjected to additional testing. (**d**) the result after applying the Sobel filter with each continuous object having a unique color. (**e**) the longest Hough Transform line segment. (**f**) a potential lead cluster, with discontinuous object being connected in the Sobel filter version of the object in panel (**d**). (**g**) a new iteration with the object from (**f**) removed from (**b**). (**h**) the final results; refer to Table 1 for the product color table.

After passing the preliminary shape tests, the next step is to identify linear features. Leads can be assumed to be polylines and for this reason a Hough Transform [36] is used to locate linear features in imagery. The Hough Transform identifies the longest linear feature within the mask of the number of times a potential lead has been detected. For example, the transform is applied to Figure 7, panel d, and the longest resulting line is shown in panel e. Processing iterations of the longest remaining line continue in a loop until the longest Hough line has three points or less (regions with less than three points are too short and flagged as a non-lead, code 53, Table 1). If the Hough line (panel f) is longer than three points, then a sub-region is defined as all continuous points connected with the longest Hough line segment (panel c). This sub-region undergoes additional testing. The length of the longest line segment within the sub-region is compared against the length of the Hough line segment (lengths of panel c points 1 to 4 compared against panel e). If the sub-region area is greater than 2 points, a few final tests classify the region as a lead or non-lead (Table 1, codes 61, 55, 101, and 56). Processing the largest remaining lead sub-region, each potential lead object is processed until no more valid Hough line segments in the region are found. Or, using the illustration, panel f is removed from panel b to form panel g, the iterative process continues with panel g replacing panel b until the only remaining features contain less than 2 continuous points (features too small to identify a linear features).

2.1.3. Step 3: Lead Branch Characterization

The last step in the processing determines lead characterization. Prior to this step a lead has been defined as a bulk object, or a continuous feature that has high thermal contrast and has passed a series of image processing tests. The previous tests derive characteristics of individual segments or branches that are within the larger, and potentially complex, multifaceted object. The input for this step is the bulk lead mask, described in previous sections. A morphological erosion function (3 by 3 square array) removes the edges from lead object regions. By essentially removing the thinnest segments from a lead, the result is a mask that contains more disconnected regions than the original mask; these discontinuous sub-regions are defined as lead branch segments. An example is shown in Figure 8. In the example, panel e represents a bulk lead object detected in Step 2, each color in panel g represents a lead branch segment. Each bulk and branch object are represented in separate text output files, the object (bulk lead or branch segment) area, length, width, orientation, and coordinates for a start and end point are listed in the respective text output file.

The branch processing loop starts by identifying the remaining branch with the largest area. For each lead branch, the first step is to isolate the edge or outline of the region. The start and end point of the lead branch are identified by finding the set of coordinates that are the furthest distance apart from each other. The great circle distance and azimuth angle are calculated between the start and end points. The segment width is another derived characteristic, found by dividing the segment area by the great circle distance. The sea-basin code for start and end points of the branch is recorded to more readily locate leads as a function of sea, while noting that some leads may span across sea boundaries.

An output text file describes lead characteristics for every (branch or bulk) lead identified in the raster mask. These files serve two purposes: the text file can provide lead location information to users with low bandwidth capabilities and the files provide the metadata of lead characteristics. For each lead branch, the start and end coordinates (x, y, longitude, and latitude), length, azimuth, width area, and the sea code that corresponds to the start and end point of the branch.

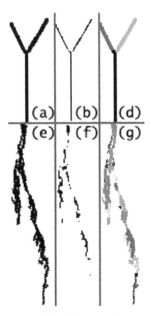

Figure 8. Left objects (**a**) and (**e**) show bulk lead; the eroded branched lead is in the center panels (**b**) and (**f**); and the right panels (**d**) and (**g**) show the restored lead with each branch shown as a different color. The top row is an idealized synthetic lead, the bottom row is an example lead detected on 15 February 2018.

2.2. Output Format

The leads product is available online (ftp://frostbite.ssec.wisc.edu). A navigation file is provided (ftp://frostbite.ssec.wisc.edu/latlon.nc) with the latitude and longitude for each point of the EASE2-Grid. The leads products are arranged in subdirectories by year. The daily product file format is NetCDF. Within the NetCDF file are four different masks, the leads mask and three ancillary arrays. The leads mask is a coded field indicating lead detection (code 100) and rejection codes (see Table 1). The ancillary data includes a cloud mask that reports the number of overpasses at each location that was cloudy (or rejected from processing due to the scan angle block-out), a coverage array that reports the number of cloud-free overpasses, and a potential lead array reports the number of overpasses a potential lead was detected. Two daily text files are also available, one for the bulk lead objects and the other for lead branches. These text files report the coordinates of the start and end point of the lead as well as characteristics such as length (great circle distance from start and end point), azimuth angle, object area, and width (derived as area divided by length). Readme files are included on the ftp site to help describe each output field.

3. Results

This paper presents a technical description of the algorithm. A later publication is anticipated with a greater focus on the results and trend analyisis. A synthetic test case and a real-world test case are presented as an assessment of the algorithm. The synthetic case is designed to assist in the development of the algorithm; tests and thresholds are set by testing performance against the synthetic scene as well as real-world case studies. Comparisons are presented against products from previously published lead detection technique. Annual lead area coverage maps over the Arctic are also included.

3.1. Synthetic Test

A synthetic test scene is used to assess algorithm performance. Using this synthetic data, in combination with test cases with real data, many of the algorithm thresholds were tested and defined. The test scene contains lead-like features as well as features that do not resemble leads and should fail a lead detection algorithm. Figure 9 shows the test scene with results from the algorithm. Black represents areas that pass as lead detections; other colors represent areas that the algorithm rejects object as a lead for reasons shown in Table 1. The top three quarters of the bulk objects resemble leads, the bottom quarter of the image represents features that were designed to not resemble leads. Several of the algorithm tests were designed and thresholds adjusted based on the detection/rejection performance with the synthetic dataset. Detection is successful for the continuous linear features in the test scene (except for some lines that are too straight and too thin). Detection failures occur sometimes when breaks in a line segment are too small or too wide. This testing provides us with the confidence to apply the algorithm to satellite imagery scenes.

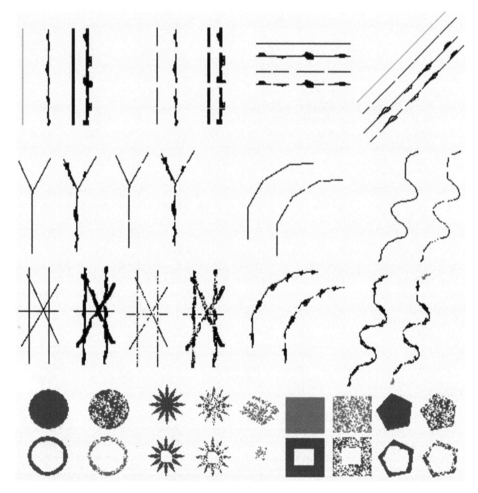

Figure 9. Synthetic algorithm test scenes and test results. Objects that pass lead detection are in black; other colors represent potential leads that fail one of the tests. Refer to Table 1 for mask code colors. Not all categories are represented in this example.

3.2. Case Studies

An example MODIS product result from 15 February 2018 is shown in Figure 10. This is the same case illustrated in previous figures. One image granule is shown on a map in Figure 1 and the potential leads identified in that single overpass are mapped in Figure 4. The majority of potential lead features found in only in one overpass (red in Figure 5) are rejected from the final leads product mask. A more detailed description of the algorithm for the small gray square was presented in Figure 7. Notice that some of features that look like leads are rejected sometimes; these objects have complex shapes that fail one of the identification tests. Also, for cases where there was only one clear overpass, a potential lead will fail our detection technique.

For comparison against other lead detection methods, a second case study from 3 March 2009 is presented in Figures 11 and 12. For a general overview, Figure 11 shows the entire polar domain. Figure 12 highlights some details enlarged on a small box region in Figure 11. Our leads products are shown in panels (A) and (a); refer to Table 1 for the color table of leads (black) and potential leads that fail the linear classification technique. MODIS lead detections from Willmes and Heinemann product [37] are shown in black and the artifact (high uncertainty) category is colored red in panels (B) and (b); lead detections from Röhrs and Kaleschke [18] (and Röhrs et al. [19]) using AMSR-E [38] are also shown in black in panels (C) and (c). These products have been remapped and resampled with a nearest neighbor technique to match the same projection and 1km resolution as our product. To highlight the product similarities and differences, the fourth panel (D) and (d) is an RGB composite of the three products. With leads from our product in the red channel, the Willmes and Heinemann product is in the blue channel, and the Röhrs AMSR-E product is in the green channel. Notice that where all three products detect a lead, the pixel is colored white; yellow, cyan, and purple result when two of the three products detect a lead. As expected, the advantage of AMSR-E is the ability to detect leads in cloud regions, the drawback is that it only detects relatively large leads and tends to over-represent lead width due to the coarser spatial resolution (see the Discussion section). The Willmes and Heinemann technique detects more leads than our product (Figures 11 and 12); many features Willmes and Heinemann classify as leads fail our tests that try to identify linear features.

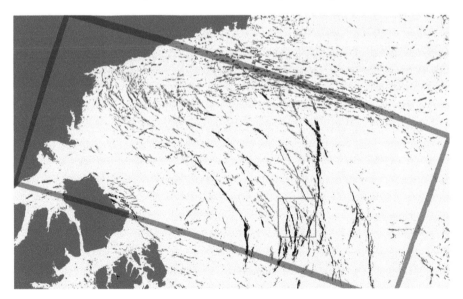

Figure 10. Leads results from 15 February 2018 in the Beaufort Sea. Accepted leads are black, refer to Table 1 for color codes. Large boxed region corresponds with the location of the 0545 UTC granule shown in Figure 1. The small square is a 200 km by 200 km box used for reference in Figure 7.

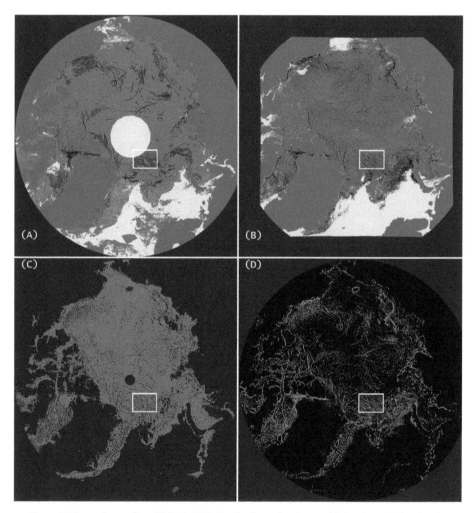

Figure 11. Example case from 3 March 2009. Our leads product in panel (**A**), refer to Table 1 for the leads color table. Leads detected from Willmes and Heinemann product [37] are shown in black and the artifact (high uncertainty) category is in red in panel (**B**). AMSR-E [38] leads are in black in panel (**C**). An RGB composite is shown in panel (**D**), here, our leads are in red, Willmes and Heinemann leads in blue, and AMSR-E in green. A 650 km by 500 km region (white box) is enlarged in Figure 12.

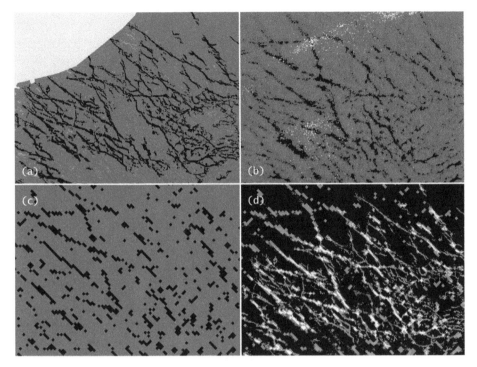

Figure 12. Enlarged region from Figure 11, Panels (**a**), (**b**), (**c**) and (**d**) correspond to the panels (**A**), (**B**), (**C**) and (**D**). The color scheme is the same as in Figure 11.

3.3. Lead Timeseries

The lead algorithm is applied to Terra and Aqua MODIS observations for the period January through April for the years 2003–2018. The reasons for limiting the study period to January through April is discussed later in the discussion section. An overview map of annual lead frequency (number of days with a lead detection from January through April) is presented in Figures 13 and 14. A daily animation of lead detections is available in Supplemental Materials S1. The variations of sea ice lead frequency from region-to-region and year-to-year is a topic to be addressed in future work.

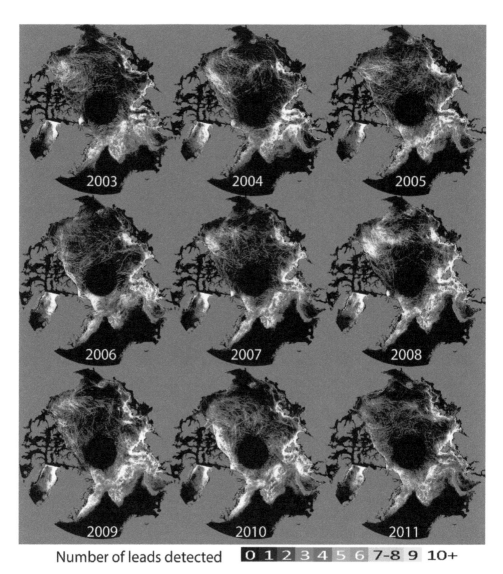

Figure 13. Lead frequency map by year 2003–2011. Color coded legend is at bottom of the figure represents number of days with a lead detected between January through April.

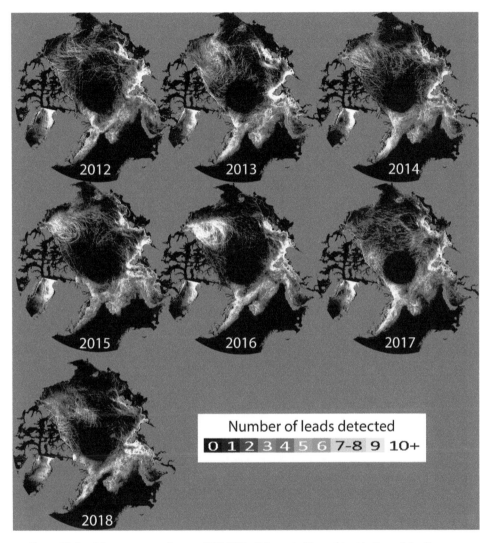

Figure 14. Lead frequency map by year 2012–2018. Color coded legend is at bottom of the figure represents number of days with a lead detected between January through April.

A comparison of the time-series of our results and the Willmes and Heinemann [37] product is presented in Table 2. The statistics are arranged in rows by year, with all years combined in the bottom row. The second through fourth columns give statistics independent of the spatial domain of other product. The remaining three columns give statistics from the subset of the domain where both products had cloud-free coverage. The area that we classify as potential leads (defined as a percentage of area with a positively identified lead or potential lead, divided by coverage area) is in better agreement with what Willmes and Heinemann classify as leads than the final lead classification ("leads"), as the linear identification techniques that our algorithm employs results in fewer positively identified leads. Comparing the statistics with and without the overlapping coverage of both products, the results are similar. The statistics tend to be only a fraction of a percent higher for the subset where both products have coverage over the same domain.

Table 2. Annual statistical difference of leads products. Area is reported as a percentage of leads area divided by coverage area. Positively identified leads and potential leads are compared against Willmes and Heinemann [37] leads (artifact category is not included), which is only available for 2003–2015. The left set of columns are for each product independently. On the right, the domains are limited to only the subset where both products have coverage.

Year	Leads	Potential Leads	Willmes and Heinemann Leads	Leads with overlapping Willmes and Heinemann coverage	Potential Leads with overlapping Willmes and Heinemann coverage	Willmes and Heinemann with overlapping coverage
2003	3.5%	11.2%	8.1%	3.7%	11.4%	8.7%
2004	3.2%	10.2%	6.6%	3.4%	10.1%	7.1%
2005	2.9%	9.8%	6.2%	3.2%	9.8%	6.7%
2006	2.9%	10.0%	7.0%	3.2%	10.2%	7.6%
2007	2.9%	10.0%	7.8%	3.3%	10.6%	8.4%
2008	2.9%	10.1%	7.8%	3.4%	10.7%	8.4%
2009	2.9%	10.2%	7.6%	3.4%	10.7%	8.3%
2010	3.0%	10.2%	7.9%	3.5%	10.8%	8.5%
2011	3.2%	11.3%	11.5%	3.7%	13.1%	12.5%
2012	2.8%	10.5%	12.0%	3.3%	12.5%	12.5%
2013	3.3%	10.9%	12.3%	3.7%	12.3%	13.3%
2014	3.1%	10.9%	6.6%	3.7%	11.8%	7.2%
2015	3.3%	10.8%	6.3%	3.5%	10.8%	6.9%
2016	2.9%	10.2%	Not available	Not available	Not available	Not available
2017	3.0%	10.5%	Not available	Not available	Not available	Not available
2018	2.8%	10.3%	Not available	Not available	Not available	Not available
All	3.0%	10.2%	7.7%	3.4%	10.7%	8.3%

Figure 15 is a map of the Arctic divided into seas as defined by Wang and Key [39]. Figure 16 shows the orientation, as an azimuth angle, of leads in the 10 seas for each year in the period 2003–2018. An azimuth of 0 degrees is north-south; azimuth is reported as degrees east of north (or degrees moving clockwise) with a range of 0–180. The distribution of azimuth angles that define orientation of the leads varies for each sea but is relatively consistent for all years within any given sea. Figure 17 shows the relative frequency of lead widths, areas, and lengths for individual years. The aggregate of all seas is used for this series of plots. Lead branch segments (described in algorithm Step 3) are on the top row and bulk lead objects (described in algorithm Step 2) are on the bottom row. Notice that the scales are different between the rows where the branches are smaller than the bulk objects; one or multiple branch segments comprise a bulk lead object. The intent here is to showcase the capabilities of the algorithm. The small inter-annual variability and analysis of the patterns in lead characteristics will be focuses of future research.

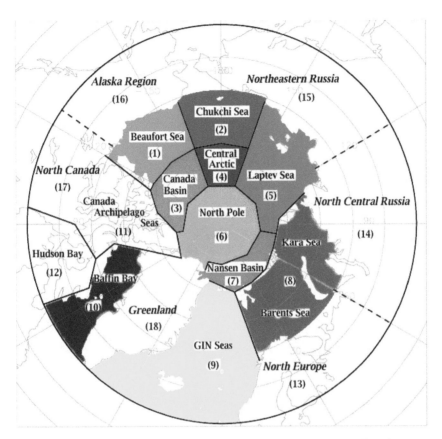

Figure 15. Color-coded Arctic map with each region assigned a sea name and number.

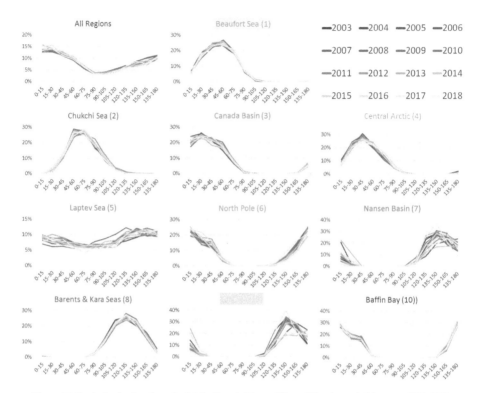

Figure 16. Sea ice lead azimuth angle distributions in 10 seas (Figure 15), color-coded for years 2003–2018.

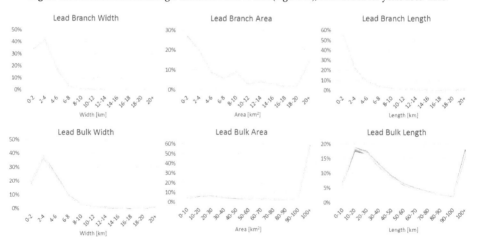

Figure 17. Relative frequency of lead widths, areas, and lengths with the same color code as Figure 16 for 2003–2018. Here, an aggregate of all seas is shown for the individual lead branch segments on the top row and for overall lead objects on the bottom row, the distinction being that a lead object is made up of one or multiple branch segments.

4. Discussion

The design philosophy of our algorithm is to minimize the errors of commission; i.e., to minimize overestimation of leads. As presented in Table 2, the area classified as potential leads is in better agreement with the Willmes and Heinemann [37] leads area than our positively identified leads. Quantitative results are presented annually; qualitative comparisons of an example day are shown in Figures 11 and 12. The quantitative results from the day of the case study are similar to the annual results (not shown). Although both products are derived from the same satellite, there are some differences in the domain. Willmes and Heinemann [16,17] does not extend as far south however, our algorithm limits the view angle (creating a coverage gap at the pole). Cloud coverage results in some difference as well. We apply a cloud screening technique that allows for some lead retrievals in areas where the ice surface temperature is not retrieved due to cloud, and therefore Willmes and Heinemann [16,17] cannot process a location for leads. Some of these differences are visible in the case study shown in Figure 11.

Furthermore, our positively identified lead area is significantly smaller. When comparing statistics where both have cloud-free coverage, our technique finds leads in approximately 3% of the domain compared with approximately 10% of the Willmes and Heinemann domain. Some of the differences can be attributed to resampling of the Willmes and Heinemann [37] product; the nominal native resolution of that product is 2 km resolution while our product resolution is 1 km. Using the coarser resolution, the contribution of sub-resolution leads may overestimate the total lead area. We are able to achieve the finer resolution by limiting the scan angle to 30°, though the trade-off is a coverage gap over the pole (north of approximately 81°N). The effects of sensor resolution on lead characteristics were examined in [11,12,40]. Using Landsat Multi-Spectral Scanner 80 m data degraded to lower resolutions, Key et al. [12] found that (a) while the manner in which widths of individual leads changes with increasing pixel size is highly variable, the mean lead width over an image changes in a more predictable way. For example, a mean lead width of 400 m derived from data with a 200 m pixel may "grow" to a mean width of 800–1000 m with a 700 m pixel (Figure 6 in [12]). (b) The change in total lead area with increasing pixel size is generally exponential. (c) Leads narrower than approximately 250 m disappear as the resolution of Landsat images is degraded to 320 and 640 m. (d) Lead orientations change if they are anisotropic (i.e., have a preferred orientation), but do not change substantially if they are isotropic. The actual effect of pixel size depends in part upon the temperature or reflectance contrast between a lead and the surrounding ice; see [12] for a quantification of the combined effect.

The difference in lead area is also a result of the different linear identification techniques that our algorithm employs. If there are applications that are less sensitive to commission error and more sensitive to omission error, users can refer back to Table 1 for a summary of lead rejection categories and they may choose to include some rejection categories in addition to our positively identified leads category. A polynya is an example of an ice feature with the thermal signature of leads, but it is irregularly shaped. [41] Our algorithm is tuned towards identifying leads as linear features within the sea ice pack.

We use Level1B 11 µm brightness temperatures [28,29] rather than the MODIS MxD29 ice surface temperature product [42–44]. The MODIS noise-equivalent temperature difference is 0.05 K at 11 µm channel, which is sufficiently accurate for lead detection. There is, of course, a strong correlation between ice surface temperature and the 11 µm brightness temperature. Also, the actual surface temperature is less important than the contrast in temperature. Leads become undetectable when the thermal contrast between leads and surrounding ice pack is small (e.g., less than 1.5 K, although the local spatial variability is also a factor, as described earlier). The primary cause of thermal contrast in the Arctic winter would be leads—the contrast between solid ice and either open water, ice and water mixed, or thin ice. In warmer seasons the contrast between ice temperature and water temperature becomes smaller. Detection capabilities decreases as the surface temperature increases. In summary, the temperature contrast becomes small as ice becomes thicker within a lead or when the surface temperature approaches the melting point of ice in the case of an unfrozen lead.

We agree with the findings of Fraser et al. [34] that cloud masks often misidentify leads as clouds. We are able to employ a similar cloud clearing method to reclassify false cloud detections as clear and detect leads using the 11 μm brightness temperature. It would not be possible to detect leads using the sea ice temperature in these locations because the sea ice temperature product would fail to provide a temperature retrieval in these areas flagged as cloudy. Another area of investigation was to use an ice concentration product as the basis for identifying potential leads, but we found that much like ice surface temperature products, cloud product omission errors (associated with leads) often prevented ice concentration retrievals just as clouds also prevent ice surface temperature retrievals.

Clouds present several obstacles in lead detection. The infrared thermal signature of clouds is often similar to that of leads, which is why cloud detection algorithms often errantly flag leads as clouds. Polar cloud detection in the winter is difficult; cloud detection algorithms are designed to detect cold and highly reflective features, which happen to also be characteristics of sea ice. Further, cloud mask performance is in general better during the day—when reflective bands contribute to better performance of cloud detection algorithms—but the polar region is predominately dark during the winter. Also, clouds in the polar winter are commonly warmer than the surface. Cloud detection algorithms use ancillary data to try to identify in advance where snow or ice may exist at the surface in an attempt to detect clouds over snow or ice. The problem with this technique is that the ancillary data might not accurately reflect the actual conditions; the product may be observation based and reflective of past conditions that may be obsolete or based on a forecast that is not reflective of current conditions. Also, the resolution of the snow/ice mask may be too coarse resolution to account for narrow features like leads.

One of the weaknesses of the algorithm is that it was not designed to work in the summer months. The thermal contrast between leads and the surrounding ice would be very different and the surface can be more complex. Melt ponds, for example, may appear in the summer and these could potentially fool the algorithm. When the ice surface temperature approaches the water temperature, the algorithm would not detect thermal contrast and therefore not detect leads. Persistent cloud coverage in the warmer months is another factor for the limited period of coverage. The January through April time period was chosen because this is generally the best season for lead detection and it is consistent with the season provided by Willmes and Heinemann [37].

One of the advantages of the polar region is the frequent coverage from polar satellites. One of the assumptions of the algorithm is that most locations in the domain will have been covered by multiple satellite overpasses. As illustrated in Figure 6, most locations have over six overpasses per day, though cloud coverage will limit the number of overpasses where lead detection is possible.

In addition to frequent repeating coverage, another assumption the algorithm makes is that leads will be stationary or slow moving in order to satisfy the requirement that a lead must be detected in multiple overpasses. To account for lead movement there is a requirement that 90% of the lead area must be detected in multiple overpasses. A fast-moving lead could fail to meet these conditions; however, the alternative has drawbacks. If a lead is moving quickly such that it appears in different locations throughout the day, a daily composite of lead area could over-report lead area because the fast-moving lead area would be counted towards the daily area in each location it was detected (versus stationary leads that would only contribute towards the daily area once). We believe the smearing effect from slow moving leads that could artificially inflate the lead width and area is more acceptable than considering each detection of a fast-moving lead as a contribution towards the daily lead area. We also believe that repeat detections are important to establishing confidence in a lead detection. The type of phenomena that would produce a thermal contrast signature similar to that of leads would either be a broken ice feature like a lead or else related to a cloud. Cloud edges, cloud shadows, and cloud mask omission errors could all produce a thermal contrast signature similar to a lead. These cloud related features are more likely to be short lived; we would expect viewing geometry and cloud movement to change overpass-to-overpass and make the reoccurrence of false lead signatures less likely than for repeat observations of a real lead. Establishing a confidence parameter would be a non-trivial undertaking. Providing the number of times a potential lead was detected can be used subjectively to help establish user confidence in lead detections. In future work a confidence may be established based on observation frequency as well as other factors such as making an attempt to track a previously detected lead or perhaps quantifying the thermal contrast signature into a confidence parameter.

5. Conclusions

This paper describes an algorithm to detect sea ice leads (fractures) over the Arctic using measurements from MODIS on the Terra and Aqua satellites. We describe the algorithm and demonstrate results by applying the algorithm over the entire MODIS Aqua/Terra period, 2003–2018, for the winter and early spring months of January–April. The algorithm consists of three main steps. The first step makes a 1-km grid composite of the potential leads based on the thermal contrast in the satellite infrared window data over the Arctic Ocean. The second step defines lead objects using a series of image analysis techniques to identify pixels that may have spatial characteristics of leads. In the third and final step, lead objects are broken into smaller individual branches and the characteristics of each branch of a potentially multi-faceted lead are calculated. This step also determines the area, length, width and orientation of the lead. The results of this algorithm are compared with the products from Willmes and Heinemann [37] who also use MODIS but employ a different methodology.

Cloud coverage constrains lead detection, optically thick clouds in particular, thus an infrared algorithm for lead detection requires an accurate cloud mask. Unfortunately, the most difficult conditions for automated cloud identification techniques—nighttime darkness, bright surfaces, and cold surfaces—are also the most prevalent conditions in the Arctic winter. Also, the thermal contrast and shape of cloud edges and cloud shadows often have a similar shape and temperature contrast as lead. Passive microwave observations overcome many problems associated with cloudy conditions but have a larger footprint. Combining collocated microwave and imager observations may be the best approach to determining changing lead characteristics in the Arctic.

Instrument resolution is another limiting factor. By limiting the sensor scan angle we are able to keep our observation resolution at a relatively consistent 1 km. We did not attempt to define the minimum detectable lead size; it would be a function of the lead size, water temperature (and sea ice concentration) within the leads, and the temperature of the surrounding sea ice.

Future work will explore the relationship between cloud cover and the lead detection, trends in lead length and width, relationships between lead azimuth angle and anomalies in wind and ocean currents, and relationships between changes in lead properties and sea ice thickness. Application of the algorithm indicates that the annual variation in lead coverage is large, both spatially and temporally. The algorithm will be adapted to the Visible Infrared Imaging Radiometer Suite (VIIRS), which has more consistent spatial resolution of 375 m across the entire 11 μm swath. With a wider swath than MODIS, the two-satellite system of NOAA-20 and the Suomi National Polar-orbiting Partnership satellites will be able to provide even greater coverage than has been available with MODIS.

Supplementary Materials: The following are available online at https://youtu.be/r7ax0uJtZX0, Video S1: 2003–2018 Daily Leads Detections.

Author Contributions: Conceptualization, J.P.H., S.A.A., Y.L. and J.R.K.; Data curation J.P.H. and Y.L.; Funding acquisition, S.A.A.; Investigation, J.P.H.; Methodology, J.P.H. and J.R.K.; Project administration, S.A.A.; Resources, J.R.K.; Software, J.P.H., Y.L. and J.R.K.; Supervision, S.A.A.; Validation, J.P.H. and Y.L.; Writing—original draft, J.P.H.; Writing—review & editing, J.P.H., S.A.A., Y.L. and J.R.K.

Funding: This research was funded by NASA, grant number NNX14AJ42G and 80NSSC18K0786.

Acknowledgments: The Aqua and Terra MODIS datasets were acquired from the Level-1 and Atmosphere Archive & Distribution System (LAADS) Distributed Active Archive Center (DAAC), located in the Goddard Space Flight Center in Greenbelt, Maryland (https://ladsweb.nascom.nasa.gov/). The Willmes and Heinemann MODIS leads product dataset acquired from https://doi.pangaea.de/10.1594/PANGAEA.854411. The Röhrs et al. AMSR-E dataset acquired from https://icdc.cen.uni-hamburg.de/1/daten/cryosphere/lead-area-fraction-amsre. html. The views, opinions, and findings contained in this report are those of the author(s) and should not be construed as an official National Oceanic and Atmospheric Administration or U.S. Government position, policy, or decision.

Remote Sens. **2019**, *11*, 521

Conflicts of Interest: The authors declare no conflict of interest.

References

1. Smith, S.D.; Muench, R.D.; Pease, C.H. Polynyas and leads: An overview of physical processes and environment. *J. Geophys. Res. Oceans* **1990**, *95*, 9461–9479. [CrossRef]
2. Alam, A.; Curry, J. Lead-induced atmospheric circulations. *J. Geophys. Res. Oceans* **1995**, *100*, 4643–4651. [CrossRef]
3. Maykut, G.A. Energy exchange over young sea ice in the central arctic. *J. Geophys. Res. Oceans* **1978**, *83*, 3646–3658. [CrossRef]
4. Lüpkes, C.; Vihma, T.; Birnbaum, G.; Wacker, U. Influence of leads in sea ice on the temperature of the atmospheric boundary layer during polar night. *Geophys. Res. Lett.* **2008**, *35*. [CrossRef]
5. Douglas, T.; Sturm, M.; Simpson, W.; Brooks, S.; Lindberg, S.; Perovich, D. Elevated mercury measured in snow and frost flowers near arctic sea ice leads. *Geophys. Res. Lett.* **2005**, *32*. [CrossRef]
6. Steiner, N.; Lee, W.; Christian, J. Enhanced gas fluxes in small sea ice leads and cracks: Effects on CO_2 exchange and ocean acidification. *J. Geophys. Res. Oceans* **2013**, *118*, 1195–1205. [CrossRef]
7. Wadhams, P.; Davis, N.R. Further evidence of ice thinning in the Arctic Ocean. *Geophys. Res. Lett.* **2000**, *27*, 3973–3975. [CrossRef]
8. Vaughan, D.G.; Comiso, J.C.; Allison, I.; Carrasco, J.; Kaser, G.; Kwok, R.; Mote, P.; Murray, T.; Paul, F.; Ren, J. Observations: Cryosphere. *Clim. Chang.* **2013**, *2103*, 317–382.
9. Laxon, S.W.; Giles, K.A.; Ridout, A.L.; Wingham, D.J.; Willatt, R.; Cullen, R.; Kwok, R.; Schweiger, A.; Zhang, J.; Haas, C. Cryosat-2 estimates of arctic sea ice thickness and volume. *Geophys. Res. Lett.* **2013**, *40*, 732–737. [CrossRef]
10. Rampal, P.; Weiss, J.; Marsan, D. Positive trend in the mean speed and deformation rate of arctic sea ice, 1979–2007. *J. Geophys. Res. Oceans* **2009**, *114*. [CrossRef]
11. Key, J.; Stone, R.; Maslanik, J.; Ellefsen, E. The detectability of sea-ice leads in satellite data as a function of atmospheric conditions and measurement scale. *Ann. Glaciol.* **1993**, *17*, 227–232. [CrossRef]
12. Key, J.; Maslanik, J.; Ellefsen, E. The effects of sensor field-of-view on the geometrical characteristics of sea ice leads and implications for large-area heat flux estimates. *Remote Sens. Environ.* **1994**, *48*, 347–357. [CrossRef]
13. Lindsay, R.; Rothrock, D. Arctic sea ice leads from advanced very high resolution radiometer images. *J. Geophys. Res. Oceans* **1995**, *100*, 4533–4544. [CrossRef]
14. Miles, M.W.; Barry, R.G. A 5-year satellite climatology of winter sea ice leads in the western arctic. *J. Geophys. Res. Oceans* **1998**, *103*, 21723–21734. [CrossRef]
15. Drüe, C.; Heinemann, G. High-resolution maps of the sea-ice concentration from MODIS satellite data. *Geophys. Res. Lett.* **2004**, *31*. [CrossRef]
16. Willmes, S.; Heinemann, G. Pan-arctic lead detection from MODIS thermal infrared imagery. *Ann. Glaciol.* **2015**, *56*, 29–37. [CrossRef]
17. Willmes, S.; Heinemann, G. Sea-ice wintertime lead frequencies and regional characteristics in the arctic, 2003–2015. *Remote Sens.* **2015**, *8*, 4. [CrossRef]
18. Röhrs, J.; Kaleschke, L. An algorithm to detect sea ice leads by using AMSR-E passive microwave imagery. *Cryosphere* **2012**, *6*, 343–352. [CrossRef]
19. Röhrs, J.; Kaleschke, L.; Bröhan, D.; Siligam, P.K. Corrigendum to "An algorithm to detect sea ice leads by using amsr-e passive microwave imagery.". *Cryosphere* **2012**, *6*, 365.
20. Bröhan, D.; Kaleschke, L. A nine-year climatology of arctic sea ice lead orientation and frequency from AMSR-E. *Remote Sens.* **2014**, *6*, 1451–1475. [CrossRef]
21. Ivanova, N.; Rampal, P.; Bouillon, S. Error assessment of satellite-derived lead fraction in the arctic. *Cryosphere* **2016**, *10*, 585–595. [CrossRef]
22. Murashkin, D.; Spreen, G.; Huntemann, M.; Dierking, W. Method for detection of leads from Sentinel-1 SAR images. *Ann. Glaciol.* **2018**, *59*, 124–136. [CrossRef]
23. Zakharova, E.A.; Fleury, S.; Guerreiro, K.; Willmes, S.; Remy, F.; Kouraev, A.V.; Heinemann, G. Sea ice leads detection using Saral/Altika altimeter. *Mar. Geod.* **2015**, *38*, 522–533. [CrossRef]
24. Wernecke, A.; Kaleschke, L. Lead detection in arctic sea ice from Cryosat-2: Quality assessment, lead area fraction and width distribution. *Cryosphere* **2015**, *9*, 1955–1968. [CrossRef]

25. Frey, R.A.; Ackerman, S.A.; Liu, Y.; Strabala, K.I.; Zhang, H.; Key, J.R.; Wang, X. Cloud detection with MODIS. Part i: Improvements in the MODIS cloud mask for Collection 5. *J. Atmos. Ocean. Technol.* **2008**, *25*, 1057–1072. [CrossRef]

26. Liu, Y.; Key, J.R.; Frey, R.A.; Ackerman, S.A.; Menzel, W.P. Nighttime polar cloud detection with MODIS. *Remote Sens. Environ.* **2004**, *92*, 181–194. [CrossRef]

27. Brodzik, M.J. EASE-Grid: A versatile set of equal-area projections and grids. In *Discrete Global Grids*; Goodchild, M., Knowles, K.W., Eds.; National Center for Geographic Information & Analysis: Santa Barbara, CA, USA, 2002.

28. MODIS Science Data Support Team (SDST). Available online: http://dx.doi.org/10.5067/MODIS/MOD021KM.006 (accessed on 29 January 2019).

29. MODIS Science Data Support Team (SDST). Available online: http://dx.doi.org/10.5067/MODIS/MYD021KM.006 (accessed on 29 January 2019).

30. MODIS Science Data Support Team (SDST). Available online: http://dx.doi.org/10.5067/MODIS/MYD03.006 (accessed on 29 January 2019).

31. MODIS Science Data Support Team (SDST). Available online: http://dx.doi.org/10.5067/MODIS/MOD03.006 (accessed on 29 January 2019).

32. Ackerman, S.A.; Strabala, K.I.; Menzel, W.P.; Frey, R.A.; Moeller, C.C.; Gumley, L.E. Discriminating clear sky from clouds with MODIS. *J. Geophys. Res. Atmos.* **1998**, *103*, 32141–32157. [CrossRef]

33. Baum, B.A. Modis cloud-top property refinements for collection 6. *J. Appl. Meteorol. Climatol.* **2012**, *51*, 1145. [CrossRef]

34. Fraser, A.D.; Massom, R.A.; Michael, K.J. A method for compositing polar MODIS satellite images to remove cloud cover for landfast sea-ice detection. *IEEE Trans. Geosci. Remote Sens.* **2009**, *47*, 3272–3282. [CrossRef]

35. Sobel, I. Camera Models and Perception. Ph.D. Thesis, California: Artificial Intelligence Lab. Stanford University, Stanford, CA, USA, 1970.

36. Ballard, D.H. Generalizing the Hough transform to detect arbitrary shapes. *Pattern Recognit.* **1981**, *13*, 111–122. [CrossRef]

37. Willmes, S.; Heinemann, G. Daily pan-arctic sea-ice lead maps for 2003–2015, with links to maps in netCDF format. In Supplement to: Willmes, S; Heinemann, G (2015): Sea-Ice Wintertime Lead Frequencies and Regional Characteristics in the Arctic, 2003–2015. *Remote Sens.* **2015**, *8*, 4. [CrossRef]

38. Integrated Climate Date Center. *AMSR-E Lead Area Fraction for the Arctic, 3 March 2009, Ed*; University of Hamburg: Hamburg, Germany.

39. Wang, X.; Key, J.R. Arctic surface, cloud, and radiation properties based on the AVHRR Polar Pathfinder dataset. Part I: Spatial and temporal characteristics. *J. Clim.* **2005**, *18*, 2558–2574. [CrossRef]

40. Key, J.R. The area coverage of geophysical fields as a function of sensor field-of-view. *Remote Sens. Environ.* **1994**, *48*, 339–346. [CrossRef]

41. Paul, S.; Willmes, S.; Heinemann, G. Long-term coastal-polynya dynamics in the southern Weddell Sea from MODIS thermal-infrared imagery. *Cryosphere* **2015**, *9*, 2027–2041. [CrossRef]

42. Hall, D.K.; Riggs, G.A.; Salomonson, V.V.; Barton, J.; Casey, K.; Chien, J.; DiGirolamo, N.; Klein, A.; Powell, H.; Tait, A. *Algorithm Theoretical Basis Document (ATBD) for the MODIS Snow and Sea Ice-Mapping Algorithms*; NASA GSFC: Greenbelt, MD, USA, 2001.

43. Riggs, G.A.; Hall, D.K.; Salomonson, V.V. *Modis Sea Ice Products User Guide to Collection 5*; NASA Goddard Space Flight Center: Greenbelt, MD, USA, 2006.

44. Riggs, G.A.; Hall, D.K.; Salomonson, V.V. *Modis Sea Ice Products User Guide to Collection 6*; NASA Goddard Space Flight Center: Greenbelt, MD, USA, 2015.

Article

Evaluation of Chlorophyll-a and POC MODIS Aqua Products in the Southern Ocean

William Moutier [1,2,*,†], Sandy J Thomalla [1,3], Stewart Bernard [2,4], Galina Wind [5,6], Thomas J Ryan-Keogh [1] and Marié E Smith [4]

1   Southern Ocean Carbon and Climate Observatory (SOCCO), CSIR, Rosebank, Cape Town 7700, South Africa
2   Department of Oceanography, University of Cape Town, Rondebosch, Cape Town 7701, South Africa
3   Marine Research Institute (MaRe), University of Cape Town, Rondebosch, Cape Town 7701, South Africa
4   Earth Observation, Council for Scientific and Industrial Research (CSIR), Rosebank,
     Cape Town 7700, South Africa
5   NASA Goddard Space Flight Center, Greenbelt, MD 20771, USA
6   SSAI, Inc., 10210 Greenbelt Road, Suite 600, Lanham, MD 20706, USA
*   Correspondence: william.moutier@meteo.be
†   Current address: Royal Meteorological Institute of Belgium, Avenue Circulaire 3, 1180 Brussels, Belgium.

Received: 13 June 2019; Accepted: 9 July 2019; Published: 31 July 2019

Abstract: The Southern Ocean (SO) is highly sensitive to climate change. Therefore, an accurate estimate of phytoplankton biomass is key to being able to predict the climate trajectory of the 21st century. In this study, MODerate resolution Imaging Spectroradiometer (MODIS), on board EOS Aqua spacecraft, Level 2 (nominal 1 km × 1 km resolution) chlorophyll-a (C^{Sat}) and Particulate Organic Carbon (POC^{sat}) products are evaluated by comparison with an in situ dataset from 11 research cruises (2008–2017) to the SO, across multiple seasons, which includes measurements of POC and chlorophyll-a ($C^{in\,situ}$) from both High Performance Liquid Chromatography (C^{HPLC}) and fluorometry (C^{Fluo}). Contrary to a number of previous studies, results highlighted good performance of the algorithm in the SO when comparing estimations with HPLC measurements. Using a time window of ±12 h and a mean satellite chlorophyll from a 5 × 5 pixel box centered on the in situ location, the median C^{Sat}:$C^{in\,situ}$ ratios were 0.89 (N = 46) and 0.49 (N = 73) for C^{HPLC} and C^{Fluo} respectively. Differences between C^{HPLC} and C^{Fluo} were associated with the presence of diatoms containing chlorophyll-c pigments, which induced an overestimation of chlorophyll-a when measured fluorometrically due to a potential overlap of the chlorophyll-a and chlorophyll-c emission spectra. An underestimation of ~0.13 mg m^{-3} was observed for the global POC algorithm. This error was likely due to an overestimate of in situ $POC^{in\,situ}$ measurements from the impact of dissolved organic carbon not accounted for in the blank correction. These results highlight the important implications of different in situ methodologies when validating ocean colour products.

Keywords: remote sensing; ocean color; algorithm; chlorophyll; HPLC; fluorometry; particulate organic carbon; southern ocean

1. Introduction

The oceans play a substantial role in mediating global climate by sequestering 25–30% of anthropogenic CO_2 from the atmosphere, with the Southern Ocean (SO) alone accounting for $\sim40\%$ of this total [1–5]. Biological production and carbon export to the deep ocean, "the biological pump" is considered a major contributor to the SO CO_2 sink while also regulating the supply of nutrients ($\sim75\%$) to thermocline waters north of 30°S, which in turn drives low latitude productivity [6]. Despite the ecological importance of this area, in situ data collection is often constrained through difficulties in accessing key areas by ship, primarily due to distance, weather and sea ice. This inability to resolve inter-annual variability and seasonal and intra-seasonal dynamics, limits our understanding of this complex system. Satellite remote sensing is one of the most effective tools available to address these spatial and temporal gaps in our knowledge. They have the added advantage of being routine, synoptic and available over decadal time scales and are in many cases, the only systematic observations available for chronically under-sampled marine systems such as the polar oceans. However, current ocean colour algorithms applied to SO data sets have been shown to perform badly, due proposedly to their typical parameterisation with low-latitude bio-optical data sets (in the absence of sufficient regional data) whose Inherent Optical Properties (IOPs) differ from those of the SO (e.g., [7]). Given the growing importance of the application of these tools in the trajectory of SO ecosystem understanding, it is necessary that we rigorously assess the quality of satellite-derived ocean colour data products such as chlorophyll-a (C^{Sat}) and particulate organic carbon (POC^{Sat}), which are currently used to study trends and trajectories of fundamental parameters such as phytoplankton biomass, primary production and the carbon cycle [8,9].

To the best of our knowledge, 20 previous studies have evaluated the ocean colour chlorophyll-a product in the SO, with most of these highlighting the poor performance of the algorithm with an average factor of 0.5 underestimate of retrieved chlorophyll-a relative to measured chlorophyll-a [7,10–26]. Holm-Hanssen et al. [18] highlighted variations in the bias associated with the range of chlorophyll-a concentration and found that in the Scotia Sea, where chlorophyll-a concentrations were <1 mg m^{-3}, ratios of satellite retrieved to in situ chlorophyll-a concentration ($C^{in\,situ}$) were as high as 0.89 ± 0.45 (N = 50), whereas for chlorophyll-a concentrations between 1–4 mg m^{-3}, lower ratios of 0.48 ± 0.18 (N = 30) were observed. Similarly, Clementson et al. [16] found a bias that depends on the concentration of chlorophyll-a with a tendency for the SeaWiFS OC4V4 algorithm to underestimate chlorophyll-a at high concentrations, while an overestimation was obtained when pigment concentrations were low (<0.15 mg m^{-3}). Dierssen & Smith, ref. [14] hypothesised that the bias in retrieving accurate chlorophyll-a concentrations for the SO was attributed to lower concentrations of bacteria, which induce a lower backscattering coefficient for similar concentrations of chlorophyll-a. Another study by Reynolds et al. [15] similarly attributed poor algorithm performance to variability in the spectral backscattering ratio, which was deduced from differentiation in algorithm performance between the Ross Sea and the Antarctic Polar Front Zone (APFZ). The majority of studies however, refer to absorption as the primary IOP responsible for the failure of the application of global chlorophyll-a algorithms to the SO. These include the following: variations in the absorption coefficient of specific detritus and colored dissolved organic matter (CDOM) absorption relative to phytoplankton absorption, large pigment packaging of the dominant phytoplankton species due to low light adaptation, species composition, low chlorophyll-a specific absorption and species-specific absorption in the 440-570 nm range (e.g., [10,12,18–20,24,26,27]). To expand on one example, Dierssen, ref. [27] used the radiative transfer model Hydrolight to analyse the impact of variations in CDOM concentration, which typically vary in the blue part of the spectrum, as a function of geographical region, age and exposure to solar degradation processes. Using a range of CDOM concentrations at 440 nm from 0 to 0.03 m^{-1} as model input, they were able to retrieve chlorophyll-a concentrations that ranged over a factor of 3 for oligotrophic waters (chlorophyll-a < 0.2 mg m^{-3}). Worth noting is that most of the studies mentioned above used $C^{in\,situ}$ measured from fluorometric methods rather than High Performance Liquid Chromatography (HPLC). In addition, many of the studies did not adhere to all the matchup procedures recommended by Bailey & Werdell, ref. [28] for their comparison between in situ and satellite data records. For instance, Johnson et al. [26] used daily, 8 days or monthly averaged NASA Aqua MODIS Level 3, 9 km data products.

On the contrary, two studies by Haëntjens et al. [29] and Marrari et al. [30] showed a good performance of global satellite algorithms in the SO over a large range of chlorophyll-a concentrations. Haëntjens et al. [29] compared chlorophyll−a and POC products of the Visible Infrared Imaging Radiometer Suite (VIIRS) on board Suomi-NPP spacecraft and the Aqua MODIS sensors with float measurements from the Southern Ocean Carbon and Climate Observations and Modeling (SOCCOM) programme. Here, chlorophyll-a and POC derived from fluorescence and backscattering sensors, were rigorously calibrated with samples collected at the time of the floats' deployment. The authors conclude that the global algorithms perform well in the SO with an average agreement to within 9% and 12%, for VIIRS and Aqua MODIS respectively. Marrari et al. [30] on the other hand compared the SeaWIFS daily chlorophyll-a (resolution: ~1 km^2/pixel) product (SeaDAS4.8, OC4v4 algorithm) with HPLC and fluorometric chlorophyll-a measurements collected during January–February of 1998–2002 around the Antarctic Peninsula. Their C^{Sat}/C^{HPLC} ratio of 1.12 ± 0.91 (N = 96) showed good agreement, while the C^{Sat}/C^{Fluo} ratio of 0.55 ± 0.63 (N = 307) suggested a ~50% underestimation, which the authors attribute to low concentrations of chlorophyll-b and high concentrations of chlorophyll-c which impact the fluorescence measurements [27,30,31]. A more recent study by Pereira & Garcia, ref. [7] however, also in the northern Antarctic Peninsula observed an underestimation of the Aqua MODIS algorithm, regardless of the method used to estimate chlorophyll-a concentrations.

At this point, an evaluation of the performance of global ocean colour algorithms applied to the SO is inconclusive, highlighting the requirement for additional in situ data with large regional and seasonal coverage to support a rigorous assessment that adheres to strict matchup criteria [28]. The aim of this study is to evaluate Aqua MODIS Level 2 (nominal 1 km × 1 km resolution) chlorophyll-a and POC products using an in situ database (>1000 data points) of POC and chlorophyll-a measured from both HPLC and fluorometric methods, from 11 research cruises (2008–2017) to the SO, spanning multiple regions and seasons.

2. Material and Methods

2.1. In Situ Dataset

A database of fluorometric (N = 1527), HPLC (N = 1010) and POC (N = 1028) measurements were assimilated from 11 research cruises to the SO conducted between 2008 and 2017 (Table 1).

Table 1. Southern Ocean research cruises during which POC, HPLC and fluorometric chlorophyll-a samples were collected.

Name	Date	N$_{POC}$ [a]	N$_{HPLC}$ [a]	N$_{Fluo}$ [a]
SANAE 48	December 2008–March 2009	0	110	198
SANAE 49	December 2009–February 2010	0	8	254
Winter 12	July 2012–August 2012	73	88	90
Expedition	January 2013–February 2013	0	117	117
SOSCEx 1	February 2013–March 2013	97	95	129
SANAE 53	November 2013–February 2014	152	147	152
Winter 15	July 2015–August 2015	76	80	83
SANAE 55	December 2015–February 2015	147	172	175
Winter 16	July 2016–July 2016	63	0	0
SANAE 56	December 2016–February 2017	100	0	0
ACE	December 2016–March 2017	320	193	329
All	December 2008–March 2017	1028	1010	1527

[a] N: the number of observations.

Sample Collection and Storage

Surface seawater samples were collected from niskin bottles attached to a conductivity–temperature–depth (CTD) rosette system and an underway intake system (nominal depth ~7 m). Fluorometric (0.25–0.5 L) and HPLC (0.5–2.0 L) samples were filtered onto GF/F filters (Whatman, diameter 25 mm, nominal pore size 0.7 μm). Fluorometric samples were measured onboard the ship whilst HPLC samples were flash frozen in liquid nitrogen and stored at −80 °C until analysis on land in Villefranche, France.

For fluorometric chlorophyll-a analysis, chlorophyll-a was extracted by placing the filters into 8 ml 90% acetone for 24 h in the dark at −20 °C. Fluorometric chlorophyll-a was measured with a fluorometer (Turner Designs 10AUTM (SANAE 48) and Trilogy® (all other cruises) Laboratory Fluorometer) following Welschmeyer, [32]. Fluorescence chlorophyll was converted to chlorophyll-a using a standard chlorophyll-a dilution calibration. Samples for HPLC were extracted at −20 °C in 3 mL of 100% Methanol, disrupted by sonication and then clarified by filtration and finally analysed on an Agilent Technologies HPLC 1200 following the methods of Ras et al. [33]. Note that total chlorophyll-c corresponds to the sum of chlorophyll-c1, c2 and c3, while total chlorophyll (C^{HPLC}) corresponds to the sum of divinyl chlorophyll-a, monovinyl chlorophyll-a, chlorophyllide a, chlorophyll-a allomers, and chlorophyll-a epimers.

POC samples (0.5–2.0 L) were filtered through GF/F filters pre-combusted at 450 °C. POC samples were dried in an oven at 40 °C for 24 h before being acid fumed with concentrated HCl for 24 h to remove inorganic carbon. Filters were pelleted into 5 × 8 mm tin capsules and analysed with a Flash EA 1112 series elemental analyser (Thermo Finnigan). Dry blanks (pre combusted GF/F filters) were interspersed every 6 to 20 samples (typically every 12) and subtracted from each sample to obtain total $POC^{in\,situ}$ (as per JGOFS POC protocols; [34]). On the most recent ACE cruise (Table 1), two additional filtered seawater (FSW) blanks were collected and analysed for POC content by filtering 2 L of seawater through a GF/F and then again through a pre-combusted GF/F, which was processed as above as an FSW-blank.

2.2. Satellite Data

Aqua MODIS Level 2 R2018.0 (nominal 1 km × 1 km resolution) data were acquired from the NASA Goddard Space Flight Center website [35]. The standard MODIS chlorophyll-a and POC products [36] (C^{Sat} and POC^{Sat} respectively) were evaluated by comparing in situ measurements with coincident satellite retrievals. A full description of the algorithm is provided in Appendix A. To ensure the quality of the matchup, an adjusted version of the Bailey & Werdell, ref. [28] procedure was applied (Figure 1). The time window to determine the closest coincidence between in situ and valid satellite data was 12 h instead of the 3 h initially advised by Bailey & Werdell, ref. [28], note however that the statistics for different time windows is presented in Table 2. Time averages were not performed, with only the closest matchup in time being recorded. For instance, for a time window of ±12 h, one matchup can be extracted at +3 h and another one at −6 h with a mean absolute time difference of 4.5 h. When a matchup was obtained, two pixel extractions were performed: a box of 5 by 5 pixels around the in situ location (named box 1) and, as recently proposed by Haëntjens et al. [29], all pixels in a radius of 8 km around the station (named box 2). Any pixel containing one or more of the following MODIS L2 quality flags ([37]) were excluded: suspect or failure in the atmospheric correction, land, sunglint, high radiance (near saturation), high sensor view zenith angle, shallow water pixels, stray light, cloud or ice contamination, coccolithophores, turbid water, high solar zenith angle, low water-leaving radiance, and moderate sun glint contamination. In addition, we masked all pixels with a solar zenith angle higher than 75° [28]. As the boxes contained several pixels, an analysis was performed to ensure the confidence of the mean value for each station. Only boxes containing >50% valid (unflagged) pixels were used for further analysis. In addition, only points within the mean ± 1.5 ×std (standard deviation) were considered. Finally, a test of homogeneity was performed by checking the coefficient of variation (CV), with the matchup being excluded if the CV was higher than 0.15 [28].

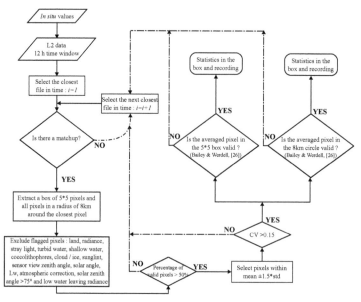

Figure 1. Flowchart of the procedures applied to extract the satellite values.

For each satellite data extraction, an iterative process was applied within the 12 h time window until a valid matchup was found. The matchup locations can be seen in Figure 2, with the matchup locations of other SO studies included for comparison. Note however that the study areas of Sullivan et al. [11], Kahru & Mitchell, ref. [23] (NOMAD and Scripps database south of 55°S latitude) and Haëntjens et al. [29] do not appear since data from these studies were distributed all around Antarctica. In addition, note that despite the apparent large distribution of data used in Johnson et al. [26] more than 95% was collected from between 140.8°E and 150.8°E. Finally, please note that an analysis of the uncertainty associated with the satellite-derived Rrs (see various sources in [38]) used to derive chlorophyll-a and POC is considered outside the scope of this study.

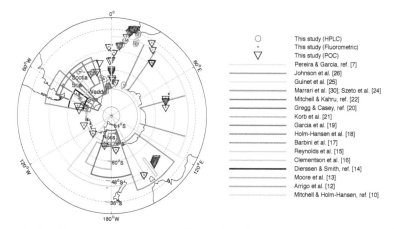

Figure 2. Localizations of matchups from this study and several research areas of previous matchup studies evaluating chlorophyll-a satellite products. The blue circles, red dots and black triangles correspond to matchups from this study for chlorophyll-a measured by HPLC, fluorometry, and POC respectively.

2.3. Statistical Metrics

The MARD (mean absolute relative difference) and the MRD (mean relative difference) were used as statistic metrics for quantifying the uncertainty associated with satellite estimations:

$$MARD(\%) = \frac{100}{N} \times \sum_{i=1}^{N} \frac{|E_i - M_i|}{M_i} \tag{1}$$

$$MRD(\%) = \frac{100}{N} \times \sum_{i=1}^{N} \frac{E_i - M_i}{M_i} \tag{2}$$

where N is the number of points, while E and M represent the retrieved and measured values respectively. Since the arithmetic mean is sensitive to potential extreme values, the median relative difference (MedRD) and the median absolute relative difference (MedRAD) have been calculated as well. As the distribution of POC and chlorophyll-a follow a lognormal curve [39–41], statistics are also shown for the logarithmically transformed (base 10) data. The formulations of Seegers et al. [41] have been followed:

$$MAD_{Log} = 10^{\frac{\sum_{i=1}^{N} |log_{10}E_i - log_{10}M_i|}{N}} \tag{3}$$

$$Bias_{Log} = 10^{\frac{\sum_{i=1}^{N} log_{10}E_i - log_{10}M_i}{N}} \tag{4}$$

Please note that Equations (5)–(6) are dimensionless. Differences between chlorophyll-a retrieved from HPLC and from fluorometry were analysed using the MRD, the root mean square difference (RMSD) and the coefficient of determination (R^2) defined as per Ricker et al. [42]:

$$R^2 = \left[\frac{\sum_{i=1}^{N}(E_i - \overline{E_i})(M_i - \overline{M_i})}{\sum_{i=1}^{N}(E_i - \overline{E_i})^2(M_i - \overline{M_i})^2} \right]^2 \tag{5}$$

$$RMSD = \sqrt{\frac{\sum_{i=1}^{N}(E_i - M_i)^2}{N}} \tag{6}$$

where E and M represent the chlorophyll-a measured from fluorometry and HPLC respectively.

3. Results

By applying all steps in the matchup procedure, a time window of ±12 h and using box 1 (mean of a 5×5 pixel box centered on the in situ location), a total of 73, 46 and 46 matchups were obtained for C^{Fluo}, C^{HPLC} and POC respectively. The concentrations of POC and chlorophyll-a (measured from HPLC) were in a typical range for the SO: between 36 and 257 mg m^{-3} with a median of 89 mg m^{-3} for POC and between 0.04 and 3.7 mg m^{-3} with a median of 0.33 mg m^{-3} for C^{HPLC} [12,29,43]. Concentrations in chlorophyll-a measured from fluorometry (C^{Fluo}) were slightly higher, ranging from 0.12 to 4.5 mg m^{-3} with a median of 0.45 mg m^{-3}. When comparing satellite retrieved values with in situ values, data points are scattered around the 1:1 line for C^{HPLC} (with a median ratio of $\sim$0.9), whereas underestimations are observed for the C^{Fluo} (median ratio of $\sim$0.5) and the POC$^{in\ situ}$ (median ratio of $\sim$0.7) (Figures 3 and 4). These observations are verified by the probability density distributions of $log_{10}(C^{in\ situ}:C^{Sat})$ from C^{HPLC}, which are normally distributed around -0.05 compared to -0.31 for C^{Fluo} and -0.13 for POC$^{in\ situ}$. Worth noting is that $C^{in\ situ}:C^{HPLC}$ and $C^{in\ situ}:C^{Fluo}$ ratios are similar to those obtained by Marrari et al. [30] in the Antarctic Peninsula region.

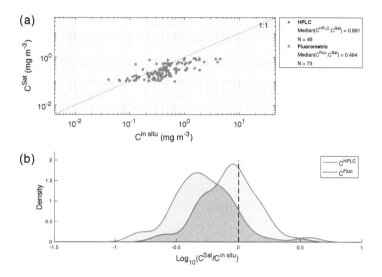

Figure 3. (**a**) Comparison between measured ($C^{in\,situ}$) and retrieved (C^{Sat}) chlorophyll-a for box 1 with a time window of 12 h. Blue and orange dots indicate samples measured from HPLC (C^{HPLC}) and fluorometry (C^{Fluo}) respectively, while the dashed line shows the 1:1 relationship. (**b**) Probability density function of the logarithm base 10 of the ratio between C^{Sat} and $C^{in\,situ}$ from fluorometry (blue line) and HPLC (orange line); with the statistics of the comparisons listed in Table 2.

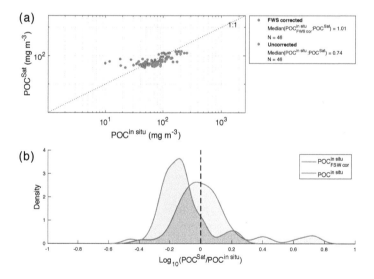

Figure 4. (**a**) Comparison between measured POC ($POC^{in\,situ}$) and retrieved (POC^{Sat}) for box 1 with a time window of 12 h. Blue and orange dots indicate dry-blank corrected samples ($POC^{in\,situ}$) and filtered seawater corrected blanks ($POC^{in\,situ}_{FSW-cor}$) respectively; the dashed line shows the 1:1 relationship. (**b**) Probability density function of the logarithm base 10 of the ratio between POC^{Sat} and $POC^{in\,situ}$ (orange line) and $POC^{in\,situ}_{FSW-cor}$ (blue line); the statistics of the comparisons are listed in Table 2.

Table 2. Statistics of the comparison between satellite and in situ values of chlorophyll-a (mg m^{-3}) and POC (mg m^{-3}) with chlorophyll-a measured from fluorometry (C^{Fluo}) and HPLC (C^{HPLC}) and POC dry-blank-corrected (POC$^{in\ situ}$), for different time windows and two spatial binning methods: a box of 5 by 5 pixel around the in situ location (box 1), and, all pixels in a radius of 8 km around the station (box 2).

Box	Parameter	C^{Fluo} vs. C^{Sat}				C^{HPLC} vs. C^{Sat}				POC$^{in\ situ}$ vs. POC^{Sat}			
	Time windows (hrs)	3	6	9	12	3	6	9	12	3	6	9	12
	N	27	41	59	73	20	29	39	46	16	25	38	46
	Ratio [a]	0.53	0.47	0.49	0.49	0.78	0.86	0.91	0.89	0.65	0.69	0.74	0.74
	MRD (%)	−33	−37	−38	−35	−8.5	−3.5	−2.2	−5.8	−32	−24	−23	−22
	MedRD (%)	−46.5	−52.5	−51.1	−50.6	−21.8	−14.4	−9.1	−10.9	−35	−31	−26	−26
Box 1	MARD (%)	58	57	53	54	43.5	40.8	37	36.2	32	32	32	32
	MedRAD (%)	52.5	53.1	52.5	52.5	33.9	34.1	33.7	31.2	35	35	31	32
	Bias_log	0.52	0.52	0.52	0.53	0.78	0.84	0.87	0.84	0.67	0.72	0.73	0.74
	MAD_log	2.16	2.17	2.1	2.1	1.63	1.56	1.49	1.5	1.51	1.48	1.47	1.46
	Median [in situ]	0.47	0.45	0.47	0.45	0.36	0.31	0.29	0.33	118	108	81	89
	Mean absolute time difference	1.8 ± 0.9	2.9 ± 1.8	4.3 ± 2.6	5.6 ± 3.5	1.9 ± 0.9	2.8 ± 1.7	3.9 ± 2.5	4.9 ± 3.4	1.8 ± 1.1	2.9 ± 1.8	4.6 ± 2.8	5.6 ± 3.5
	Time windows (hrs)	3	6	9	12	3	6	9	12	3	6	9	12
	N	36	54	74	91	27	40	52	63	25	38	55	66
	Ratio [a]	0.48	0.47	0.48	0.48	0.83	0.82	0.87	0.85	0.68	0.72	0.73	0.74
	MRD (%)	−38	−42	−42	−39	−9.8	−10.6	−8.4	−10	−31	−23	−22	−19
	MedRD (%)	−52.2	−52.6	−52	−52	−17.5	−18.1	−13.2	−14.5	−32	−28	−27	−26
Box 2	MARD (%)	57	57	54	54	35.4	37.5	35.1	34.3	31	33	32	31
	MedRAD (%)	52.6	55.7	52.6	52.5	26.4	30.5	30	29.6	32	32	29	30
	Bias_log	0.49	0.47	0.48	0.5	0.8	0.79	0.82	0.81	0.68	0.73	0.74	0.76
	MAD_log	2.27	2.32	2.22	2.18	1.49	1.54	1.49	1.48	1.47	1.47	1.45	1.43
	Median [in situ]	0.44	0.44	0.46	0.44	0.31	0.34	0.32	0.33	115	109	90	89
	Mean absolute time difference	1.6 ± 1	2.7 ± 1.8	3.9 ± 2.6	5.2 ± 3.6	1.7 ± 1	2.6 ± 1.7	3.6 ± 2.5	4.9 ± 3.5	1.6 ± 1.1	2.6 ± 1.8	4.2 ± 2.8	5.3 ± 3.6

[a] Median of the $C^{Sat}/C^{in\ situ}$ ratio.

Similar results are obtained when statistics are performed for different averaging products in box 1 and 2 (see Section 2.2) and different time windows (±3 h, ±6 h, ±9 h and ±12 h) (Table 2). For example, using box 1 and a time window of ±12 h, the MRD and MARD between retrieved and in situ data were −6% and 36% for C^{HPLC}, whereas higher differences were observed for C^{Fluo} (MRD = −35%, MARD = 54%) and POC (MRD = −22%, MARD = 32%). Marrari et al. [30] similarly noted a larger bias using fluorometric measurements when comparing SeaWIFS daily chlorophyll-a data (resolution: ∼1 km^2/pixel from SeaDAS4.8, OC4v4 algorithm) with in situ chlorophyll-a from HPLC and fluorometric measurements in the Antarctic Peninsula, where the authors obtained a MRD of 12% for C^{HPLC} and −45.2% for C^{Fluo}. For the different time windows tested, the errors remained relatively constant for all different algorithms. Using a radius of 8 km centered on the in situ location (box 2) enabled an increase in the number of matchups without necessarily impacting the accuracy or bias. To elaborate, the number of matchups between box 1 and box 2 increased by an average factor of 1.4, while the medians of the ratio of the MARD and the MRD between box 2 and box 1 were 1 and 1.1 respectively. Surprisingly, in some instances there was a tendency for the accuracy of the algorithm to be slightly better using box 2. For instance, the MARD for C^{HPLC} was 44% for box 1 and 35% for box 2 when using a time window of ±3 h. Note however that the difference between in situ concentrations of C^{HPLC} between box 1 and box 2 were the highest as opposed to inter box comparisons of C^{Fluo} and POC$^{in\ situ}$. To elaborate, the MARD between in situ concentrations for C^{HPLC} in box 1 and box 2 was 8.4% while these diffrences were low for C^{Fluo} (MARD = 3.2%) and POC$^{in\ situ}$ (MARD = 3.6%). Thus, differences in the results observed between box 1 and 2 for C^{HPLC} were likely due to variations in in situ concentrations. Indeed, algorithm performance has been known to differ according to variations in the range of concentration. To test this, an evaluation of the errors according to different concentration ranges was performed utilising a time window of ±12 h and box 1 (Table 3). For chlorophyll-a, the limits were chosen to separate the Color Index (<0.2 mg m^{-3}; [44]) and OC3M algorithms (see Appendix A). For POC, the concentration limits correspond to the first and third quartile (Q1 = 67 mg m^{-3} and Q3 = 124 mg m^{-3}). Results confirm that errors in the chlorophyll-a product are more sensitive to the range of concentration than the POC product. The coefficient of variation of the MARD was 12% for C^{HPLC} and 32% for C^{Fluo}, whereas it was of 3% for POC. However, we note that no significant variation in the statistics was observed when using only the OC3M algorithm instead of the blended version. For instance, the MARD increased by ~1% using exclusively the OCx model ($C^{HPLC} < 0.2$; N = 13).

Table 3. Statistics of the comparison between satellite and in situ values for different ranges in concentration. A time window of 12 h and box 1 was used for the extraction procedure.

Parameter	Fluo			HPLC			POC$^{in\ situ}$			
Range	0.2<	≥0.2	All	0.2<	≥0.2	All	[0,67[	[67,124[	≥124	All
N	5	68	73	13	33	46	10	24	12	46
Ratio [a]	1.04	0.48	0.49	1.34	0.8	0.9	0.93	0.67	0.65	0.74
MRD (%)	68.2	−42.6	−35	39.1	−23.4	−5.6	13	−30	−33	−22
MedRD (%)	4.1	−52.5	−50.6	33.6	−23.3	−10.9	−6.6	−33.4	−35.4	−26
MARD (%)	87.6	51	53.5	42.6	33.7	36.2	33	31	33	32
MedRAD (%)	36.7	52.6	52.5	33.6	28.7	31.2	25.3	33.4	35.4	31.9
Bias_log	1.34	0.49	0.53	1.32	0.7	0.84	1.06	0.68	0.65	0.74
MAD_log	1.69	2.13	2.1	1.37	1.55	1.5	2.14	1.35	1.47	1.54
Median [in situ]	0.15	0.47	0.45	0.11	0.44	0.33	51	85	159	89

[a] Median of the $C^{Sat}/C^{in\ situ}$ ratio.

4. Discussion

4.1. Satellite Versus In Situ Chlorophyll-a Comparison

There is a recognised need in the user community for ocean colour products to be regionally optimised and their uncertainties well characterised [45]. Despite previous conclusions of a poor performance of the ocean colour chlorophyll-a product (with a typical underestimate of 50%) from the majority of validation studies performed in the SO, results from this study show good agreement between chlorophyll-a retrievals from MODIS and in situ HPLC derived chlorophyll-a. These results are in agreement with only two other studies in the literature [29,30]. The more recent study from Haëntjens et al. [29] focused on in situ data derived from floats, which themselves have significant uncertainties in estimating chlorophyll-a, primarily due to variability in the chlorophyll-a to fluorescence yield that changes during the floats life time in response to adjustments in photophysiology, nutrients, temperature and species composition [46–48]. They however tested the float bias using an independent data set of 97 matchups of HPLC derived chlorophyll-a (from NASA's SeaBASS database) with MODIS OCI, which showed similar results to ours, supporting their conclusion that the default algorithm to estimate chlorophyll-a from NASA performs well in the SO, and that a regional specific algorithm is not required. Our results from an extensive in situ database of HPLC derived chlorophyll (using consistent methods and analysis) covering a broad regional and seasonal range and stricter matchup criteria, supports this conclusion. In addition, our dataset of co-located fluorescence and HPLC derived chlorophyll allows us to go one step further and interrogate possible reasons for the average factor of 0.5 underestimate typical of previous validation studies.

4.2. Comparison of HPLC and the Fluorometric Chlorophyll-a Methods

HPLC and fluorometry are two distinct methods currently used to measure the concentration of chlorophyll-a in marine environments. The HPLC method separates phytoplankton pigments in order of polarity upon passage through a column [33], with the most polar pigments removed earlier than the less polar pigments [34,49]. The fluorometric method uses the capacity of chlorophyll-a pigments to fluoresce in the red part of the spectrum when they are excited by blue light. Briefly, fluorescence of in vivo chlorophyll-a is measured by irradiating a water sample in the blue-green region of the spectrum (~440 nm), following which, the amount of energy fluoresced in the red region (~685 nm) as a result of light interactions with the chlorophyll molecules, is measured by a detector to produce raw fluorescence units (RFU). The RFU are then converted into chlorophyll-a concentration by means of a calibration curve pre-established with a range of chlorophyll-a standards, where the curve represents an ideal case of fluorescence measurements being linearly related to chlorophyll-a concentrations. Two conditions have to be encountered to satisfy this case: (i) the excitation energy (μmol photon $m^{-2} s^{-1} nm^{-1}$) is saturating and constant among measurements and (ii) the spectral chlorophyll-a specific absorption coefficient (m^2 mg Chl^{-1}) and fluorescence quantum yield product (μmol photons fluoresced μmol photons absorbed^{-1}) have to be linearly related to fluorescence. It is however well known that the second postulate is not always satisfied [46,48,50]. Firstly, because the chlorophyll-a specific absorption coefficient depends on cell size, pigment concentration, the package effect and pigment composition [51,52]; and secondly, because fluorescence quantum yield changes as a function of species, nutritional status, ambient light and light history [46,48,53]. In addition, the integrity of fluorescence measurements may suffer from interference from the presence of significant amounts of chlorophyll-b, chlorophyll-c and degradation products (i.e., phaeopigments phaeophytin-a and phaeophorbide-a), which fluoresce in a similar spectral region as chlorophyll-a; this overlap may result in overestimates of the chlorophyll-a concentration (as in our case when using the non-acidification technique) and/or an underestimation (if the acidification technique is used) [30–32,48,49,54–57]. To elaborate, The acidification technique follows the method of Holm-Hansen et al. [58], where the impacts on fluorescence by phaeopigments are quantified by acidifying the sample, which converts all of the chlorophyll-a to phaeopigments. The difference between the two fluorescnece measurements (before and after acidification) thus reflects the total amount of chlorophyll-a in the sample. However, the acidification process results in an underestimate of chlorophyll-a in the presence of chlorophyll-b as the acidification step converts all chllorophyll-b to pheophytin b which has an overlapping emission spectra with pheophytin a [32]. To improve the accuracy of fluorescence measurements the Welschmeyer, [32] non-acidification method (used here) was optimised to avoid the overlapping phenomena by implementing a narrow band width optical approach to measured fluorescence that provides maximum sensitivity to chlorophyll-a while maintaining desensitized responses to chlorophyll-c, chlorophyll-b and phaeopigments. However, biases are still present making the HPLC method the more reliable method for quantifying chlorophyll-a concentration [32,48,55].

4.3. C^{HPLC} vs. C^{Fluo}

Coincident HPLC and fluorometric measurements were collected from different locations and time periods in the SO (Table 1). Results show a typical overestimation of C^{Fluo} when compared to C^{HPLC}, with median C^{Fluo}:C^{HPLC} ratios that range from 1.1 (SANAE 48) to 2.8 (ACE) (all data median = 1.5) (Table 4; Figure 5). Linear regression slopes between C^{HPLC} and C^{Fluo} ranged from 0.85 to 2.21 while the MRD ranged from 8% to 284%. Such discrepancies between methods have been reported previously in the literature [7,30–32,48,49,54–57], which could be explained by concentrations of chlorophyll-b or chlorophyll-c that vary regionally with different dominant species composition [30,54–57,59,60]. Worth noting is that ACE is the only research cruise that collected samples from the continental margin region around Antarctica (excluding ACE data puts the "all cruise" median C^{Fluo} / C^{HPLC} at 1.37 instead of 1.5). In an attempt to determine the drivers of this discrepancy, the data have been analysed according to latitude. Results indicate that the MRD between C^{HPLC} and C^{Fluo} is linked to progression in latitude, total biomass (estimated from C^{HPLC}) and the proportion of chlorophyll-c and fucoxanthin (Table 5). Such pigments are found in the taxa Bacillariophytes (diatoms) and Haptophytes (e.g., *Phaeocystis antarctica*), which typically dominate the SO [61–63]. The role of chlorophyll-c and fucoxanthin pigments are crucial in marine phytoplankton environments as their absorption covers the blue-green part of the light spectrum [64], which is prevalent in marine environments. However, while chlorophyll-c and fucoxanthin are found in both taxa, fucoxanthin is considered a marker pigment for the diatoms, with much lower concentrations being typical of the SO strain of Phaeocystis [62,65]. Results therefore suggest that diatoms are the dominant source of chlorophyll-c and that their presence is primarily responsible for the observed range in C^{Fluo}:C^{HPLC} ratios presented in Table 5. The dominance of specific species in certain regions is mainly driven by sea surface temperature, sea ice presence, grazing pressure and water column structure which influences both nutrient and light availability ([63,66] and references therein). For diatom species, favorable growth conditions are associated with high nutrient concentrations (nitrate, silicate and iron) [63,67,68]. In the SO, the Polar Front with a mean position in the Atlantic SO of 54°S [69] forms an important transitional boundary, south of which high silicate concentrations prevail and diatoms typically dominate [68,70]. In addition, diatoms typically dominate in regions with shallow mixed layers that are characteristic of stratified waters (e.g., from ice melt) that provide a high light environment by allowing dense cells to remain suspended in the illuminated surface waters ([71] and references therein). It is thus recommended that in regions where diatoms dominate and a high proportion of chlorophyll-c is anticipated (e.g., the SO), that HPLC be the method implemented for chlorophyll-a analysis.

Table 4. Statistics of the relation between chlorophyll-a measured from HPLC and fluorometry for each expedition. The number of observations (N), the slope (a), the intercept (b) and their standard deviations (Δa and Δb) and coefficients of determination (R^2) of the linear regression are indicated. RMSD is the root mean square difference and MRD is the mean relative difference (see Section 2.3 for formulations).

Expedition	N	a ± Δa	b ± Δb	R^2	RMSD	$\frac{C^{Fluo}}{C^{HPLC}}$	MRD
SANAE 48	107	1.1 ± 0.06	0.09 ± 0.05	0.8	0.35	1.1	26
SANAE 49	8	2.02 ± 0	−0.12 ± 0	0.9	0.69	1.8	74
Winter 12	88	0.85 ± 0.16	0.12 ± 0.04	0.2	0.16	1.3	39
Expedition	117	1.39 ± 0.07	0.02 ± 0.04	0.8	0.38	1.4	47
SOSCEx 1	75	1.07 ± 0.1	0.02 ± 0.03	0.6	0.12	1.2	8
SANAE 53	142	0.7 ± 0.08	0.16 ± 0.05	0.3	0.36	1.2	8
Winter 15	80	1.48 ± 0.13	0 ± 0.04	0.6	0.16	1.4	47
SANAE 55	172	1.99 ± 0.05	−0.04 ± 0.04	0.9	0.77	1.9	91
ACE	192	2.21 ± 0.08	0.23 ± 0.05	0.8	1.15	2.8	283
All	981	1.66 ± 0.04	0	0.7	0.65	1.5	90

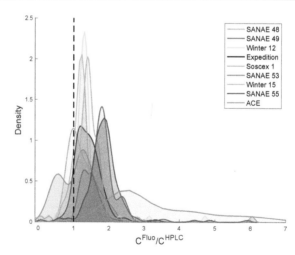

Figure 5. Probability density function of the ratio between chlorophyll-a measured from fluorometry (C^{Fluo}) and HPLC (C^{HPLC}). Each color corresponds to a specific expedition.

Table 5. Statistics of the relationship between chlorophyll-a measured from HPLC and fluorometry for different ranges of latitude, the proportion of chlorophyll-c and Fucoxanthin is also indicated. The number of observations (N), the slope (a), the intercept (b) and their standard deviations (Δa and Δb) and the coefficient of determination (R^2) of the linear regression are indicated. RMSD is the root mean square difference and MRD is the mean relative difference (see Section 2.3 for formulations).

Latitude Range	N	$a \pm \Delta a$	$b \pm \Delta b$	R^2	RMSD	$\frac{C^{Fluo}}{C_{HPLC}}$	MRD	$\frac{C_c}{C_{HPLC}}$ [a]	$\frac{Fuco}{C_{HPLC}}$	C^{HPLC} (mg m^{-3})
$[-80°, -70°]$	28	2.4 ± 0.32	0.06 ± 0.38	0.68	2.03	2.13	146	0.41	0.47	0.87
$[-70°, -60°]$	196	1.47 ± 0.08	0.12 ± 0.06	0.64	0.7	1.69	144	0.31	0.44	0.37
$[-60°, -50°]$	320	1.52 ± 0.06	0.07 ± 0.04	0.66	0.7	1.62	104	0.29	0.42	0.32
$[-50°, -40°]$	317	1.6 ± 0.06	-0.04 ± 0.03	0.67	0.34	1.38	54	0.25	0.11	0.3
$[-40°, -30°]$	120	1.15 ± 0.08	0.08 ± 0.03	0.6	0.2	1.38	45	0.17	0.07	0.27

[a] C_c: chlorophyll-c concentration.

4.4. Satellite Versus In Situ POC Comparison

An intercomparison and validation study of various ocean colour POC algorithms by Evers-King et al. [45] found that the Stramski et al. [72] algorithm (applied here), performed well and consistently across a broad range in POC concentration (2.7–8 097 mg m^{-3}) and across different water types, with the majority of pixels falling within an error range of 30%. The SO specific matchup performed here, across a lower range in POC$^{in\,situ}$ (36–257 mg m^{-3}) found a 30% underestimate in the Stramski et al. [72] algorithm, which was verified by the probability density distribution offset in POC$^{in\,situ}$:POC^{Sat} ratios of -0.13 (Figure 4). However, a number of studies have highlighted the issue of different methodologies for treating POC blanks as a possible source of bias, particularly at low POC concentrations. Cetinic et al. [73] (and references therein) show that the effect of dissolved organic carbon (DOC) adsorption onto filters (if not accounted for with an adequate blank correction) can result in an overestimate in POC of between 11 and 25 mg m^{-3}. On the most recent SO cruise (ACE, Table 1), DOC adsorption onto filters was tested by passing filtered seawater through an unused, pre combusted GF/F filter to obtain a filtered seawater blank of 25.7 mg m^{-3}. On ACE, the difference between the FSW-blank and the mean dry-filter-blank (i.e., the DOC contribution) was 18.92 mg m^{-3}. If this is assumed to be representative of DOC adsorption and subtracted from all SO cruises in Table 1 (in addition to the dry-filter-blank), the revised matchup with satellite POC results in a median POC^{Sat}:POC$^{in\,situ}$ ratio of 1.01 (Figure 4). These results suggest a robust performance of the Stramski et al. [72] POC algorithm in the SO and highlight the requirement for revised JGOFS POC protocols to include a blank exposed to filtered seawater [74].

5. Conclusions

The standard MODIS L2 chlorophyll-a and POC products were evaluated by comparing satellite retrievals with an in situ dataset encompassing a broad geographical and seasonal range. The database included POC concentrations (mg m^{-3}) and chlorophyll-a concentrations (mg m^{-3}) measured from both fluorometric and HPLC methods. Using strict matchup criteria of a time window of ± 12 h and the mean of a 5 × 5 pixel box centered on the in situ location. The median of the C^{Sat}:$C^{in\,situ}$ ratios were 0.89 for C^{HPLC} (N = 46) and 0.49 for C^{Fluo} (N = 73). The mean relative difference and mean relative absolute difference were -5.8% and 36.2% for C^{HPLC} whereas they were -35% and 54% for C^{Fluo}. Note that an increase in the spatio-temporal resolutions to a time window of ± 12 h and to a radius of 8 km around the in situ sample localization did not impact the results.

The consensus observed in our study between C^{HPLC} and C^{Sat} suggests that the MODIS global chlorophyll-a algorithm performs well in the SO, which agrees with a recent publication by Haëntjens et al. [29]. The comparatively poor performance of C^{Fluo} supports the more common factor of ~0.5 difference between retrieved and measured values of chlorophyll-a obtained by most previous validation studies in the SO, that were typically done with chlorophyll-a measured from fluorometry rather than HPLC. Similarly to Marrari et al. [30], our results suggest that the typical overestimation of

C^{Fluo} was due to the presence of chlorophyll-c pigments from diatom species. The median relative difference of -26% observed for the POC^{Sat} to $POC^{in\ situ}$ comparison has to be interpreted carefully as this underestimate is likely due to a methodological bias from an inadequate blank correction applied to in situ POC samples. If an assumption is made on the representativeness of the contribution of DOC adsorption onto POC filters (18.92 mg m^{-3}), then the performance of the satellite algorithm improves to a median of 1.01 (median relative difference and median relative absolute difference of 1% and 23% respectively).

These results highlight the importance of accurately calibrating fluorescence measurements and promotes the use of the HPLC method of chlorophyll-a analysis, particularly in regions where diatoms are known to dominate and despite the high costs involved. In addition, as noted by Boss et al. [53], the best calibration should be more often and as close as possible in time and space to the sampled area. Similarly, results highlight the need for community consensus for a standard protocol for POC analysis that includes an FSW-blank.

Author Contributions: Conceptualization, W.M., S.J.T. and S.B.; Data Curation, W.M., S.J.T. and T.J.R.-K.; Formal Analysis, W.M.; Funding Acquisition, S.J.T.; Investigation, W.M., S.J.T. and S.B.; Methodology, W.M., S.J.T., S.B., G.W. and M.E.S.; Project Administration, S.J.T.; Resources, S.J.T.; Software, W.M. and G.W.; Supervision, S.J.T. and S.B.; Validation, W.M. and T.J.R.-K.; Visualization, W.M.; Writing—Original Draft, W.M.and S.J.T.; Writing—Review & Editing, W.M., S.J.T., S.B., G.W., T.J.R.-K. and M.E.S.

Funding: This work was undertaken as part of the CSIR's Southern Ocean Carbon and Climate Observatory (SOCCO) (http://socco.org.za). This work was supported by CSIR's Parliamentary Grant (SNA2011112600001), the NRF South African National Antarctic Programme (SANAP) grants (SNA2007051100001, SNA2011120800004, SNA14071475720), the European Commission 7th Framework program through the GreenSeas Collaborative Project, FP7-ENV-2010 contract No. 265294 and the ACE scientific expedition carried out under the auspices of the Swiss Polar Institute, supported by funding from the ACE Foundation and Ferring Pharmaceuticals.

Acknowledgments: We would like to thank our funders and the captains and crews of the S.A. Agulhas, S.A. Agulhas II and RV Akademic Tryoshnikov for their professional support throughout the cruises. This work was undertaken as part of the CSIR's Southern Ocean Carbon and Climate Observatory (SOCCO) (http://socco.org.za). We thank the anonymous reviewers for their valuable comments and suggestions.

Conflicts of Interest: The authors declare no conflict of interest.

Abbreviations

The following abbreviations are used in this manuscript:

APFZ	Antarctic Polar Front Zone
C_c	Chlorophyll-c concentration
C^{Fluo}	Chlorophyll-a concentration measured by Fluorometry
C^{HPLC}	Chlorophyll-a concentration measured by High Performance Liquid Chromatography
C^{Sat}	Chlorophyll-a concentration retrieved by satellite
CDOM	Colored Dissolved Organic Matter
CTD	Conductivity-Temperature-Depth
CV	Coefficient of Variation
DOC	Dissolved Organic Carbon
EOS	Earth Observing System
FSW	filtered seawater
HPLC	High Performance Liquid Chromatography
IOP	Inherent Optical Properties
JGOFS	Joint Global Ocean Flux Studies
MLD	Mixed Layer Depth
MODIS	MODerate resolution Imaging Spectroradiometer
MRAD	Mean Relative Absolute Difference
MRD	Mean Relative Difference

NASA	National Aeronautics and Space Administration
POC	Particulate Organic Carbon
$POC^{in\,situ}$	in situ concentration in Particulate Organic Carbon
POC^{Sat}	Particulate Organic Carbon concentration retrieved by satellite
RFU	Raw Fluorescence Units
RMSD	Root Mean Square Difference
SeaWIFS	Sea-viewing Wide Field-of-view Sensor
SO	Southern Ocean
SOCCOM	Southern Ocean Carbon and Climate Observations and Modeling
VIIRS	Visible Infrared Imager Radiometer Suite

Appendix A

Algorithms Description

The current chlorophyll-a algorithm is a blend between the standard OCx band ratio algorithm (named OC3M for Aqua MODIS) and the Color Index (CI) of Hu et al. [44]:

$$Chla_{OC3M} = 10^{0.2424 - 2.7423 \times R + 1.8017 \times R^2 + 0.0015 \times R^3 - 1.2280 \times R^4} \tag{A1}$$

with:

$$R = log_{10}\left(\frac{R_{rs}(\lambda_{blue})}{R_{rs}(\lambda_{green})}\right) \tag{A2}$$

where $R_{rs}(\lambda_{green})$ is the remote sensing reflectance (R_{rs}) at 547 nm, and $R_{rs}(\lambda_{blue})$ is the greatest R_{rs} between 443 and 488 nm. The CI algorithm is defined as follows:

$$Chla_{CI} = 10^{-0.4909 + 191.6590 \times CI} \tag{A3}$$

with:

$$CI = R_{rs}(\lambda_{green}) - [R_{rs}(\lambda_{blue}) + \frac{\lambda_{green} - \lambda_{blue}}{\lambda_{red} - \lambda_{blue}} \times (R_{rs}(\lambda_{red}) - R_{rs}(\lambda_{blue}))] \tag{A4}$$

where λ_{blue}, λ_{green} and λ_{red} represent the closest wavelength to 443, 555 and 670 nm respectively. A weighted model (WM) is used to blend between the CI and OC3M algorithms at chlorophyll-a concentrations between 0.15 and 0.20 mg m^{-3}, while only the CI and OC3M are used at concentrations of <0.15 and >0.20 mg m^{-3} respectively. The weighted model is defined as follows:

$$Chla_{WM} = \left(Chla_{CI} \times \frac{0.2 - Chla_{CI}}{0.05}\right) + \left(Chla_{OC3M} \times \frac{Chla_{CI} - 0.15}{0.05}\right) \tag{A5}$$

The current POC algorithm was developed by Stramski et al. [72] and is defined as follows:

$$POC = 203.2 \times \left[\frac{R_{rs}(443)}{R_{rs}(547)}\right]^{-1.034} \tag{A6}$$

References

1. Raven, J.A.; Falkowski, P.G. Oceanic sinks for atmospheric CO2. *Plant Cell Environ.* **1999**, *22*, 741–755. [CrossRef]
2. Sabine, C.L.; Feely, R.A.; Gruber, N.; Key, R.M.; Lee, K.; Bullister, J.L.; Wanninkhof, R.; Wong, C.S.L.; Wallace, D.W.R.; Tilbrook, B.; et al. The oceanic sink for anthropogenic CO2. *Science* **2004**, *305*, 367–371. [CrossRef] [PubMed]
3. Khatiwala, S.; Primeau, F.; Hall, T. Reconstruction of the history of anthropogenic CO2 concentrations in the ocean. *Nature* **2009**, *346–349*, 554–577. [CrossRef] [PubMed]

4. Takahashi, T.; Sutherland, S.C.; Wanninkhof, R.; Sweeney, C.; Feely, R.A.; Chipman, D.W.; Hales, B.; Friederich, G.; Chavez, F.; Sabine, C. Climatological mean and decadal change in surface ocean pCO2, and net sea–air CO2 flux over the global oceans. *Deep Sea Res. Part 2 Top. Stud. Oceanogr.* **2009**, *56*, 554–577. [CrossRef]

5. Frölicher, T.L.; Sarmiento, J.L.; Paynter, D.J.; Dunne, J.P.; Krasting, J.P.; Winton, M. Dominance of the Southern Ocean in Anthropogenic Carbon and Heat Uptake in CMIP5 Models. *J. Clim.* **2015**, *28*, 862–886. [CrossRef]

6. Sigman, D.M.; Boyle, E.A. Glacial/interglacial variations in atmospheric carbon dioxide. *Nature* **2000**, *859–869*, 554–577. [CrossRef]

7. Pereira, E.S.; Garcia, C.A. Evaluation of satellite-derived MODIS chlorophyll algorithms in the northern Antarctic Peninsula. *Deep Sea Res. Part 2 Top. Stud. Oceanogr.* **2018**, *149*, 124–137. [CrossRef]

8. Muller-Karger, F.; Varela, R.; Thunell, R.; Astor, Y.; Zhang, H.; Luerssen, R.; Hu, C. Processes of coastal upwelling and carbon flux in the Cariaco Basin. *Deep Sea Res. Part 2 Top. Stud. Oceanogr.* **2004**, *51*, 927–943. [CrossRef]

9. Hu, C.; Muller-Karger, F.E.; Taylor, C.J.; Carder, K.L.; Kelble, C.; Johns, E.; Heil, C.A. Red tide detection and tracing using MODIS fluorescence data: A regional example in SW Florida coastal waters. *Remote Sens. Environ.* **2005**, *97*, 311–321. [CrossRef]

10. Mitchell, B.G.; Holm-Hansen, O. Bio-optical properties of Antarctic Peninsula waters: Differentiation from temperate ocean models. *Deep Sea Res. A* **1991**, *38*, 1009–1028. [CrossRef]

11. Sullivan, C.W.; Arrigo, K.R.; McClain, C.R.; Comiso, J.C.; Firestone, J. Distributions of phytoplankton blooms in the Southern Ocean. *Science* **1993**, *262*, 1832–1837. [CrossRef] [PubMed]

12. Arrigo, K.R.; Worthen, D.; Schnell, A.; Lizotte, M.P. Primary production in Southern Ocean waters. *J. Geophys. Res.* **2008**, *103*, 15587–15600. [CrossRef]

13. Moore, J.K.; Abbott, M.R.; Richman, J.G.; Smith, W.O.; Cowles, T.J.; Coale, K.H.; Gardner, W.D.; Barber, R.T. SeaWiFS satellite ocean color data from the Southern Ocean. *Geophys. Res. Lett.* **1999**, *26*, 1465–1468. [CrossRef]

14. Dierssen, H.M.; Smith, R.C. Bio-optical properties and remote sensing ocean color algorithms for Antarctic Peninsula waters. *J. Geophys. Res. Oceans* **2000**, *105*, 26301–26312. [CrossRef]

15. Reynolds, R.A.; Stramski, D.; Mitchell, B.G. A chlorophyll-dependent semianalytical reflectance model derived from field measurements of absorption and backscattering coefficients within the Southern Ocean. *J. Geophys. Res. Oceans* **2001**, *106*, 7125–7138. [CrossRef]

16. Clementson, L.A.; Parslow, J.S.; Turnbull, A.R.; McKenzie, D.C.; Rathbone, C.E. Optical properties of waters in the Australasian sector of the Southern Ocean. *J. Geophys. Res. Oceans* **2001**, *106*, 31611–31625. [CrossRef]

17. Barbini, R.; Colao, F.; Fantoni, R.; Fiorani, L.; Palucci, A.; Artamonov, E.S.; Galli, M. Remotely sensed primary production in the western Ross Sea: Results of in situ tuned models. *Remote Sens. Environ.* **2003**, *15*, 77–84. [CrossRef]

18. Holm-Hansen, O.; Kahru, M.; Hewes, C.D.; Kawaguchi, S.; Kameda, T.; Sushin, V.A.; Krasovski, I.; Priddle, J.; Korb, R.; Hewitt, R.P.; et al. Temporal and spatial distribution of chlorophyll-a in surface waters of the Scotia Sea as determined by both shipboard measurements and satellite data. *Deep Sea Res. Part 2 Top. Stud. Oceanogr.* **2004**, *51*, 1323–1331. [CrossRef]

19. Garcia, C.A.E.; Garcia, V.M.T.; McClain, C.R. Evaluation of SeaWiFS chlorophyll algorithms in the Southwestern Atlantic and Southern Oceans. *Remote Sens. Environ.* **2005**, *95*, 125–137. [CrossRef]

20. Gregg, W.W.; Casey, N.W. Global and regional evaluation of the SeaWiFS chlorophyll data set. *Remote Sens. Environ.* **2004**, *93*, 463–479. [CrossRef]

21. Korb, R.E.; Whitehouse, M.J.; Ward, P. SeaWiFS in the southern ocean: Spatial and temporal variability in phytoplankton biomass around South Georgia. *Deep Sea Res. Part 2 Top. Stud. Oceanogr.* **2004**, *51*, 99–116. [CrossRef]

22. Mitchell, B.G.; Kahru, M. Bio-optical algorithms for ADEOS-2 GLI. *J. Remote Sens. Soc. Jpn.* **2009**, *29*, 80–85. [CrossRef]

23. Kahru, M.; Mitchell, B.G. Blending of ocean colour algorithms applied to the Southern Ocean. *Remote Sens. Lett.* **2010**, *1*, 119–124. [CrossRef]

24. Szeto, M.; Werdell, P.J.; Moore, T.S.; Campbell, J.W. Are the world's oceans optically different?. *J. Geophys. Res. Oceans* **2011**, *116*, 1–14. [CrossRef]

25. Guinet, C.; Xing, X.; Walker, E.; Monestiez, P.; Marchand, S.; Picard, B.; Jaud, T.; Authier, M.; Cotté, C.; Dragon, A; et al. Calibration procedures and first data set of Southern Ocean chlorophyll a profiles collected by elephant seals equipped with a newly developed CTD-fluorescence tags. *Earth Syst. Sci. Data* **2013**, *5*, 15–29. [CrossRef]
26. Johnson, R.; Strutton, P.G.; Wright, S.W.; McMinn, A.; Meiners, K.M. Three improved satellite chlorophyll algorithms for the Southern Ocean. *J. Geophys. Res. Oceans* **2013**, *118*, 3694–3703. [CrossRef]
27. Dierssen, H.M. Perspectives on empirical approaches for ocean color remote sensing of chlorophyll in a changing climate. *Proc. Natl. Acad. Sci. USA* **2010**, *107*, 17073–17078. [CrossRef]
28. Bailey, S.W.; Werdell, P.J. A multi-sensor approach for the on-orbit validation of ocean color satellite data products. *Remote Sens. Environ.* **2006**, *102*, 12–23. [CrossRef]
29. Haëntjens, N.; Boss, E.; Talley, L.D. Revisiting Ocean Color algorithms for chlorophyll a and particulate organic carbon in the Southern Ocean using biogeochemical floats. *J. Geophys. Res. Oceans* **2017**, *122*, 6583–6593. [CrossRef]
30. Marrari, M.; Hu, C.; Daly, K. Validation of SeaWiFS chlorophyll a concentrations in the Southern Ocean: A revisit. *Remote Sens. Environ.* **2006**, *105*, 367–375. [CrossRef]
31. Gibbs, C.F. Chlorophyll b interference in the fluorometric determination of chlorophyll a and 'phaeo-pigments'. *Mar. Freshw. Res.* **1979**, *30*, 597–606. [CrossRef]
32. Welschmeyer, N.A. Fluorometric analysis of chlorophyll a in the presence of chlorophyll b and pheopigments. *Limnol. Oceanog.* **1994**, *39*, 1985–1992. [CrossRef]
33. Ras, J.; Claustre, H.; Uitz, J. Spatial variability of phytoplankton pigment distributions in the Subtropical South Pacific Ocean: Comparison between in situ and predicted data. *Biogeosciences* **2008**, *5*, 353–369. [CrossRef]
34. Knap, A.H.; Michaels, A.; Close, A.R.; Ducklow, H.; Dickson, A.G. *Protocols for the Joint Global Ocean Flux Study (JGOFS) Core Measurements*; Reprint of Intergovernmental Oceanographic Commission Manuals and Guides, No. 29; UNESCO: Paris, France, 1994; p. 170.
35. Ocean Color Feature. Available online: http://oceancolor.gsfc.nasa.gov (accessed on 4 July 2019).
36. NASA Goddard Space Flight Center, Ocean Ecology Laboratory, Ocean Biology Processing Group. *Moderate-resolution Imaging Spectroradiometer (MODIS) Aqua Ocean Color Data*; 2018 Reprocessing; NASA OB.DAAC: Greenbelt, MD, USA, 2018. [CrossRef]
37. Level 2 Ocean Color Flags. Available online: https://oceancolor.gsfc.nasa.gov/atbd/ocl2flags/ (accessed on 4 July 2019).
38. Zheng, G.; DiGiacomo, P.M. Uncertainties and applications of satellite-derived coastal water quality products. *Prog. Oceanogr.* **2017**, *159*, 45–72. [CrossRef]
39. Campbell, J.W. The lognormal distribution as a model for bio-optical variability in the sea. *J. Geophys. Res. Oceans* **1995**, *100*, 13237–13254. [CrossRef]
40. Campbell, J.W.; O'Reilly, J.E. Metrics for Quantifying the Uncertainty in a Chlorophyll Algorithm: Explicit equations and examples using the OC4. v4 algorithm and NOMAD data. In Proceedings of the Ocean Color Bio-Optical Algorithm Mini (OCBAM) Workshop, New England Center, Southborough, MA, USA, 27–29 September 2005; pp. 1–15.
41. Seegers, B.N.; Stumpf, R.P.; Schaeffer, B.A.; Loftin, K.A.; Werdell, P.J. Performance metrics for the assessment of satellite data products: An ocean color case study. *Opt. Express* **2018**, *26*, 7404–7422. [CrossRef] [PubMed]
42. Ricker, W.E. Linear regressions in fishery research. *J. Fish. Res. Board Can.* **1973**, *30*, 409–434. [CrossRef]
43. Stramski, D.; Reynolds, R.A.; Kahru, M.; Mitchell, B.G. Estimation of particulate organic carbon in the ocean from satellite remote sensing. *Sciences* **1999**, *285*, 239–242. 285.5425.239. [CrossRef]
44. Hu, C.; Lee, Z.; Franz, B. Chlorophyll a algorithms for oligotrophic oceans: A novel approach based on three-band reflectance difference. *J. Geophys. Res. Oceans* **2012**, *117*. [CrossRef]
45. Evers-King, H.; Martinez-Vicente, V.; Brewin, R.J.W.; Dall'Olmo, G.; Hickman, A.E.; Jackson, T.; Kostadinov, T.S.; Krasemann, H.; Loisel, H.; Röttgers, R.; et al. Validation and intercomparison of ocean color algorithms for estimating particulate organic carbon in the oceans. *Front. Mar. Sci.* **2017**, *4*, 1–19. [CrossRef]
46. Cullen, J.J. The deep chlorophyll maximum: Comparing vertical profiles of chlorophyll a. *Can. J. Fish. Aquat. Sci.* **1982**, *39*, 791–803. [CrossRef]
47. Proctor, C.W.; Roesler, C.S. New insights on obtaining phytoplankton concentration and composition from in situ multispectral Chlorophyll fluorescence. *Limnol. Oceanogr. Methods* **2010**, *8*, 695–708. [CrossRef]

48. Roesler, C.; Uitz, J.; Claustre, H.; Boss, E.; Xing, X.; Organelli, E.; Briggs, N.; Bricaud, A.; Schmechtig, C.; Poteau, A.; et al. Recommendations for obtaining unbiased chlorophyll estimates from in situ chlorophyll fluorometers: A global analysis of WET Labs ECO sensors. *Limnol. Oceanogr. Methods* **2017**, *15*, 572–585. [CrossRef]

49. Kumari, B. Comparison of high performance liquid chromatography and fluorometric ocean colour pigments. *J. Indian Soc. Remote* **2005**, *33*, 541–546. [CrossRef]

50. Kahru, M.; Mitchell, B.G. Chlorophyll a fluorescence in marine centric diatoms: Responses of chloroplasts to light and nutrient stress. *Mar. Biol.* **1973**, *23*, 39–46. [CrossRef]

51. Morel, A.; Bricaud, A. Theoretical results concerning light absorption in a discrete medium, and application to specific absorption of phytoplankton. *Deep Sea Res. A* **1981**, *28*, 1375–1393. [CrossRef]

52. Bricaud, A.; Morel, A.; Prieur, L. Optical efficiency factors of some phytoplankters. *Limnol. Oceanogr.* **1983**, *28*, 816–832. [CrossRef]

53. Boss, E.; Swift, D.; Taylor, L.; Brickley, P.; Zaneveld, R.; Riser, S.; Perry, M.J.; Strutton, P.G. Observations of pigment and particle distributions in the western North Atlantic from an autonomous float and ocean color satellite. *Limnol. Oceanogr.* **2008**, *53*, 2112–2122._part_2.2112. [CrossRef]

54. Lorenzen, C.J. Chlorophyll b in the eastern North Pacific Ocean. *Deep Sea Res. A* **1981**, *28*, 1049–1056. [CrossRef]

55. Trees, C.C.; Kennicutt, M.C., II; Brooks, J.M. Errors associated with the standard fluorimetric determination of chlorophylls and phaeopigments. *Mar. Chem.* **1985**, *17*, 1–12. [CrossRef]

56. Bianchi, T.S.; Lambert, C.; Biggs, D.C. Distribution of chlorophyll a and phaeopigments in the northwestern Gulf of Mexico: A comparison between fluorometric and high-performance liquid chromatography measurements. *Bull. Mar. Sci.* **1995**, *56*, 25–32.

57. Dos Santos, A.C.A.; Calijuri, M.D.C.; Moraes, E.M.; Adorno, M.A.T.; Falco, P.B.; Carvalho, D.P.; Deberdt, G.L.B.; Benassi, S.F. Comparison of three methods for Chlorophyll determination: Spectrophotometry and Fluorimetry in samples containing pigment mixtures and spectrophotometry in samples with separate pigments through High Performance Liquid Chromatography. *Acta Limnol. Bras.* **2003**, *15*, 7–18.

58. Holm-Hansen, O.; Lorenzen, C.J.; Holmes, R.W.; Strickland, J.D. Fluorometric determination of chlorophyll. *ICES J. Mar. Sci.* **1965**, *30*, 3–15. [CrossRef]

59. Jeffrey, S.W. A report of green algal pigments in the central North Pacific Ocean. *Mar. Biol.* **1976**, *37*, 33–37. [CrossRef]

60. Bidigare, R.R.; Frank, T.J.; Zastrow, C.; Brooks, J.M. The distribution of algal chlorophylls and their degradation products in the Southern Ocean. *Deep Sea Res. A* **1986**, *33*, 923–937. [CrossRef]

61. Parsons, T.R.; Takahashi, M.; Hargrave, B. *Biological Oceanographic Processes*, 3rd ed.; Oxford Pergamon Press: Oxford, UK, 1984; pp. 40–50, ISBN 0-08-030766-3.

62. Arrigo, K.R.; Mills, M.M.; Kropuenske, L.R.; van Dijken, G.L.; Alderkamp, A.C.; Robinson, D.H. Photophysiology in two major Southern Ocean phytoplankton taxa: Photosynthesis and growth of *Phaeocystis antarctica* and *Fragilariopsis cylindrus* under different irradiance levels. *Integr. Comp. Biol.* **2010**, *50*, 950–966. [CrossRef] [PubMed]

63. Mendes, C.R.B.; de Souza, M.S.; Garcia, V.M.T.; Leal, M.C.; Brotas, V.; Garcia, C.A.E. Dynamics of phytoplankton communities during late summer around the tip of the Antarctic Peninsula. *Deep Sea Res. Part 1 Oceanogr. Res. Pap.* **2012**, *65*, 1–14. [CrossRef]

64. Papagiannakis, E.; van Stokkum, I.H.M.; Fey, H.; Buchel, C.; van Grondelle, R. Spectroscopic characterization of the excitation energy transfer in the fucoxanthin–chlorophyll protein of diatoms. *Photosynth. Res.* **2005**, *86*, 241–250. [CrossRef] [PubMed]

65. Vaulot, D.; Birrien, J.L.; Marie, D.; Casotti, R.; Veldhuis, M.J.W.; Kraay, G.W.; Chrétiennot-Dinet, MJ. Morphology, ploidy, pigment composition, and genome size of cultured strains of *Phaeocystis* (Prymnesiophycea). *J. Phycol.* **1994**, *30*, 1022–1035. [CrossRef]

66. Crosta, X.; Romero, O.; Armand, L.K.; Pichon, J.J. The biogeography of major diatom taxa in Southern Ocean sediments: 2. Open ocean related species. *Palaeogeogr. Palaeoclimatol. Palaeoecol.* **2005**, *223*, 66–92. [CrossRef]

67. Boyd, P.W. Environmental factors controlling phytoplankton processes in the Southern Ocean. *J. Phycol.* **2002**, *38*, 844–861. [CrossRef]

68. Coale, K.H.; Johnson, K.S.; Chavez, F.P.; Buesseler, K.O.; Barber, R.T.; Brzezinski, M.A.; Cochlan, W.P.M.; Millero, F.J.; Falkowski, P.G.; Bauer, J.E.; et al. Southern Ocean iron enrichment experiment: Carbon cycling in high-and low-Si waters. *Science* **2004**, *304*, 408–414. [CrossRef] [PubMed]

69. Swart, S.; Speich, S. An altimetry-based gravest empirical mode south of Africa: 2. Dynamic nature of the Antarctic Circumpolar Current fronts. *J. Geophys. Res. Oceans* **2010**, *115*. [CrossRef]

70. Trull, T.W.; Bray, S.G.; Manganini, S.J.; Honjo, S.; François, R. Moored sediment trap measurements of carbon export in the Subantarctic and Polar Frontal zones of the Southern Ocean, south of Australia. *J. Geophys. Res. Oceans* **2001**, *106*, 31489–31509. [CrossRef]

71. Lindenschmidt, K.E.; Chorus, I. The effect of water column mixing on phytoplankton succession, diversity and similarity. *J. Plankton Res.* **1998**, *20*, 1927–1951. [CrossRef]

72. Stramski, D.; Reynolds, R.A.; Babin, M.; Kaczmarek, S.; Lewis, M.R.; Röttgers, R.; Sciandra, A.; Stramska, M.; Twardowski, M.S.; Franz, B.A.; et al. Relationships between the surface concentration of particulate organic carbon and optical properties in the eastern South Pacific and eastern Atlantic Oceans. *Biogeosciences* **2008**, *5*, 171–201. [CrossRef]

73. Cetinić, I.; Perry, M.J.; Briggs, N.T.; Kallin, E.; D'Asaro, E.A.; Lee, C.M. Particulate organic carbon and inherent optical properties during 2008 North Atlantic Bloom Experiment. *J. Geophys. Res. Oceans* **2012**, *117*, 13237–13254. [CrossRef]

74. Gardner, W.D.; Richardson, M.J.; Carlson, C.A.; Hansell, D.; Mishonov, A.V. Determining true particulate organic carbon: Bottles, pumps and methodologies. *Deep Sea Res. Part 2 Top. Stud. Oceanogr.* **2003**, *50*, 655–674. [CrossRef]

Article

Evaluation of Satellite-Based Algorithms to Retrieve Chlorophyll-a Concentration in the Canadian Atlantic and Pacific Oceans

Stephanie Clay [1,†], Angelica Peña [2], Brendan DeTracey [3] and Emmanuel Devred [3,*,†]

1   Takuvik Joint International Laboratory, Laval University-CNRS, Quebec City, QC G1V 0A6, Canada;
    Stephanie.Clay.1@ulaval.ca
2   Institute of Ocean Sciences, Fisheries and Oceans Canada, Sidney, BC V8L 5T5, Canada;
    Angelica.Pena@dfo-mpo.gc.ca
3   Bedford Institute of Oceanography, Fisheries and Oceans Canada, Dartmouth, NS B2Y 4A2, Canada;
    Brendan.DeTracey@dfo-mpo.gc.ca
*   Correspondence: emmanuel.devred@dfo-mpo.gc.ca; Tel.: +1-902-426-4681
†   These authors contributed equally to this work.

Received: 3 October 2019; Accepted: 1 November 2019; Published: 7 November 2019

Abstract: Remote-sensing reflectance data collected by ocean colour satellites are processed using bio-optical algorithms to retrieve biogeochemical properties of the ocean. One such important property is the concentration of chlorophyll-a, an indicator of phytoplankton biomass that serves a multitude of purposes in various ocean science studies. Here, the performance of two generic chlorophyll-a algorithms (i.e., a band ratio one, Ocean Colour X (OCx), and a semi-analytical one, Garver–Siegel Maritorena (GSM)) was assessed against two large *in situ* datasets of chlorophyll-a concentration collected between 1999 and 2016 in the Northeast Pacific (NEP) and Northwest Atlantic (NWA) for three ocean colour sensors: Sea-viewing Wide Field-of-view Sensor (SeaWiFS), Moderate Resolution Imaging Spectroradiometer (MODIS), and Visible Infrared Imaging Radiometer Suite (VIIRS). In addition, new regionally-tuned versions of these two algorithms are presented, which reduced the mean error (mg m^{-3}) of chlorophyll-a concentration modelled by OCx in the NWA from -0.40, -0.58 and -0.45 to 0.037, -0.087 and -0.018 for MODIS, SeaWiFS, and VIIRS respectively, and -0.34 and -0.36 to -0.0055 and -0.17 for SeaWiFS and VIIRS in the NEP. An analysis of the uncertainties in chlorophyll-a concentration retrieval showed a strong seasonal pattern in the NWA, which could be attributed to changes in phytoplankton community composition, but no long-term trends were found for all sensors and regions. It was also found that removing the 443 nm waveband for the OCx algorithms significantly improved the results in the NWA. Overall, GSM performed better than the OCx algorithms in both regions for all three sensors but generated fewer chlorophyll-a retrievals than the OCx algorithms.

Keywords: ocean colour; satellite-derived chlorophyll-a concentration; algorithm evaluation; Northwest Atlantic; Northeast Pacific

1. Introduction

The product derived from satellite ocean colour that is the most used is undoubtedly chlorophyll-a (*chla*) concentration, an index of phytoplankton biomass, which has numerous applications in biogeochemical oceanography, such as phytoplankton ecology and phenology [1,2], carbon cycles [3], climate change, transfer of energy to higher trophic levels, and water quality [4]. Satellite-derived *chla* concentration is a mature product of ocean colour that is used not only by experts in the field of bio-optics but also by the entire oceanographic community in various ways, such as modelling and data assimilation [5], fisheries applications [6], and ecosystem management [7]. For instance,

European programs such as Copernicus (https://www.copernicus.eu/) or NOAA's COASTWATCH (https://coastwatch.noaa.gov/cw/index.html) provide daily *chla* data that are accessible to the public for fisheries and water quality applications. In Canada, Fisheries and Oceans have been relying on ocean colour for several decades to monitor the state of the marine ecosystem [8,9], and in particular on *chla* concentration, and its derived product primary production, to infer ecosystem indicators that are at the foundation of ocean management.

Chla concentration is derived from remote sensing reflectance (R_{rs}), which is the ratio of water-leaving radiance to downwelling irradiance corrected for sun geometry to allow for comparisons independent of locations, times and dates. Several approaches to infer *chla* from R_{rs} have been developed and embedded by space agencies in their data processing software, including band ratio [10] and semi-analytical models [11]. As its name suggests, band ratio algorithms exploit the ratio of wavebands in the blue and green to retrieve *chla*; the Ocean Colour X (OCx) (x stands for 2, 3, or 4 and indicates the number of bands that were used in the algorithm) suite of empirical algorithms have been developed by the National Aeronautics and Space Administration (NASA) using a global dataset of *in situ* measurements of *chla* concentration fitted to remote sensing reflectance (NOMAD, the NASA bio-Optical Marine Algorithm Dataset [12]) and a fourth-degree polynomial expression. On the other hand, semi-analytical algorithms (e.g., Garver–Siegel Maritorena (GSM)) consist of optimizing bio-optical parameters (including *chla* concentration) in an approximate solution of the radiative transfer equation to match modelled reflectance to the reflectance measured by the satellite. This type of algorithm has the advantage of decoupling the contribution of the optically active components (i.e., phytoplankton, non-algal particles and coloured dissolved organic carbon) such that *chla* concentration should, in theory, be retrieved with higher accuracy than the band ratio algorithms. These two types of approaches are well-suited for the so-called case-1 waters (i.e., waters where *chla* concentration drives the optical characteristics of bulk seawater) but do not perform as well in case-2 waters (i.e., where optical signals of marine components are uncorrelated). For case-2 waters, models that use fluorescence [13] or more advanced statistical methods, such as neural networks [14] or principal component analysis [15], have been demonstrated to perform better than band ratio or semi-analytical approaches. Note that the performance of algorithms that use remote sensing reflectance will be inherently dependent on the performance of the atmospheric correction procedure, which will not be addressed here.

It has been shown that OCx [15,16] and GSM [17] algorithms contain regional biases, such that even if its overall performance is satisfying, application to a given region results in systematic bias. Here, we assessed the performance of the OCx and GSM algorithms in Canadian waters (Northwest Atlantic (NWA) and Northeast Pacific (NEP)) using a dataset of *in situ chla* concentration that was collected by Fisheries and Oceans Canada as part of their monitoring of the marine ecosystem. Performance of these algorithms was evaluated for NASA's three sensors, namely the Sea-viewing Wide-field-of-View Sensor (SeaWiFS, 1998–2010), the Moderate Resolution Imaging Spectrometer on the Aqua platform (MODIS-Aqua, 2002–current) and the Visible and Infrared Imaging Spectrometer on the NPP platform (VIIRS, 2012–current). Uncertainties of the algorithms with respect to environmental variables were also discussed.

2. Material and Methods

2.1. Regions of Interest

Fisheries and Oceans Canada (DFO) carries out sea-going expeditions as part of its oceanographic monitoring programs to obtain information on the health of the marine ecosystem to support decision-making in the management of the ocean. In the Northwest Atlantic (NWA) Ocean, more than a hundred stations are visited twice a year in spring and fall on the Scotian Shelf and once a year (in spring) in the Labrador Sea as part of the fieldwork of the Atlantic Zone Monitoring Program (AZMP) and Atlantic Zone Offshore Monitoring Program (AZOMP) respectively (Figure 1).

During these missions, a large number of physical, chemical and biological oceanographic parameters are collected [8]. As part of its remote sensing operations, DFO archives satellite data of ocean colour and sea-surface temperature in a region bounded by 39° to 82° north and 95° to 42° west. These data were used to assess the performance of ocean colour *chla* data products in the NWA. In a similar fashion, DFO runs monitoring programs in the NEP along Line P three times a year (in winter, spring and summer), off the west coast of Vancouver Island twice a year (in spring and summer) and in the Strait of Georgia three times a year (in spring, summer and fall), where physical, chemical and biological data are collected [18]. The in situ data from this region used in this study are bounded by 47° to 57° north, 148° to 123° west.

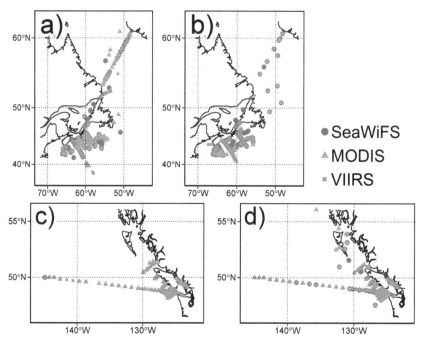

Figure 1. Locations of satellite matchups (**a**) in the Northwest Atlantic (NWA) from January to June, (**b**) from July to December, (**c**) in the Northeast Pacific (NEP) from January to June, and (**d**) from July to December.

2.2. In Situ Samples

DFO collects data on *chla* concentration using two methods, namely, Turner fluorescence (TF) and high-performance liquid chromatography (HPLC). While the most comprehensive dataset is the one obtained using the TF method (tens of thousands of data), here we used the dataset containing HPLC-derived *chla* concentration since it provides the most accurate estimates of *chla* and it is also the method recommended by space agencies for validation activities. A large dataset was compiled for the purpose of this study, consisting of the *in situ* samples gathered in the NWA and NEP which were analyzed in their respective DFO regional laboratories. In brief, water samples were rapidly filtered after collection through 25 mm GF/F glass fiber filters, immediately flash-frozen using liquid nitrogen, and stored in a −80 °C freezer until analysis in the laboratory. The NWA samples were analyzed using a Beckman–Coulter Gold HPLC system (1998–2013) and more recently an Agilent 1200 (2013–2014) HPLC instrument following the protocol of Head and Horne [19] as modified by Stuart and Head [20]. Similarly, the NEP samples were analyzed using a Waters Alliance System HPLC following the protocol of Zapata et al. [21]. The analysis was performed for 23 pigments for both locations, but only *chla* concentrations from both regions and fucoxanthin (*fucox*) concentrations from the NWA were used in this study. Note that chlorophyll-a concentration includes chlorophyll-a epimer and allomer pigments.

The complete dataset consisted of 1857 samples from the NWA between 1999 and 2014, and 1231 samples from the NEP between 2006 and 2016, all collected within ten metres from the surface. Both datasets showed a skewed normal distribution with a peak around 0.43 (0.42) mg m^{-3} in the NWA (NEP) Ocean and a similar range of variation of *chla* from 0.03 to 29.41 mg m^{-3}, but samples collected in the NEP had a mean concentration (2.26 mg m^{-3}) greater than in the NWA, where the mean concentration was 1.68 mg m^{-3} (Figure 2a). In the NWA, samples were collected mainly in spring (April to June) and fall (September to November) with a few exceptions of data collected in July and August, whereas the NEP dataset consisted of samples collected in winter (February to March), late spring-early summer (June to July) and late summer to early fall (August through October) (Figure 2b). Both datasets captured the seasonal cycle of phytoplankton, which was consistent with the wide range of biomass concentrations. The dataset was reduced to measurements collected within a day of satellite passes and ten kilometres from a valid pixel.

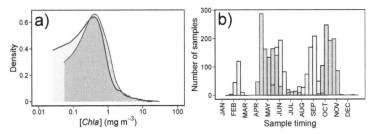

Figure 2. (a) Density function distribution of *in situ* chlorophyll-a concentration for the Northwest Atlantic (grey) and Northeast Pacific (lavender), and (b) total number of *in situ* samples available in the entire dataset for each month.

2.3. Satellite Matchups

The 2014 reprocessed datasets of SeaWiFS merged local area coverage (MLAC), MODIS-Aqua local area coverage (LAC), and VIIRS-Suomi NPP level-2 satellite scenes were downloaded from NASA's ocean colour L1/L2 browser (https://oceancolor.gsfc.nasa.gov/cgi/browse.pl?sen=am). For each image, a mask was applied to remove pixels with invalid data, defined as any pixel where one or more of the following criteria were met:

1. The pixel was marked by any of the following level 2 flags: atmospheric correction failure (ATMFAIL), deleted overlapping pixels (BOWTIEDEL), pixel overland (LAND), high sun glint (HIGLINT), pixel contains cloud or ice (CLDICE), radiance too high (HILT), high solar zenith (HISOLZEN), or high sensor zenith angle (HISATZEN).
2. The pixel contained a no-data (NA) value.
3. The pixel had more than one negative remote sensing reflectance (R_{rs}) value, implying the data might be flawed by the atmospheric correction procedure.

A search was performed for potential matches in any satellite image taken within a day of the *in situ* sample, as exact sampling times were not always available in the database. In order to match a satellite value to an *in situ* measurement, the satellite image was projected using a Gnomonic projection centred on the *in situ* point, which stretched great circles to straight lines. In other words, the image is projected onto a plane tangent to the *in situ* point, from which Cartesian coordinates (x, y) can be extracted for each corresponding (longitude, latitude) pair. For the sake of speed and simplicity, the Pythagorean theorem was then used to compute the distance between the *in situ* point and each satellite pixel. Though this method did not account for Earth's curvature, the computed distances at the maximum allowed radius around the *in situ* point (i.e., 10 km) were found to be within ±23 m of the more accurate geodetic distance as calculated by the *distGeo*() function using R Statistical Software [22]. A successful match was defined as the closest pixel, within a 10 km radius, that contained at least three valid pixels in a 3 × 3 pixel box extracted around the matching point. The distance between the *in situ* sample and the closest satellite pixel was recorded in the database. The median of the 3 × 3 pixel matrix was computed for each of the extracted remote sensing reflectance (R_{rs}) wavelengths, and the resulting dataset was considered as the total number of matches, mapped in Figure 1.

From the initial dataset of 1857 and 1231 samples for the NWA and NEP respectively, only 416 (SeaWiFS), 530 (MODIS) and 176 (VIIRS) *in situ* samples were matched to valid satellite pixels in the NWA, and 45 (SeaWiFS), 487 (MODIS) and 342 (VIIRS) in the NEP (Table 1). The small number of matchups compared to the initial dataset attests to the difficulty of compiling archives of satellite data that match *in situ* samples for validation purposes and emphasizes the need for recurrent measurements such as monitoring programs and moored automatic sensors.

Table 1. Specifications of sensors and matchup datasets. Bold and underlined numbers correspond to blue (λ_{blue}) and green (λ_{green}) bands used in the band ratio Ocean Colour X (OCx) algorithms respectively (see Section 2.4.1).

Sensor	Resolution (km)	Wavebands (nm)	Period (Year)	N	Period (Year)	N
			NWA		NEP	
MODIS	1.1	412, **443**, 469, **488**, 531, <u>547</u>, 555, 645, 667, 678	2002–2014	530	2006–2016	487
SeaWiFS	1.0	412, **443**, **490**, **510**, <u>555</u>, 670	1999–2010	416	2006–2010	45
VIIRS	0.75	410, **443**, **486**, <u>551</u>, 671	2012–2014	176	2012–2016	342

Finally, *chla* concentration was derived using the OCx algorithm (Section 2.4.1) and pixels with a coefficient of variation (computed as the ratio of the standard deviation to the mean of all pixels available in the 3 × 3 matrix) exceeding 0.5 were removed from the matchup dataset.

2.4. Chlorophyll-a Algorithms

Two main algorithms were tested: The Ocean Colour (OCx) [10] and the Garver–Siegel Maritorena (GSM) [23] algorithms, as well as variations of these original versions that were regionally-tuned. The variations consist of four new versions for OCx, following the same polynomial format in degrees 1–4 with regional coefficients, and three optimal versions of GSM, with different combinations of three modifications to the algorithm, described in Section 2.4.4. For the sake of simplicity, results from only six algorithms were analyzed in this study: the two original algorithms (OCx and GSM), two regionally-tuned counterparts for OCx, and two for GSM. The results of the remaining algorithms have been included in Appendix A. The algorithms were implemented using version 3.3.2 of R Statistical Software [24] and several functions and packages referenced in the text.

2.4.1. OCx-Type Algorithms

These empirical algorithms use a polynomial equation to express the log-10 transformed *chla* as a function of the log10-transformed ratio of blue to green remote sensing reflectance (Table 1). The blue band is selected to maximise the blue-to-green ratio, R_{OCx}:

$$R_{OCx} = \log_{10}\left(\max\left(\frac{R_{rs}(\lambda_{blue})}{R_{rs}(\lambda_{green})} \right) \right). \tag{1}$$

The ratio of blue to green R_{rs} values is used as opposed to a single R_{rs} value in an attempt to normalize the data, as R_{rs} values in the blue part of the spectrum vary greatly with *chla*, whereas the green wavebands have a more limited range of variation. As *chla* increases, R_{rs}(443) will decrease and become close to zero, reaching the noise level of the sensor. To avoid this issue, the algorithm switches to longer wavebands where the signal still correlates with *chla* and remains detectable. Additionally, longer wavelengths (e.g., 490 and 510) are less affected than the 443 nm band by variations in the concentration of coloured dissolved organic matter (i.e., yellow substances), and are therefore selected in the numerator of the ratio. Log10 *chla* is calculated using the following polynomial algorithm:

$$\log_{10}(chla) = a + b\,R_{OCx} + c\,R_{OCx}^2 + d\,R_{OCx}^3 + e\,R_{OCx}^4. \tag{2}$$

NASA uses the method prescribed in O'Reilly et al. [10] to retrieve the optimal coefficients *a*, *b*, *c*, *d*, and *e*. This is an iterative method that constrains the slope and intercept of the linear regression between model and *in situ chla* to 1 and 0 respectively, minimizing the *RMSLE* (root mean squared logarithmic error, see Equation (12) in Section 2.5.1) and maximizing the coefficient of determination, r^2. The comparison between band ratio models is then simplified since the slope, intercept, and consequently, the bias are all equal, and only the *RMSLE* and r^2 need to be evaluated. The *in situ* data used for the optimization of the original OCx algorithm is the global NASA Bio-Optical Marine Algorithm Dataset (NOMAD, version 2). Each sensor has a different set of wavelengths in the blue and green part of the spectrum (Table 1), such that the band ratio algorithm parameters are optimized for each sensor and are referred to as OC3M for MODIS, OC4 for SeaWiFS, and OC3V for VIIRS.

2.4.2. Regional Tuning of the OCx Algorithm

To remain consistent with the matchup exercise, the median value of the 3 × 3 pixel matrix of each waveband was computed to represent the $R_{rs}(\lambda)$ of the matchup. Note that the 443 nm waveband was removed from the potential blue wavebands in the band ratio as it was found to negatively affect the retrieval of *chla* (see discussion in Section 3.1). Several polynomial formulations from degrees one to four were tested and are hereafter referred to as "POLY1", "POLY2", "POLY3", and "POLY4" respectively. The optimal coefficients were retrieved using an iterative procedure to force the slope and intercept of the linear regression between satellite-derived and *in situ chla* to one and zero respectively as in O'Reilly et al. [10]. A 95% confidence interval for each optimal coefficient was computed using the *boot()* function in R with 2000 iterations [25,26]. For the sake of clarity and to investigate any potential benefits of adding higher-degree terms to the polynomial, the scores were computed for every polynomial degree, but only the results of the first and fourth-degree polynomials are discussed in the study. The parametrization and full results for the second and third polynomial are reported in Tables A1, A2 and A7 of Appendix A.

2.4.3. Original GSM

The GSM semi-analytical algorithm was developed by Garver and Siegel [27], and modified by Maritorena et al. [23]. The model uses a quadratic formulation (Equation (3)) to relate the underwater remote-sensing reflectance values to the ratio of total absorption, *a* (m^{-1}), and backscattering, b_b (m^{-1}) [28]. Note that the wavelength dependence in the following equations has been removed for the sake of legibility:

$$R_{rs,model}(0^-) = g_1\left(\frac{b_b}{a+b_b}\right) + g_2\left(\frac{b_b}{a+b_b}\right)^2, \tag{3}$$

where $g_1 = 0.0949$ and $g_2 = 0.0794$. Total absorption and backscattering are defined as follows:

$$
\begin{aligned}
a(\lambda) &= a_w(\lambda) + a_{ph}(\lambda) + a_{dg}(\lambda), & (4)\\
&= a_w(\lambda) + chla \times a_{ph}^{\star}(\lambda) + a_{dg}(443)\exp(-S(\lambda - 443)), & (5)
\end{aligned}
$$

and

$$b_b(\lambda) \quad = \quad b_{bw}(\lambda) + b_{bp}(\lambda) \tag{6}$$

$$= \quad b_{bw}(\lambda) + b_{bp}(443) \left(\frac{443}{\lambda}\right)^Y, \tag{7}$$

where λ is the wavelength (nm). The term a corresponds to the total absorption term and is the sum of the absorption terms of pure seawater (a_w), phytoplankton (a_{ph}), and coloured detrital and dissolved organic materials (a_{dg}), which are combined in a single term given their similar spectral shapes; S expresses the exponential decrease of a_{dg} with increasing wavelength using the reference wavelength at 443 nm (i.e., $a_{dg}(443)$). The phytoplankton absorption term is defined as the concentration of *chla* multiplied by its specific absorption coefficient, $a_{ph}^\star$ (i.e., absorption per unit *chla*), to account for its spectral variation. The total backscattering term is the sum of the backscattering terms of pure seawater (b_{bw}) and particulate matter (b_{bp}), the latter expressed as the product between a reference value (i.e., $b_{bp}(443)$) and the power-law-like decrease with increasing wavelength by the exponent Y. The coefficient $a_{ph}^\star$, given in Table A10 of Appendix A, was determined using *in situ* measurements of *chla* and phytoplankton absorption at the wavelengths of interest, and computing the mean value of absorption divided by *chla* across all available records within the AZMP database that spans from 1998 to 2014. The a_w coefficients were derived from tabulated values with a 2.5 nm resolution from 400 to 700 nm [29], and when necessary, interpolated to the corresponding wavelengths of each sensor. Similarly, the b_{bw} coefficients were derived from a 10 nm resolution table of values generated by Smith and Baker [30]. The spectral slope of yellow substances absorption S and the power-law decrease of particulate backscattering Y were optimized for use in the original algorithm to 0.02061 and 1.03373 respectively using a large simulated dataset [31]. The updated GSM models in this study tuned S and Y using the dataset collected from the NWA and NEP (see Section 2.4.4 for details).

$R_{rs,satellite}(0^-)$ was retrieved by taking the median $R_{rs,satellite}(0^+)$ of the 3×3 matrix at each wavelength for a given match, then converting it from surface to underwater reflectance using Lee et al. [32]:

$$R_{rs,satellite}(0^-) = \frac{R_{rs}(0^+)}{0.52 + 1.7 R_{rs}(0^+)}. \tag{8}$$

Non-linear least squares regression, using the Gauss–Newton algorithm of the *nls()* function [24] to minimize the difference between modelled and satellite-derived R_{rs}, was implemented with a maximum of 30 iterations to derive the remaining three unknowns in Equation (3): *chla*, the absorption coefficient of coloured detrital and dissolved material at the reference wavelength ($a_{dg}(443)$), and the backscattering coefficient of particulate matter at the reference wavelength ($b_{bp}(443)$). After retrieving these unknowns, a correction factor of 0.754188 was applied to the $a_{dg}(443)$ value according to Maritorena et al. [23]. The following constraints were then placed on each to ensure realistic values:

$$\begin{cases} 0.01 \leq chla \leq 64 \, \text{mg m}^{-3} \\ 0.0001 \leq a_{dg}(443) \leq 2 \, \text{m}^{-1} \\ 0.0001 \leq b_{bp}(443)) \leq 0.1 \, \text{m}^{-1} \end{cases} \tag{9}$$

The original version of the model is abbreviated here as GSM_ORIG.

2.4.4. Regional Tuning of GSM

Similarly to the optimization of the OCx algorithm, $R_{rs}(\lambda)$ for a given satellite matchup was computed as the median of the corresponding 3×3 pixel matrix centered on the matchup pixel. Three modifications to the original GSM algorithm were tested. First, a new optimized exponent, P, was added to the *chla* term to correct for systematic underestimation at high *chla* and overestimation at low *chla* (see Sections 3.2 and 3.3) that were potentially caused by the change of a_{ph}^{*} with increasing *chla* (i.e., packaging effect [33]). The expression of the total absorption became:

$$a(\lambda) = a_w(\lambda) + chla^P \times a_{ph}^{*}(\lambda) + a_{dg}(443) \times \exp(-S(\lambda - 443)). \tag{10}$$

Second, the *g* parameters (see Equation (3) and Table A8) were optimized and spectral dependency was included as well as a new term g_3 (Equation (11)) using the synthetic dataset from the IOCCG working group on remote sensing of inherent optical properties [31]. Coefficients were retrieved at 10 nm intervals from 400 nm to 700 nm and interpolated to sensor-specific wavelengths when necessary.

$$R_{rs,model}(0^-) = g_1 \left(\frac{b_b}{a + b_b} \right) + g_2 \left(\frac{b_b}{a + b_b} \right)^{g_3}. \tag{11}$$

Third, a sensitivity analysis was performed to compute the optimal exponents *S* (in the a_{dg} term), *Y* (in the b_{bp} term), and the new exponent *P* (in the a_{ph} term), which provided the best agreement between *in situ* and satellite-derived *chla* for the NWA and NEP regions. A total of 5355 (17 × 35 × 9) possible combinations of the three exponents were tested, ranging from 0.008 to 0.04 with an increasing step of 0.002 for *S*, 0.5 to 2.2 with an increasing step of 0.05 for *Y*, and 0.4 to 0.8 with an increasing step of 0.05 for *P*. These combinations of exponents were evaluated for all sensors and regions, as well as for both constant *g* values (see Equation (3)) and the new spectral *g* values (see Equation (11)). The best set of exponents was determined using a similar scoring method as that used for the algorithm performance evaluation (see Section 2.5.2), where each set of exponents was ranked according to a selected set of metrics. To remain consistent with the methods of optimizing band ratio coefficients, the slope and intercept, as well as *RMSLE* and r^2, were selected for this test, and a score was assigned to each statistic based on its mean across all possible combinations and its value relative to other combinations. The optimal set of exponents was defined as the median of all sets with the highest sum of scores. At least 50% valid retrievals were required for each potential combination in the test, in order to avoid sets with low error values but with a poor predictive capacity. The optimal exponents are given in Table A9.

The modified versions of GSM were divided into three separate algorithms according to the changes that were made. All new versions included the exponent *P* on the *chla* term: 1) GSM_GC used the original *g* coefficients and regionally-tuned *P*, *S*, and *Y*, 2) GSM_GCGS used the spectral *g* coefficients in combination with the optimized *P*, *S*, and *Y* parameters from GSM_GC; and 3) GSM_GS used the spectral *g* coefficients optimized simultaneously with *P*, *S*, and *Y*. The results of GSM_GC and GSM_GS are discussed in this study to demonstrate improvement in *chla* retrieval when using the new spectrally-dependent *g* coefficients and the *P* exponent in the phytoplankton-dependent term, however only the scores of GSM_GCGS are included in the main text for simplicity's sake, and the full results in Appendix A (see Tables A1 and A2).

2.5. Performance Metrics

2.5.1. Statistical Models

Several sets of statistics were used for the evaluation of the algorithm. First, the total number of matchups (*N*) and the number of valid retrievals (*n*), were used to calculate the percentage of retrievals. Second, several formulations were used to express the error between *in situ* and satellite-derived *chla*, including the mean error (μ_{error}, units of mg m^{-3}), as well as the root mean squared logarithmic error (*RMSLE*):

$$RMSLE = \sqrt{\frac{1}{n} \sum_{i=1}^{n} \left(\log_{10}(C_i^*) - \log_{10}(C_i) \right)^2}, \tag{12}$$

the mean log-transformed error (*MLE*):

$$MLE = 10^{\left(\frac{1}{n}\sum_{i=1}^{n}\left(\log_{10}(C_i^*) - \log_{10}(C_i)\right)\right)}, \tag{13}$$

and the mean magnitude of log-transformed error (*MMLE*),

$$MMLE = 10^{\left(\frac{1}{n}\sum_{i=1}^{n}\left|\log_{10}(C_i^*) - \log_{10}(C_i)\right|\right)}, \tag{14}$$

where C_i^* corresponds to satellite-derived *chla* of matchup i and C_i corresponds to *in situ chla* of matchup i. These three metrics are unitless as a result of the log transformation, and the latter two metrics have an ideal value of one, stemming from the reverse log transformation that converts them from linear to multiplicative space. Third, the intercept, slope, and coefficient of determination (r^2) were computed for the linear regression of log-transformed satellite-derived *chla* on log-transformed *in situ chla* using the Standard Major Axis method of Type 2 regression in the *lmodel2*() function [34]. Finally, a metric referred to as the "win ratio" based on Seegers et. al. [35], was also computed. The full dataset of *in situ*/satellite matchups for a region/sensor was pared down to a new total value, T_{valid}, which only included matchups with valid satellite-derived *chla* for all algorithms. This facilitated comparisons between the errors of individual matchups for each algorithm. For a single *in situ*/satellite pair, the algorithm that generated the lowest magnitude of error was the "winner". Each model's total number of "wins" in the set, $wins_{model}$, was then divided by T_{valid}:

$$win\ ratio_{model} = \frac{wins_{model}}{T_{valid}}. \tag{15}$$

For example, if a dataset contains 500 matchups for which every algorithm generates a valid satellite *chla* concentration, and one of the algorithms (e.g., POLY1) produces the lowest error out of all algorithms for 70 of the 500 matchups, then the win ratio of POLY1 would be 70/500 = 0.14.

2.5.2. Scoring Method

The performance of all algorithms was evaluated using a modified version of the ranking approach developed by Brewin et al. [36], which relies on a slightly different set of metrics. Five statistics were selected as suggested in Seegers et al. [35] in an attempt to summarize the bias, accuracy, and precision of the model, as described below:

1. *MLE*, to account for the possible bias in the algorithm,
2. *MMLE*, to determine the accuracy of the algorithm,
3. r^2 of the linear regression, to test the precision or "goodness of fit" of the model,
4. percentage of valid retrievals, to ensure that a high score and low error are not reported as a result of a small number of retrievals, and
5. win ratio, to judge algorithm performance based on individual matches rather than a summary statistic.

The *RMSLE* was excluded from the scoring method in order to avoid redundant metrics that could skew the results by adding more emphasis to one element of the assessment (for example, bias) instead of giving all elements equal weight, however, its result is presented here for comparison with the literature. In our evaluation approach, all the metrics used were considered of equal weight. Each statistic was transformed to its magnitude of difference from the ideal value (for example, $r^2 = 0.6$ would be transformed to 0.4, as the ideal r^2 value is one) to give each the same reference point of zero so that lower statistics get higher scores. The mean of these transformed statistics was calculated across the results for all six algorithms. For a given algorithm, each of the five statistics was then scored as follows, taking into account their value relative to the mean and to the individual values of other algorithms:

- two points if the algorithm's transformed statistic was in the lowest 20% of the statistic's values across all algorithms (i.e., closer to the ideal value of zero) and below the mean,
- zero points if the algorithm's statistic was in the highest 20% of the statistic's values across all algorithms (i.e., further from the ideal) and above the mean, and
- one point otherwise

Statistical scores for each algorithm were then summed to give the algorithm's overall performance in comparison to others (Tables 3 and 4).

2.6. Temporal, Geographic and Phytoplankton Composition Influences on Algorithm Performance

The performance of the algorithms was also evaluated against environmental properties to detect any systematic bias or increase in uncertainties due to external factors. We studied the variability in the error (E_{chla}) as a function of time (defined as year + day of year/365) using boxplots generated by *ggplot2()* and *geom_boxplot()* in R [37], given that the validation exercises span over several years for all three sensors and both regions. To examine the relationships between the uncertainties and other environment variables, a correlation analysis was performed on the magnitude of error ($|E_{chla}|$) between *in situ* and satellite-derived *chla* against the variables listed below, using the *shapiro.test()* function to check that each subset followed a normal distribution, and the *cor.test()* function to retrieve the correlation coefficient and *p*-value [24]:

1. time, as defined above (units in decimal year),
2. day of year (units in days),
3. chlorophyll-a concentration (mg m^{-3}),
4. fucoxanthin to chlorophyll-a ratio (unitless),
5. latitude (decimal degrees),
6. bathymetry in coastal ocean ($\leq$200 m depth, units in metres),
7. bathymetry in open ocean (>200 m depth, units in metres), and
8. distance (in metres) from central pixel to *in situ* samples, as we relax the constraint to 10 km.

The ratio of *fucox* to *chla* was selected to provide the ability to detect any bias due to change in phytoplankton composition, as the *fucox* pigment is an indicator of the presence of diatoms [38,39].

To test if the seasonal cycle induced different uncertainties we performed a Student's *t*-test on μ_{error} for each sensor and algorithm between spring and fall in the NWA, and similarly a One-Way ANOVA test between spring, summer and fall in the NEP. A separate *t*-test was also performed on μ_{error} between coastal and open ocean bathymetry subsets (as defined above). Again, the assumption of normality of each subset was checked using the *shapiro.test()* function and assumptions of equal variance between subsets were checked using the *var.test()* and *fligner.test()* functions for comparisons of two groups and three groups respectively. The tests for differences in means between subsets were conducted using the functions *t.test()* (for two groups) and *aov()* (three groups). Tukey's honest significant difference was performed on the ANOVA results from the NEP to determine which seasons generated mean errors that were statistically different from each other, using the *TukeyHSD()* function [24].

3. Results

3.1. Parameters of the Optimized Band Ratio Algorithms

Regionalization of the OCx and GSM algorithms was carried out using satellite-derived and *in situ* *chla*, which differs from the original algorithms where the coefficients of the models were derived using remote sensing reflectances and *chla* measured *in situ*. The regionally-tuned coefficients of the band ratio models remained consistent with the original parameters and notably, the first two coefficients (i.e., *a* and *b*, Table 2) were within the same order of magnitude and of the same sign as the original coefficients. In general, the first term *a* was higher in the regional versions of the algorithms. This can be explained by the overall underestimation of *chla* by the original OCx algorithms. The *a* and *b* coefficients for the NEP for all sensors were closer to the original ones than for the NWA, which was consistent with the better performance of the original algorithms in the NEP than in the NWA (see Section 3.2 and 3.3). Regarding the coefficients for high degree polynomials (i.e., $>= 2$), there were no distinct patterns and the coefficients of the regional algorithms tend to be different and often opposite in sign from the original coefficients. The removal of the 443 nm band in the new band ratio algorithms may explain the differences between the original and regional coefficients of the algorithms. The numerator of the band ratio in the OCx algorithm varies between two or three possible wavebands depending on the sensor, where one of the options is the 443 nm band. Our analysis revealed that this band produced more outliers in the polynomial relationship between log-transformed *in situ chla* and the log-transformed band ratio than other numerator wavebands, which is particularly obvious in the NWA MODIS dataset (Figure 3). The blue boxes in Figure 3a,b show how the polynomial fit translates to a small range of modelled *chla* corresponding to a wide range of *in situ chla*. Removing the 443 nm waveband from the potential blue wavebands produced a tighter fit with the polynomial (Figure 3c,d). SeaWiFS and VIIRS NWA matchups datasets showed similar improvements after removing the 443 nm band, while the NEP datasets did not show significant changes after removing this band.

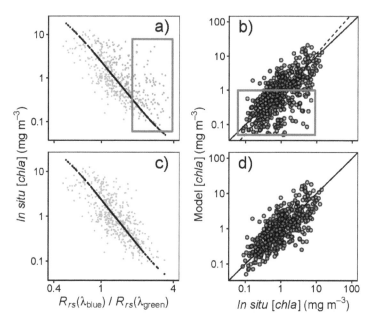

Figure 3. (**a**) *In situ chla* as a function of R_{rs} ratio for POLY4 including the 443 nm band, where the blue dots represent the band ratios that used the 488 nm band in the numerator and the pink dots used the 443 nm band in the numerator, (**b**) Satellite-derived versus *in situ chla* when including the 443 nm band in the ratio, where green and red circles represent spring and fall matchups respectively. (**c**) same as (**a**) but without including $R_{rs}(443)$ in the POLY4 algorithm, (**d**) same as (**b**) but without including $R_{rs}(443)$ in the POLY4 algorithm.

Table 2. Coefficients of the OCx algorithms for each sensor and region.

Coefficients	MODIS			SeaWiFS			VIIRS		
	OC3M	POLY1	POLY4	OC4	POLY1	POLY4	OC3V	POLY1	POLY4
					NWA				
a	0.2424	0.36695	0.37925	0.3272	0.51664	0.51824	0.2228	0.43399	0.44786
b	−2.7423	−3.27757	−3.28487	−2.9940	−3.84589	−3.68431	−2.4683	−3.09652	−3.11091
c	1.8017	-	−0.75830	2.7218	-	−0.97401	1.5867	-	−0.77987
d	0.0015	-	1.49122	−1.2259	-	0.84875	−0.4275	-	1.42500
e	−1.2280	-	0.80020	−0.5683	-	0.77874	−0.7768	-	0.90445
					NEP				
a		0.24947	0.26575		0.41867	0.42516		0.31886	0.33055
b		−2.84152	−2.84142		−3.14708	−3.14271		−2.65010	−2.76455
c		-	−0.57938		-	−0.70269		-	−0.39595
d		-	0.74974		-	1.21802		-	1.52198
e		-	0.47743		-	1.59686		-	0.46509

3.2. Algorithms performance in the NWA

The statistic metrics remain consistent across the three sensors, with the regionally-tuned algorithms providing the best performance, as expected. In general, the polynomial-based approach performed better than the semi-analytical algorithms (Figure 4a and Table 3, see scores). It is noteworthy that the polynomial approaches yielded more matchups (96, 81 and 98% of possible matchups for MODIS, SeaWiFS and VIIRS respectively) than the semi-analytical methods (between 57 and 87% of possible matchups depending on sensors, global and regional methods). For the semi-analytical approach, the regional versions systematically provided more valid pixels that the original one.

Table 3. Statistics computed for OCx, POLY1, POLY4, Garver–Siegel Maritorena (GSM)_ORIG, GSM_GC, and GSM_GS algorithms for the Northwest Atlantic. Note that other versions of the algorithms OCx and GSM were tested but only included in Appendix A, so the "win ratio" column does not add up to one for each sensor as it does not include all values. Also recall that as a result of the log and reverse log transformations, the ideal value of *MLE* and *MMLE* is one. *N* is the total number of matchups as defined in Section 2.3, and *n* is the number of matchups that returned valid *chla* for the particular algorithm.

Algorithm	N	n	Intercept	Slope	r^2	μ_{error}	RMSLE	MLE	MMLE	Win Ratio	Score
					MODIS						
OC3M	530	508	−0.067	0.839	0.45	−0.40	0.37	0.857	1.95	0.15	6
POLY1	530	508	-1.0×10^{-5}	1.00	0.57	0.081	0.34	1.00	1.85	0.037	6
POLY4	530	508	-4.2×10^{-6}	1.00	0.57	0.037	0.33	1.00	1.85	0.083	9
GSM_ORIG	530	354	−0.21	0.745	0.22	−0.84	0.47	0.644	2.16	0.32	2
GSM_GC	530	439	5.7×10^{-3}	0.995	0.56	0.13	0.34	1.01	1.88	0.10	5
GSM_GS	530	439	7.3×10^{-3}	0.963	0.57	0.10	0.33	1.02	1.86	0.023	5
					SeaWiFS						
OC4	416	336	−0.035	0.674	0.55	−0.58	0.32	0.928	1.75	0.25	7
POLY1	416	336	1.1×10^{-5}	1.00	0.62	−0.062	0.31	1.00	1.72	0.093	6
POLY4	416	336	-1.2×10^{-5}	1.00	0.62	−0.087	0.31	1.00	1.73	4.2×10^{-3}	5
GSM_ORIG	416	239	−0.18	0.734	0.28	−0.87	0.44	0.681	2.15	0.23	1
GSM_GC	416	282	0.025	0.957	0.64	0.057	0.29	1.06	1.69	0.11	5
GSM_GS	416	282	0.044	0.938	0.66	0.11	0.29	1.11	1.68	0.034	6
					VIIRS						
OC3V	176	172	−0.12	0.728	0.38	−0.45	0.37	0.831	1.90	0.13	3
POLY1	176	172	3.1×10^{-5}	1.00	0.55	-8.7×10^{-3}	0.33	1.00	1.82	0.035	6
POLY4	176	172	1.1×10^{-5}	1.00	0.55	−0.018	0.33	1.00	1.81	0.15	7
GSM_ORIG	176	144	−0.12	0.600	0.55	−0.45	0.31	0.906	1.78	0.24	5
GSM_GC	176	153	−0.036	0.959	0.56	−0.013	0.32	0.934	1.74	0.069	5
GSM_GS	176	153	0.018	1.04	0.58	0.054	0.32	1.02	1.75	0.21	6

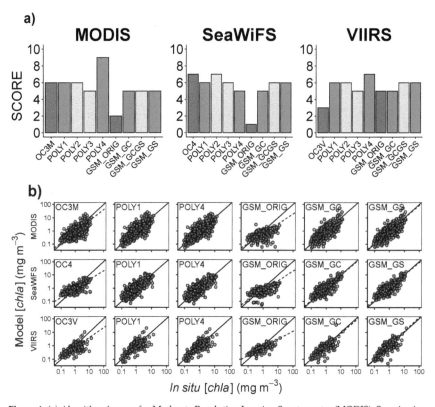

Figure 4. (**a**) Algorithms' scores for Moderate Resolution Imaging Spectrometer (MODIS), Sea-viewing Wide Field-of-view Sensor (SeaWiFS) and Visible Infrared Imaging Radiometer Suite (VIIRS) in the NWA based on the mean log-transformed error (*MLE*), mean magnitude of log-transformed error (*MMLE*), and r^2 from the linear regression of satellite-derived on *in situ chla*, the percentage of retrievals, and the win ratio metric. Red bars correspond to the original algorithms, green bars correspond to regionally-tuned algorithms discussed in the text, grey bars corresponds to regionally-tuned algorithms presented only in Tables A1 and A2 of Appendix A. (**b**) Satellite-derived versus *in situ chla*, green solid circles correspond to the spring season and red solid circles correspond to the fall.

As shown in Table 3, all the generic algorithms (i.e., OCx and GSM_ORIG) had a slope of the linear regression of satellite-derived *chla* on *in situ chla* lower than 0.84 (0.839 for OC3M) and as low as 0.600 for VIIRS GSM_ORIG. Note that the slope of the linear regression of the tuned algorithms was forced to one. The regional algorithms also had a higher correlation coefficient than the original methods for all sensors, all algorithms. The *RMSLE* exhibited values (from 0.29 to 0.47) that were in agreement with values reported in the literature for regional algorithms for the retrieval of *chla* [40–42]. Tuned algorithms showed smaller *RMSLE* than original algorithms for both semi-analytical and band ratio approaches except for the GSM approaches for VIIRS. The *MLE* (Equation (13)) was lower than one for the original algorithms for all three sensors, as they have an overall negative bias, as seen in Figure 4b. After correcting the bias in the regional algorithms, the *MLE* was closer to one (particularly in the band ratio models) and the *MMLE* was reduced. In general, the improvement in the *MMLE* was greater for the GSM-type approaches than for the OCx-type approaches. The better performance of the regional algorithms is most noticeable in the mean error, which generally improved by a full order of magnitude. When the algorithms were applied to the identical subset of retrievals, GSM_ORIG showed the highest win ratio for MODIS and VIIRS, while narrowly losing to OC4 in the SeaWiFS dataset. The regional linear and fourth-degree polynomial fits (POLY1 and POLY4) provided very similar results for all sensors, which is consistent with the study of Laliberté et al. [15] in the Gulf of Saint-Lawrence (Canada).

3.3. Algorithm Performance in the NEP

For the NEP, the original OCx algorithms exhibited a better linear fit than for the NWA (Table 4 and Figure 5b). For instance, OC3M showed a slope and intercept of the regression of satellite versus *in situ* data of 0.966 and 0.021 respectively against 0.839 and -0.067 for the NWA. The good performance of the original algorithms is also reflected in the correlation coefficients ($r^2 \geq 0.55$) for both original algorithms (OCx and GSM_ORIG) and all sensors, as they have higher values than for the NWA, with the exception of GSM_ORIG applied to VIIRS data where r^2 is 0.48 (Table 4). The regional algorithms, which aimed at reducing the *RMSLE* and forcing the slope and intercept to one and zero respectively, also improved the *MLE*. The *RMSLE* for all algorithms and sensors in the NEP are within the same order of magnitude as in the NWA except for the regional GSM models for SeaWiFS (i.e., GSM_GC and GSM_GS), which exhibited low *RMSLE* of 0.16 and 0.17 for GSM_GC and GSM_GS, respectively. As in the NWA, regional tuning generally improved the mean error, with the exception of the OCx and band ratio models in the MODIS dataset, where the magnitude of mean error increased, attesting to the good performance of the ocean colour model currently in use for this region and sensor. The semi-analytical algorithms (GSM-type) exhibited higher win ratios in general across all sensors, with GSM_ORIG showing the highest win ratio for MODIS and VIIRS (0.28 and 0.25 respectively), while GSM_GS showed the highest win ratio for SeaWiFS. Ultimately, the simple linear band ratio model (POLY1) obtained the highest score across the three sensors but tied with POLY4 and the regional GSM models in the SeaWiFS dataset. It is worth noting that other variations of the band ratio models (POLY2 and POLY3) also performed well, with the third-degree polynomial (POLY3) outperforming POLY1 in the SeaWiFS set, and POLY2 obtaining a higher score than other polynomials in the NWA dataset (Table A1). Similarly, the MODIS NEP dataset contained an exception to the pattern of high-scoring GSM_GC and GSM_GS, where GSM_GCGS outperformed the two.

Table 4. Same as Table 3, for the NEP.

Algorithm	N	n	Intercept	Slope	r^2	μ_{error}	RMSLE	MLE	MMLE	Win Ratio	Score
					MODIS						
OC3M	487	461	0.021	0.966	0.59	−0.016	0.37	1.05	1.87	0.12	5
POLY1	487	461	-5.3×10^{-6}	1.00	0.66	−0.21	0.34	1.00	1.78	0.11	7
POLY4	487	461	1.2×10^{-5}	1.00	0.67	−0.27	0.33	1.00	1.79	0.040	5
GSM_ORIG	487	279	−0.18	0.729	0.55	−0.96	0.37	0.708	1.81	0.28	3
GSM_GC	487	356	0.011	1.02	0.69	0.14	0.30	1.02	1.67	0.11	5
GSM_GS	487	355	2.1×10^{-3}	0.968	0.69	−0.013	0.29	1.01	1.65	0.040	5
					SeaWiFS						
OC4	45	40	−0.042	0.851	0.63	−0.34	0.30	0.897	1.65	0.15	3
POLY1	45	40	-1.3×10^{-5}	1.00	0.75	-3.8×10^{-4}	0.25	1.00	1.55	0.030	6
POLY4	45	40	5.3×10^{-6}	1.00	0.75	-5.5×10^{-3}	0.25	1.00	1.54	0.091	6
GSM_ORIG	45	33	−0.028	0.634	0.81	−0.42	0.23	0.959	1.56	0.15	4
GSM_GC	45	34	0.017	0.958	0.88	9.0×10^{-3}	0.16	1.04	1.30	0	6
GSM_GS	45	34	4.4×10^{-3}	0.972	0.87	−0.010	0.17	1.01	1.32	0.21	6
					VIIRS						
OC3V	342	332	−0.042	0.930	0.63	−0.36	0.34	0.902	1.78	0.17	6
POLY1	342	332	7.7×10^{-6}	1.00	0.69	−0.13	0.32	1.00	1.74	0.045	7
POLY4	342	332	1.4×10^{-5}	1.00	0.69	−0.17	0.31	1.00	1.74	0.063	6
GSM_ORIG	342	230	−0.10	0.725	0.48	−0.88	0.38	0.831	1.80	0.25	2
GSM_GC	342	265	0.021	1.01	0.68	0.095	0.29	1.05	1.60	0.13	6
GSM_GS	342	266	-3.4×10^{-3}	1.01	0.66	0.21	0.31	0.991	1.62	0.063	5

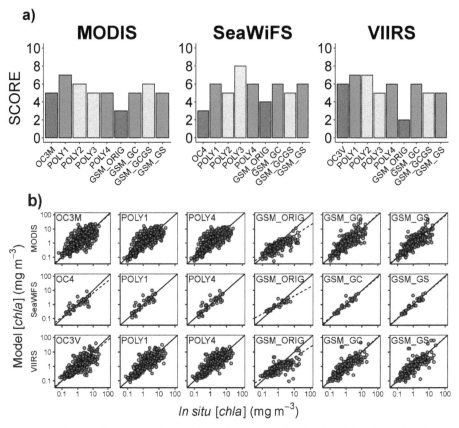

Figure 5. Same as Figure 4, for the NEP. Note the colour-coding for the solid circles in (**b**): blue corresponds to winter, green to spring, purple to summer, and red to fall.

3.4. Variation of Satellite-Derived Chla with Environmental Factors

3.4.1. Patterns in *Chla* Uncertainties with Phytoplankton Composition, Time, and Number of Retrievals

The correlation analysis described in Section 2.6 revealed that the magnitude of uncertainty in the retrieved *chla* ($|E_{chla}|$) was strongly positively correlated to *chla* concentration for all sensors, regions, and algorithms, with r^2 varying from 0.40 (VIIRS, GSM_GC, NWA) to 0.97 (GSM_ORIG, NWA, all sensors). GSM_GC presented an exception in the NEP SeaWiFS dataset with a weaker correlation between $|E_{chla}|$ and *chla* that can be explained by the very low number of matchups (Tables 5 and A2). Similarly, we found a relationship between $|E_{chla}|$ and the ratio of *fucox* to *chla* concentration for all algorithms, which was weaker in the regionally-tuned algorithms (Table 5) indicating that phytoplankton taxonomic composition has an impact on these types of algorithms, which can be reduced by the use of the regional models.

Table 5. Pearson's correlation coefficients (*r*) from the test for correlation between $|E_{chla}|$ and phytoplankton composition and time (see Section 2.6). * = statistically significant (*p*-value < 0.05).

	MODIS		SeaWiFS		VIIRS		MODIS		SeaWiFS		VIIRS	
			NWA						NEP			
Algorithm	N	r	N	r	N	r	N	r	N	r	N	r
						Time						
OCx	508	−0.022	336	0.18 *	172	−0.10	461	0.17 *	40	0.20	332	−0.031
POLY1	508	1.1×10^{-3}	336	0.10	172	−0.087	461	0.17 *	40	0.14	332	6.7×10^{-4}
POLY4	508	3.4×10^{-3}	336	0.11 *	172	−0.088	461	0.17 *	40	0.15	332	0.020
GSM_ORIG	354	−0.058	239	0.27 *	144	−0.094	279	0.076	33	0.26	230	−0.083
GSM_GC	439	−0.14	282	0.16 *	153	−0.043	356	6.5×10^{-4}	34	0.25	265	−0.033
GSM_GS	439	−0.14	282	0.15 *	153	−0.012	355	-8.1×10^{-3}	34	0.21	266	−0.10
						Day of year						
OCx	508	−0.41 *	336	−0.35 *	172	−0.46 *	461	0.14 *	40	0.15	332	0.14 *
POLY1	508	−0.33 *	336	−0.26 *	172	−0.39 *	461	0.15 *	40	0.11	332	0.15 *
POLY4	508	−0.35 *	336	−0.26 *	172	−0.41 *	461	0.15 *	40	0.11	332	0.16 *
GSM_ORIG	354	−0.48 *	239	−0.40 *	144	−0.35 *	279	0.065	33	0.19	230	0.13 *
GSM_GC	439	−0.22 *	282	−0.21 *	153	−0.25 *	356	0.12 *	34	0.039	265	0.088
GSM_GS	439	−0.22 *	282	−0.22 *	153	−0.39 *	355	0.12 *	34	0.087	266	0.095
						[Chla]						
OCx	508	0.78 *	336	0.91 *	172	0.96 *	461	0.71 *	40	0.68 *	332	0.82 *
POLY1	508	0.56 *	336	0.70 *	172	0.70 *	461	0.79 *	40	0.46 *	332	0.81 *
POLY4	508	0.61 *	336	0.72 *	172	0.73 *	461	0.80 *	40	0.46 *	332	0.82 *
GSM_ORIG	354	0.97 *	239	0.97 *	144	0.97 *	279	0.95 *	33	0.87 *	230	0.91 *
GSM_GC	439	0.52 *	282	0.64 *	153	0.40 *	356	0.59 *	34	0.28 *	265	0.57 *
GSM_GS	439	0.52 *	282	0.66 *	153	0.73 *	355	0.64 *	34	0.42 *	266	0.63 *
						[Fucox]/[chla]						
OCx	508	0.37 *	336	0.36 *	172	0.53 *						
POLY1	508	0.27 *	336	0.34 *	172	0.43 *						
POLY4	508	0.29 *	336	0.34 *	172	0.45 *						
GSM_ORIG	354	0.48 *	239	0.38 *	144	0.54 *						
GSM_GC	439	0.26 *	282	0.34 *	153	0.27 *						
GSM_GS	439	0.27 *	282	0.35 *	153	0.38 *						

There were no significant long-term trends of error magnitude in the time series for all sensors and both regions (Table 5). However, there was a negative correlation between the magnitude of error and the day of year across all sensors in the NWA, which is also apparent in the larger spread of error (the interquartile range) occurring during the first half of each year (1.5 mg m^{-3} on average across sensors for both POLY4 and GSM_GS) than during the last half (0.59 and 0.63 for POLY4 and GSM_GS respectively), with a general negative bias in the spring across the whole time series (Figure 6 and Table A6). The differences in the spread of error between the first and last halves of the year is generally less noticeable in the NEP (Figure 7), where there is a smaller positive correlation between error magnitude and day of year, corresponding to a larger average spread of error in the second half of each year (1.0 and 0.57 mg m^{-3} for POLY4 and GSM_GS respectively) than in the first half (0.51 and 0.34). This can possibly be attributed in part to the timing of the samples, which were spread out over the year instead of concentrated in the spring and fall as in the NWA.

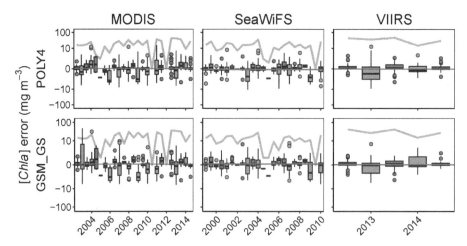

Figure 6. Model residuals across the full-time series for the POLY4 and GSM_GS algorithms in the NWA for MODIS, SeaWiFS and VIIRS. Blue lines represent μ_{error}, and the grey lines indicate the number of valid retrievals. Green boxes represent the first half of the year (January to June) and red boxes represent the second half (July to December). Note the axis is asinh-transformed.

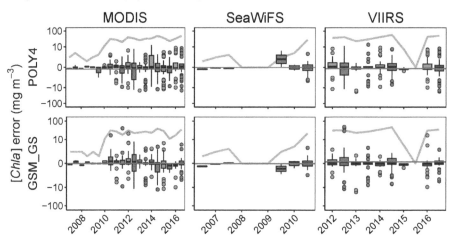

Figure 7. Same as Figure 6, for the NEP.

A *t*-test and ANOVA also revealed that in the NWA, the average μ_{error} across algorithms in the spring (−0.51, −0.60, and −0.60 mg m^{-3} for MODIS, SeaWiFS, and VIIRS respectively) were generally significantly different (opposite in sign and higher in magnitude) than in the fall (0.20, 0.19, and 0.032) (see Table A3), but there were no significant differences in the mean error between seasons in the NEP (Table A4). We did not find any correlation between the number of matchups and the magnitude of the uncertainties for a given season/sensor/region except when only a very low number of matchups was available (e.g., MODIS in spring 2005 in the NWA, 2 matchups, Figure 6) and one bad matchup could negatively weight the entire matchup dataset. In the NEP (Figure 7), there was a large number of outliers spread evenly across spring and fall seasons, unlike the NWA, where spring blooms generated the highest spread of error. Only a few data were available for SeaWiFS in the NEP, which were concentrated mainly in 2009 and 2010, and therefore the correlation and ANOVA results for this region/sensor dataset were less reliable. All variations of GSM also returned fewer valid retrievals than the band ratio algorithms, mainly due to negative remote sensing reflectances at either end of the spectrum, which prevented the model from generating a valid fit (Table 6).

Table 6. Missing retrievals by GSM_GS algorithm for each region/sensor combination, subdivided by the cause of lost data, with the percentage of lost retrievals given for each cause. $R_{rs}(41X)$ is the remote sensing reflectance value at 412 nm for MODIS and SeaWiFS, and 410 nm for VIIRS. $R_{rs}(6XX)$ varies from 645 to 671 nm depending on sensor. Note that there were no instances of negative model *chla*, unlike $a_{dg}(443)$ and $b_{bp}(443)$. "Unrealistic unknowns" occur when *chla*, $a_{dg}(443)$, or $b_{bp}(443)$ are outside their acceptable ranges, given in Equation (9). Multiple issues were a combination of $a_{dg}(443) < 0$ and *chla* above the acceptable range, with two instances where $b_{bp}(443)$ was also too high (>0.18).

	MODIS		SeaWiFS		VIIRS		MODIS		SeaWiFS		VIIRS	
Cause of Missing Pixel	**NWA**						**NEP**					
	#	%	#	%	#	%	#	%	#	%	#	%
$R_{rs}(41X) < 0$	34	49.3	28	51.9	7	36.8	66	62.3	0	0	26	39.4
$R_{rs}(6XX) < 0$	5	7.25	3	5.56	1	5.26	8	7.55	2	33.3	3	4.55
$a_{dg}(443) < 0$	27	39.1	20	37.0	6	31.6	17	16.0	4	66.7	18	27.3
$b_{bp}(443) < 0$	0	0	0	0	0	0	1	0.940	0	0	0	0
Unrealistic unknowns	3	4.35	2	3.70	2	10.5	7	6.60	0	0	3	4.55
Multiple issues	0	0	0	0	0	0	7	6.60	0	0	2	3.03
Unexplained	0	0	1	1.85	3	15.8	0	0	0	0	14	21.2
Total	69		54		19		106		6		66	

3.4.2. Uncertainties Related to Geographic Considerations

The correlation analysis of $|E_{chla}|$ on latitude, bathymetry and distance from *in situ* to satellite measurements was carried out to investigate if these variables could explain some of the biases and discrepancies observed between the *in situ* and satellite-derived *chla* (Table 7). There were no correlations observed between $|E_{chla}|$ and latitude, which is an interesting result given the range of latitudes in the NWA dataset from approximately 38° to 61°N. Possible exceptions exist in the NWA VIIRS and NEP SeaWiFS datasets, but as mentioned in the previous section, the SeaWiFS dataset had very few data, and the high latitude data in the VIIRS dataset were all retrieved during the spring bloom (see Figure 1a,b), when *chla* concentration would typically be higher and thus $|E_{chla}|$ would likely also be higher as the two were collinear (see Section 3.4.1).

There were low negative correlations between $|E_{chla}|$ and open ocean bathymetry in the NWA MODIS dataset, and larger negative correlations in the NEP, which again can be explained by the collinearity of *chla* concentration and $|E_{chla}|$ and the fact that in the open ocean subset, bathymetry was also negatively correlated with *chla* concentration ($r = -0.47$). As for seasonal bias, the t-test on the coastal and offshore datasets in the NWA revealed that the mean uncertainties for both datasets were statistically different in the regional band ratio models for MODIS and SeaWiFS, with μ_{error} ranging from 0.14 to 0.42 mg m^{-3} for the coastal regions and -0.31 to -0.20 in open ocean, indicating that this family of algorithms systematically underestimated *chla* offshore and overestimates in shallower water, while the difference was less significant with the GSM-type algorithms (see Table A5). The pattern reversed in the NEP, where the regional band ratio underestimated coastal *chla* (-0.90 to -0.77 mg m^{-3}) and overestimated *chla* in the open ocean (0.076 to 0.27), again ignoring the SeaWiFS dataset due to insufficient data.

Distance between a satellite matchup and the *in situ* measurement was up to 10 km to maximize the number of matchups, though over half the matchups in each region/sensor subset were within 2 km of the measurement. Regression of $|E_{chla}|$ on the distance did not reveal a relationship ($|r| < 0.1$) for any datasets except NWA VIIRS GSM_ORIG, and the small NEP SeaWiFS dataset, suggesting that in open waters, where spatial variations in *chla* are small (mesoscale) compared to the coastal environment, the stringent criteria of pixel to *in situ* collocation can be relaxed to increase the number of matchups.

Table 7. Pearson's correlation coefficients (*r*) for the test for correlation between $|E_{chla}|$ and geographic environmental factors (see Section 2.6). * = statistically significant (*p*-value < 0.05).

	MODIS		SeaWiFS		VIIRS		MODIS		SeaWiFS		VIIRS	
			NWA						NEP			
Algorithm	N	r	N	r	N	r	N	r	N	r	N	r
Latitude												
OCx	508	0.095 *	336	−0.077	172	0.37 *	461	0.011	40	0.23	332	0.022
POLY1	508	0.076	336	−0.022	172	0.33 *	461	-4.3×10^{-3}	40	0.19	332	0.019
POLY4	508	0.079	336	−0.031	172	0.34 *	461	−0.010	40	0.20	332	0.016
GSM_ORIG	354	0.049	239	−0.12	144	0.30 *	279	−0.076	33	0.13	230	−0.075
GSM_GC	439	−0.017	282	−0.014	153	0.21 *	356	3.2×10^{-3}	34	0.45 *	265	−0.076
GSM_GS	439	−0.018	282	−0.021	153	0.35 *	355	4.2×10^{-3}	34	0.38 *	266	−0.084
Open ocean												
OCx	276	−0.16 *	171	−0.13	98	−0.029	278	−0.45 *	30	−0.41 *	197	−0.41 *
POLY1	276	−0.18 *	171	-0.13	98	-0.059	278	−0.44 *	30	−0.38 *	197	−0.45 *
POLY4	276	−0.18 *	171	-0.13	98	-0.059	278	−0.45 *	30	−0.39 *	197	−0.46 *
GSM_ORIG	212	−0.17 *	132	−0.14	87	−0.033	202	−0.32 *	25	−0.53 *	161	−0.28 *
GSM_GC	244	−0.15 *	148	−0.10	89	3.9×10^{-3}	238	−0.33 *	26	−0.54 *	177	−0.32 *
GSM_GS	244	−0.15 *	148	−0.10	89	0.074	237	−0.35 *	26	−0.47 *	177	−0.33 *
Coastal ocean												
OCx	232	-9.3×10^{-3}	165	−0.051	74	0.13	186	−0.083	10	−0.59	137	−0.090
POLY1	232	−0.011	165	−0.11	74	0.12	186	−0.10	10	−0.37	137	−0.090
POLY4	232	−0.024	165	−0.10	74	0.11	186	−0.11	10	−0.36	137	−0.079
GSM_ORIG	142	0.032	107	-1.5×10^{-3}	57	0.18	78	−0.091	8	−0.05	69	−0.051
GSM_GC	195	−0.019	134	0.052	64	0.22	119	-1.9×10^{-3}	8	−0.48	88	0.037
GSM_GS	195	−0.017	134	0.054	64	0.18	119	-8.3×10^{-3}	8	−0.36	89	0.015
Distance to in situ measurement												
OCx	508	−0.033	336	0.037	172	−0.038	461	0.036	40	−0.11	332	0.020
POLY1	508	−0.054	336	0.020	172	−0.085	461	0.046	40	−0.14	332	0.037
POLY4	508	−0.049	336	0.026	172	−0.085	461	0.050	40	−0.13	332	0.052
GSM_ORIG	354	−0.057	239	−0.15 *	144	−0.026	279	-6.3×10^{-3}	33	−0.32	230	−0.010
GSM_GC	439	−0.055	282	−0.077	153	−0.045	356	0.047	34	−0.32	265	0.076
GSM_GS	439	−0.059	282	−0.075	153	−0.072	355	0.057	34	−0.35 *	266	0.012

4. Discussion

Satellite-derived *chla* concentration remains the most used product to infer global scale information on the status of the marine ecosystem, and non-specialists represent an important fraction of ocean colour data users. Here we have assessed the performance of two algorithms currently implemented in NASA's SeaDAS software (https://seadas.gsfc.nasa.gov) in Canadian waters (Atlantic and Pacific Oceans) to inform on biases associated with these algorithms in those regions. In addition, we carried out modifications of these original models by optimizing their parameterization and formulation to correct for regional bias in the NWA and NEP. Both OCx and GSM exhibited a negative overall bias particularly in the NWA, where low *chla* were overestimated and high concentrations were underestimated, which resulted in a linear fit of satellite-derived on *in situ chla* with a slope lower than one. This bias was partially corrected in the modified versions of the models, which also improved the tightness of the relationship as indicated by higher r^2 than for the original algorithms, but at the expense of the number of individual retrievals with the lowest magnitude of error between different algorithms ("win ratio", Tables 3 and 4). Significantly different means in *chla* concentration between spring and fall in the NWA (see Section 3.4.1) suggest that the dataset could be further subdivided by season, but at the expense of a synoptical approach. It would also require a larger matchup dataset to achieve statistical significance.

Our approach differed from NASA's in that the algorithms were fitted directly between satellite-derived and *in situ chla*, under the assumption that the satellite R_{rs} were reliable, while the original algorithms were developed using both *in situ* radiometric and *chla* measurements. Furthermore, the regional parameters were optimized, tested, and compared to the original algorithms using the entire dataset for a given region and sensor, rather than extracting a subset for training and using the remainder for testing. Confidence intervals for the polynomial algorithm coefficients were retrieved using a bootstrap method that subsamples the original dataset to compute 2000 different sets of coefficients (see Section 2.4.2), revealing interval widths ranging from 0.08 to 1.26 for the first three coefficients and 0.38 to 2.39 for the last two, which had the smallest impact on the shape of the polynomial in the region defined by the band ratio values. This suggests that the coefficients derived from the full dataset are representative of any subsets. For the GSM algorithms, the sensitivity study on combinations of spectral variations in *P*, *S*, and *Y* (see Section 2.4.4) was performed for each region, sensor, and set of *g* coefficients (spectral or constant), and the median of the subset of highest-scoring combinations was selected. For this reason, we can assume that using the entire dataset or a subset for each combination would have led to similar results after retrieving the median of the optimal combinations. Finally, the small size of some of the datasets, particularly the NEP SeaWiFS set, presented a challenge in developing a reliable set of parameters spanning a wide range of values for *in situ chla* and the environmental variables that were tested for correlations with algorithm error. For these reasons, the regional tuning of the algorithms was carried out on the full datasets rather than subsets.

One of the main results of our studies was the identification of the 443 nm waveband as a strong contributor to the overall uncertainty in the OCx algorithms. Removing this waveband from the OCx algorithms and iteratively forcing the slope and intercept of the linear model between satellite-derived and *in situ chla* to one and zero respectively provided better results than the original algorithms (Table 3). Increasing the degree of the polynomial improved the results for some combinations of region and sensor (see scores in Tables 3 and 4), but overall offered minimal improvements. The negative bias of the original GSM model was corrected by introducing a new exponent, *P*, on the chlorophyll term, which lowered satellite-derived *chla* concentrations <1 mg m^{-3} and increased concentrations >1 mg m^{-3}. This exponent *P* was derived in combination with the exponents on the a_{dg} and b_{bp} terms, giving the best set of spectral slopes for each region and sensor. Inclusion of spectral dependence in the *g* coefficients in Equation (11) slightly improved the results, however, optimization of the spectral slopes of the phytoplankton and yellow substances absorption terms, as well as the particulate backscattering term, had a more positive impact. This finding highlights the importance of accounting for regional properties of absorption and backscattering terms, and particularly their spectral dependence. Overall, the GSM-type algorithms provided the highest win ratio (i.e., the highest number of matchups with the smallest *chla* uncertainties), however, this good performance was hampered by the lower number of retrievals compared to the polynomial formulations (Tables 3 and 4). This limitation has to be considered according to the end-user's applications of the *chla* product.

Uncertainties in satellite-derived products are currently an active field of research (IOCCG report on Uncertainties in Ocean Colour Remote Sensing, in preparation) that remains highly complex given the possible sources of uncertainties associated with satellite observations that include, among others, uncertainties in radiometry (and calibration), atmospheric corrections, data binning, and bio-optical algorithms. For instance, it has been shown that at the global scale, the OCx algorithm contains an inherent uncertainty of about 35% [43]. The difference in scales between satellite and *in situ* measurements also creates another source of uncertainty, as *in situ* data refer to about two litres or less of seawater at a discrete depth while satellites integrate a volume of tens of millions of litres of seawater. Note that uncertainties in HPLC-derived *chla* were partially accounted for in the linear model comparing the results by using Type 2 linear regression as described in Section 2.5.1, which minimizes the sum of areas of triangles between each point and the regression line, assuming that both variables contain errors. Here, we have addressed discrepancies between satellite-derived *chla* and *in situ chla* and included temporal and spatial effects. For the two areas of interest, namely the Northwest Atlantic and the Northeast Pacific, we did not find significant patterns of temporal drift, thanks to regular reprocessing of ocean colour products by NASA following a vicarious calibration exercise [44] that ensures stable measurements of the radiative field with time. We did not find spatial patterns, with the exception of minor positive correlations between latitude and magnitude of error in the NWA VIIRS dataset, and negative correlations between open ocean bathymetry and error magnitude in the NEP datasets, which are likely due to changes in *chla* concentration that affect the degree of error. Seasonal biases were observed with an overall underestimation of *chla* in the spring and a small overestimation in the fall. This could be explained by the change in phytoplankton community composition (Table 5), such that when the ratio of fucoxanthin, a marker pigment for diatoms, to *chla* concentration increased, the magnitude of error in *chla* retrieval also increased. This pattern was observed across all algorithms, but weaker in the regionally-tuned algorithms. GSM-type algorithms offered the greatest improvements, in particular, GSM_GC in the VIIRS dataset where the correlation coefficient of $|E_{chla}|$ and the ratio of fucox to *chla* was reduced from 0.54 to 0.27. This demonstrates the effectiveness of regional algorithms at addressing changes in phytoplankton absorption properties, and notably the modified GSM algorithms, which include an exponent on the phytoplankton absorption term to account for changes in absorption properties with changes in phytoplankton community composition in the Northwest Atlantic [45–47]. However, the fact that the $fucox$ to *chla* ratio impacted the regional algorithms for SeaWiFS (i.e., both OCx-like and GSM-like) to a similar degree as the original models suggests that the 510 nm band, which is used across all models for the SeaWiFS sensor, can contain information on the presence of diatoms, as highlighted in Sathyendranath et al. [48] who developed an algorithm to discriminate diatoms from other phytoplankton types for SeaWiFS.

Finally, another notable source of uncertainty is the atmospheric correction method applied to NASA's level 2 images, based on the "black pixel assumption", which assumes that scattering in the near-infrared bands from approximately 670 nm to 865 nm [49] is negligible, thus the ocean appears black. This was a reasonable assumption for case 1 waters [50], but optically-complex case 2 waters can contain elements that contribute to the scattering in those bands, giving an inaccurate correction of remote sensing reflectance values used in *chla* algorithms [51]. Given that many matchups in this report were near coastal or shelf areas, there could have been a degree of error associated with poor atmospheric correction. As discussed in Section 3.4.1, many satellite matchups did contain negative R_{rs} in a waveband near one end of the visible spectrum. These matchups were incapable of computing valid *chla* concentrations for the GSM algorithms but were included in the band ratio algorithms.

5. Conclusions

This comparison exercise between satellite-derived and *in situ chla* concentration measurements for three ocean colour sensors (i.e., SeaWiFS, MODIS and VIIRS), two widely-used algorithms based on different approaches (i.e., band ratio and semi-analytical) and for two regions (i.e., Northwest Atlantic and Northeast Pacific) revealed systematic bias (slopes varying from 0.60 to 0.84 in the NWA and from 0.63 to 0.97 in the NEP) and relatively large uncertainties. While regional tuning permitted correction for the biases, the uncertainties remained substantial, attesting to the fact that natural variations of phytoplankton optical properties, the presence of other optically active components, and accuracy of the atmospheric correction will challenge accurate retrieval of *chla* concentration. In the Northwest Atlantic, the removal of the 443 nm band in the regional band ratio algorithm improved the performance of the OCx-type algorithms. It is noteworthy that while band ratio algorithms provided the most matchups between satellite and *in situ* data, the GSM-type approach appeared to be most accurate thanks to its ability to decipher the phytoplankton signal from yellow substances, but its performance was hindered by the significantly lower number of retrievals, mainly due to negative radiances present in the blue or red part of the visible spectrum, perhaps due to poor atmospheric corrections.

Author Contributions: E.D. conceptualized and led the project, E.D. and S.C. developed the methodology. S.C. wrote most of the code and carried out all the statistical analyses. B.D. contributed to the coding and provided technical support. A.P. provided the dataset for the Northeast Pacific. S.C. and E.D. drafted the manuscript, which was edited by B.D. and A.P.

Funding: This research was funded by MEOPAR under the Observation Core Project and the Strategic Program for Ecosystem-Based Research and Advice (SPERA) from Fisheries and Oceans Canada.

Acknowledgments: The authors would like to thank NASA's Ocean Biology Processing Group (OBPG) for supplying a large up-to-date satellite ocean colour dataset available for public use (https://oceancolor.gsfc.nasa.gov/), as well as detailed information on the current ocean colour algorithms.

Conflicts of Interest: The authors declare no conflict of interest.

Abbreviations

The following abbreviations are used in this manuscript:

NWA	Northwest Atlantic
NEP	Northeast Pacific
SeaWiFS	Sea-viewing Wide Field-of-view Sensor
MODIS	Moderate Resolution Imaging Spectrometer
VIIRS	Visible Infrared Imaging Radiometer Suite
$RMSLE$	Root mean squared logarithmic error
MLE	Mean log-transformed error
$MMLE$	Mean magnitude of log-transformed error
OCx	Ocean Colour X
GSM	Garver–Siegel Maritorena
N	total number of matchups
n	number of valid retrievals
μ_{error}	mean error
r^2	coefficient of determination
E_{chla}	error in chlorophyll-a concentration retrieval
$|E_{chla}|$	magnitude of the error in chlorophyll-a concentration retrieval

Appendix A. Results Omitted from the Main Text

Table A1. Statistics of extra algorithms not presented in the text, as in Tables 3 and 4.

Algorithm	N	n	Intercept	Slope	r^2	μ_{error}	$RMSLE$	MLE	$MMLE$	Win Ratio	Score
					NWA—MODIS						
POLY2	530	508	-1.2×10^{-5}	1.00	0.57	1.3×10^{-3}	0.33	1.00	1.85	0.10	6
POLY3	530	508	1.4×10^{-5}	1.00	0.57	0.036	0.33	1.00	1.85	0.011	5
GSM_GCGS	530	439	0.037	0.959	0.57	0.23	0.33	1.09	1.86	0.17	5
					NWA—SeaWiFS						
POLY2	416	336	-5.5×10^{-7}	1.00	0.62	-0.10	0.31	1.00	1.74	0.068	7
POLY3	416	336	1.9×10^{-5}	1.00	0.62	-0.077	0.31	1.00	1.73	0.076	6
GSM_GCGS	416	283	0.065	0.931	0.66	0.19	0.29	1.17	1.69	0.14	6
					NWA—VIIRS						
POLY2	176	172	5.8×10^{-5}	1.00	0.55	-0.11	0.33	1.00	1.83	0.090	6
POLY3	176	172	3.9×10^{-5}	1.00	0.55	-0.029	0.33	1.00	1.81	6.9×10^{-3}	5
GSM_GCGS	176	154	3.7×10^{-3}	0.919	0.57	0.15	0.30	1.04	1.72	0.069	6

Table A1. *Cont.*

Algorithm	N	n	Intercept	Slope	r^2	μ_{error}	RMSLE	MLE	MMLE	Win Ratio	Score
				NEP—MODIS							
POLY2	487	461	-1.9×10^{-5}	1.00	0.67	-0.36	0.33	1.00	1.80	0.098	6
POLY3	487	461	1.4×10^{-5}	1.00	0.67	-0.33	0.33	1.00	1.79	0.040	5
GSM_GCGS	487	355	0.039	0.975	0.69	0.16	0.29	1.10	1.66	0.16	6
				NEP—SeaWiFS							
POLY2	45	40	2.9×10^{-5}	1.00	0.75	-0.036	0.25	1.00	1.55	0	5
POLY3	45	40	2.2×10^{-7}	1.00	0.75	-0.071	0.25	1.00	1.55	0.21	8
GSM_GCGS	45	34	0.038	0.941	0.88	0.083	0.16	1.09	1.32	0.15	5
				NEP—VIIRS							
POLY2	342	332	-4.9×10^{-5}	1.00	0.69	-0.23	0.31	1.00	1.74	0.080	7
POLY3	342	332	1.7×10^{-5}	1.00	0.69	-0.18	0.31	1.00	1.73	0.022	5
GSM_GCGS	342	270	0.026	0.997	0.56	0.028	0.35	1.06	1.69	0.17	5

Table A2. Pearson's correlation coefficients (r) from the test for correlation between $|E_{chla}|$ and environmental factors (see Section 2.6) for model versions not presented in the text. * = statistically significant (p-value < 0.05).

	MODIS		SeaWiFS		VIIRS		MODIS		SeaWiFS		VIIRS	
			NWA						NEP			
Algorithm	N	r	N	r	N	r	N	r	N	r	N	r
						Latitude						
POLY2	508	0.081	336	-0.035	172	0.38 *	461	-0.014	40	0.21	332	0.010
POLY3	508	0.079	336	-0.028	172	0.34 *	461	-0.011	40	0.22	332	0.016
GSM_GCGS	439	-0.018	283	-0.020	155	0.19 *	355	2.9×10^{-3}	34	0.40 *	270	-0.092
						Open ocean						
POLY2	276	-0.18 *	171	-0.13	98	-0.068	278	-0.44 *	30	-0.41 *	197	-0.46 *
POLY3	276	-0.18 *	171	-0.13	98	-0.059	278	-0.44 *	30	-0.44 *	197	-0.46 *
GSM_GCGS	244	-0.14 *	149	-0.13	89	-0.016	237	-0.36 *	26	-0.45 *	179	-0.35 *
						Coastal ocean						
POLY2	232	-0.024	165	-0.097	74	0.067	186	-0.090	10	-0.39	137	-0.076
POLY3	232	-0.022	165	-0.10	74	0.11	186	-0.091	10	-0.41	137	-0.078
GSM_GCGS	195	-0.025	134	0.055	66	0.20	119	8.3×10^{-3}	8	-0.48	91	0.013
						Distance to *in situ* measurement						
POLY2	508	-0.047	336	0.027	172	-0.080	461	0.048	40	-0.14	332	0.053
POLY3	508	-0.049	336	0.024	172	-0.085	461	0.049	40	-0.13	332	0.053
GSM_GCGS	439	-0.060	283	-0.077	155	-0.050	355	0.056	34	-0.35 *	270	0.057
						Time						
POLY2	508	1.9×10^{-3}	336	0.11 *	172	-0.094	461	0.17 *	40	0.17	332	0.015
POLY3	508	2.8×10^{-3}	336	0.11	172	-0.089	461	0.17 *	40	0.21	332	0.020
GSM_GCGS	439	-0.14 *	283	0.14 *	155	-0.040	355	-0.020	34	0.22	270	-0.031
						Day of year						
POLY2	508	-0.35 *	336	-0.26 *	172	-0.42 *	461	0.15 *	40	0.12	332	0.16 *
POLY3	508	-0.35 *	336	-0.25 *	172	-0.41 *	461	0.15 *	40	0.12	332	0.16 *
GSM_GCGS	439	-0.20 *	283	-0.21 *	155	-0.20 *	355	0.11 *	34	0.088	270	0.094
						[Chla]						
POLY2	508	0.64 *	336	0.74 *	172	0.87 *	461	0.84 *	40	0.49 *	332	0.84 *
POLY3	508	0.61 *	336	0.72 *	172	0.75 *	461	0.83 *	40	0.54 *	332	0.82 *
GSM_GCGS	439	0.49 *	283	0.65 *	155	0.33 *	355	0.58 *	34	0.39 *	270	0.59 *
						[Fucox] / [chla]						
POLY2	508	0.30 *	336	0.34 *	172	0.47 *						
POLY3	508	0.29 *	336	0.34 *	172	0.45 *						
GSM_GCGS	439	0.25 *	283	0.34 *	155	0.22 *						

Table A3. Results of the *t*-test performed on μ_{error} of seasonal subsets in the NWA to test for statistically significant differences in mean error between spring and fall. N = size of subset. * = statistically significant (*p*-value < 0.05).

Algorithm	N_{spring}	N_{fall}	μ_{error}^{spring}	μ_{error}^{fall}	*p*-Value
			MODIS		
OCx	254	239	−0.86	0.074	3.3×10^{-7} *
POLY1	254	239	−0.077	0.26	0.14
POLY4	254	239	−0.18	0.27	0.024 *
GSM_ORIG	161	179	−1.7	−0.11	5.5×10^{-20} *
GSM_GC	220	204	−0.10	0.38	0.13
GSM_GS	221	203	−0.13	0.35	0.14
			SeaWiFS		
OCx	183	124	−1.1	0.040	2.1×10^{-8} *
POLY1	183	124	−0.32	0.29	2.7×10^{-3} *
POLY4	183	124	−0.37	0.30	5.8×10^{-4} *
GSM_ORIG	112	100	−1.7	−0.14	3.8×10^{-9} *
GSM_GC	144	109	−0.17	0.32	3.7×10^{-3} *
GSM_GS	144	109	−0.082	0.35	0.011 *
			VIIRS		
OCx	51	121	−1.4	−0.061	9.0×10^{-7} *
POLY1	51	121	−0.42	0.16	0.083
POLY4	51	121	−0.48	0.17	0.038 *
GSM_ORIG	39	105	−1.2	−0.16	1.0×10^{-5} *
GSM_GC	42	111	0.047	−0.036	0.87
GSM_GS	42	111	−0.098	0.11	0.33

Table A4. Results of the ANOVA on μ_{error} of seasonal subsets in the NEP, to check for statistically significant differences in means between seasons. The SeaWiFS subset has been excluded from this test, as all SeaWiFS matchups were collected during the summer, with the exception of a single matchup from fall and one from winter. Tukey Honest Significant Difference *p*-values between each pair of seasons are included to determine which seasons, if any, are statistically different from each other.

Algorithm	N_{spring}	N_{summer}	N_{fall}	μ_{error}^{spring}	μ_{error}^{summer}	μ_{error}^{fall}	*p*-Value	p_{spring}^{summer}	p_{spring}^{fall}	p_{summer}^{fall}
					MODIS					
OCx	87	194	135	−0.28	0.23	−0.20	0.47	0.56	0.99	0.58
POLY1	87	194	135	−0.37	−0.054	−0.38	0.63	0.75	1.0	0.67
POLY4	87	194	135	−0.40	−0.14	−0.44	0.69	0.82	1.00	0.71
GSM_ORIG	47	139	58	−1.5	−0.89	−1.3	0.39	0.41	0.91	0.66
GSM_GC	65	155	97	0.35	0.13	0.065	0.89	0.92	0.89	0.99
GSM_GS	64	154	97	0.24	−0.020	−0.20	0.75	0.87	0.73	0.92
					VIIRS					
OCx	67	115	116	−0.89	0.014	−0.51	0.34	0.34	0.83	0.60
POLY1	67	115	116	−0.70	0.17	−0.093	0.32	0.29	0.55	0.85
POLY4	67	115	116	−0.80	0.12	−0.11	0.23	0.21	0.41	0.87
GSM_ORIG	36	93	74	−1.5	−0.48	−1.4	0.085	0.19	0.98	0.14
GSM_GC	52	99	83	−0.52	0.30	0.29	0.42	0.44	0.47	1.0
GSM_GS	52	101	83	−0.68	0.97	−0.042	0.16	0.17	0.78	0.41

Table A5. Results of the *t*-test to check for differences in μ_{error} between coastal (bathymetry <= 200 m) and open ocean (>200 m) subsets. * = statistically significant (*p*-value < 0.05).

Algorithm	N Coastal	N Open	μ_{error} Coastal	μ_{error} Open	*p*-Value	N Coastal	N Open	μ_{error} Coastal	μ_{error} Open	*p*-Value
			NWA					NEP		
			MODIS							
OCx	232	276	−0.27	−0.51	0.20	183	275	−0.46	0.18	0.064
POLY1	232	276	0.42	−0.20	4.9×10^{-3} *	183	275	−0.78	0.083	4.6×10^{-3} *
POLY4	232	276	0.32	−0.20	6.7×10^{-3} *	183	275	−0.90	0.076	8.4×10^{-4} *
GSM_ORIG	142	212	−1.0	−0.71	0.077	77	201	−2.4	−0.41	7.3×10^{-9} *
GSM_GC	195	244	0.36	−0.066	0.17	118	237	0.17	0.12	0.91
GSM_GS	195	244	0.33	−0.090	0.19	118	236	−0.093	0.016	0.77
			SeaWiFS							
OCx	165	171	−0.53	−0.64	0.55	10	30	−0.17	−0.40	0.69
POLY1	165	171	0.19	−0.30	6.9×10^{-3} *	10	30	0.36	−0.12	0.39
POLY4	165	171	0.14	−0.31	0.011 *	10	30	0.37	−0.13	0.35
GSM_ORIG	107	132	−1.1	−0.72	0.17	8	25	−1.1	−0.20	0.032 *
GSM_GC	134	148	0.24	−0.11	0.029 *	8	26	0.19	−0.047	0.44
GSM_GS	134	148	0.34	−0.091	7.1×10^{-3} *	8	26	0.36	−0.12	0.17
			VIIRS							
OCx	74	98	−0.23	−0.61	0.13	135	195	−1.1	0.12	4.5×10^{-3} *
POLY1	74	98	0.40	−0.32	0.021	135	195	−0.77	0.27	9.0×10^{-3} *
POLY4	74	98	0.36	−0.31	0.020	135	195	−0.87	0.26	2.4×10^{-3} *
GSM_ORIG	57	87	−0.47	−0.43	0.87	69	161	−2.3	−0.28	5.5×10^{-7} *
GSM_GC	64	89	0.26	−0.21	0.29	88	177	0.29	3.7×10^{-4}	0.56
GSM_GS	64	89	0.010	0.085	0.70	89	177	0.77	−0.069	0.20

Table A6. Average difference between first and third quartiles presented in the boxplots of Figures 6 and 7 across boxes representing the first half of the year, and across the second half of the year.

	POLY4	GSM_GS	POLY4	GSM_GS	POLY4	GSM_GS
	MODIS		SeaWiFS		VIIRS	
	NWA					
January–June	1.7	2.0	1.1	0.94	1.7	1.5
July–December	0.74	0.84	0.57	0.60	0.46	0.44
	NEP					
January–June	0.65	0.37	0.33	0.27	0.54	0.37
July–December	1.0	0.72	0.94	0.38	1.1	0.62

Table A7. Coefficients of the remaining OCx algorithms for each sensor and region.

Coefficients	MODIS POLY2	MODIS POLY3	SeaWiFS POLY2	SeaWiFS POLY3	VIIRS POLY2	VIIRS POLY3
	NWA					
a	0.37539	0.37657	0.51424	0.52039	0.41461	0.44156
b	−3.12409	−3.26173	−3.59265	−3.75269	−2.54637	−3.05795
c	−0.75408	−0.60435	−0.95058	−0.92392	−1.47087	−0.65894
d	-	1.1404	-	1.71524	-	1.21248
e	-	-	-	-	-	-
	NEP					
a	0.28424	0.2805	0.42171	0.42506	0.33771	0.3303
b	−2.66996	−2.77728	−2.95509	−2.74285	−2.56462	−2.74252
c	−1.09915	−1.01747	−0.68104	−1.48743	−0.5314	−0.34545
d	-	0.92282	-	0.17624	-	1.35569
e	-	-	-	-	-	-

Table A8. g coefficients used in Equation (11) after performing linear interpolation to the necessary wavelengths.

λ	g_1	g_2	g_3
400	0.0742	0.0805	1.4839
410	0.0716	0.0820	1.4520
420	0.0697	0.0841	1.4353
430	0.0685	0.0862	1.4300
440	0.0697	0.0890	1.4595
450	0.0773	0.1009	1.6387
460	0.0801	0.1142	1.7489
470	0.0832	0.1394	1.9055
480	0.0869	0.2095	2.1890
490	0.0878	0.2621	2.3091
500	0.0875	0.2820	2.3212
510	0.0861	0.2568	2.2215
520	0.0844	0.2233	2.1058
530	0.0821	0.1967	1.9920
540	0.0800	0.1811	1.9097
550	0.0781	0.1717	1.8464
560	0.0763	0.1651	1.7968
570	0.0754	0.1624	1.7722
580	0.0757	0.1640	1.7816
590	0.0768	0.1712	1.8211
600	0.0781	0.1864	1.8879
610	0.0784	0.1939	1.9143
620	0.0783	0.1956	1.9172
630	0.0782	0.1969	1.9186
640	0.0780	0.1973	1.9160
650	0.0782	0.2009	1.9283
660	0.0789	0.2227	1.9923
670	0.0798	0.2513	2.0663
680	0.0795	0.2465	2.0510
690	0.0789	0.2270	2.0000
700	0.0791	0.2323	2.0137

Table A9. Exponents used in Equations (6) and (10), where the g_c columns give exponents optimized to the original constant g_1 and g_2 coefficients, and g_s columns give the exponents optimized for the use of the spectral g coefficients g_1, g_2, and g_3.

	MODIS		SeaWiFS		VIIRS		MODIS		SeaWiFS		VIIRS	
	NWA						NEP					
	g_c	g_s	g_c	g_s	g_c	g_s	g_c	g_s	g_c	g_s	g_c	g_s
median P	0.500	0.500	0.500	0.500	0.600	0.500	0.600	0.600	0.700	0.650	0.600	0.600
median S	0.038	0.036	0.035	0.034	0.026	0.026	0.038	0.036	0.028	0.026	0.034	0.030
median Y	0.800	0.750	0.600	0.525	1.400	1.750	0.900	0.750	0.750	0.650	0.800	0.750

Table A10. Spectrally-dependent $a^\star_{ph}$ coefficients used in Equation (10) for the wavebands used in MODIS, SeaWiFS, and VIIRS.

λ	$a^\star_{ph}$
410	0.054343
412	0.055765
443	0.063252
469	0.051276
486	0.04165
488	0.040648
490	0.039546
510	0.025105
531	0.015745
547	0.011477
551	0.010425
555	0.009382
645	0.008967
667	0.019878
670	0.022861
671	0.023646
678	0.024389

References

1. Vargas, M.; Brown, C.W.; Sapiano, M.R.P. Phenology of marine phytoplankton from satellite ocean color measurements. *Geophys. Res. Lett.* **2009**, *36*. [CrossRef]
2. Racault, M.F.; Quéré, C.L.; Buitenhuis, E.; Sathyendranath, S.; Platt, T. Phytoplankton phenology in the global ocean. *Ecol. Indic.* **2012**, *14*, 152–163. [CrossRef]
3. Siegel, D.A.; Buesseler, K.O.; Doney, S.C.; Sailley, S.F.; Behrenfeld, M.J.; Boyd, P.W. Global assessment of ocean carbon export by combining satellite observations and food-web models. *Glob. Biogeochem. Cycles* **2014**, *28*, 181–196. [CrossRef]
4. Toming, K.; Kutser, T.; Uiboupin, R.; Arikas, A.; Vahter, K.; Paavel, B. Mapping Water Quality Parameters with Sentinel-3 Ocean and Land Colour Instrument imagery in the Baltic Sea. *Remote Sens.* **2017**, *9*, 1070. [CrossRef]
5. Natvik, L.J.; Evensen, G. Assimilation of ocean colour data into a biochemical model of the North Atlantic: Part 1. Data assimilation experiments. *J. Mar. Syst.* **2003**, *40-41*, 127–153. [CrossRef]
6. IOCCG. *Remote Sensing in Fisheries and Aquaculture*; Reports of the International Ocean Colour Coordinating Group; IOCCG: Dartmouth, NS, Canada, 2009; Volume 8. [CrossRef]
7. McIver, R.; Breeze, H.; Devred, E. Satellite remote-sensing observations for definitions of areas for marine conservation: Case study of the Scotian Slope, Eastern Canada. *Remote Sens. Environ.* **2018**, *214*. [CrossRef]
8. Johnson, C.; Devred, E.; Casault, B.; Head, E.; Cogswell, A.; Spry, J. Optical, Chemical, and Biological Oceanographic Conditions on the Scotian Shelf and in the Eastern Gulf of Maine during 2015. Available online: http://publications.gc.ca/site/eng/9.833512/publication.html (accessed on 5 November 2019).
9. Hannah, C.G.; McKinnell, S. (Eds.) *Applying Remote Sensing Data to Fisheries Management in BC*; Technical Report; Department of Fisheries and Oceans Canada: Ottawa, ON, Canada, 2016.
10. O'Reilly, J.E.; Maritorena, S.; Mitchell, B.G.; Siegel, D.A.; Carder, K.L.; Garver, S.A.; Kahru, M.; McClain, C. Ocean color chlorophyll algorithm for SeaWiFS. *J. Geophys. Res.* **1998**, *103*, 24937–24953. [CrossRef]
11. Werdell, J.; Franz, B.; Bailey, S.; Feldman, G.; Boss, E.; Brando, V.; Dowell, M.; Hirata, T.; Lavender, S.; Lee, Z.; et al. Generalized ocean color inversion model for retrieving marine inherent optical properties. *Appl. Opt.* **2013**, *52*, 2019–2037. [CrossRef]
12. Werdell, P.J.; Bailey, S.W. An improved *in situ* bio-optical dataset for ocean colour algorithm development and satellite data production validation. *Remote Sens. Environ.* **2005**, *98*, 122–140. [CrossRef]
13. Gower, J. On the use of satellite-measured chlorophyll fluorescence for monitoring coastal waters. *Int. J. Remote Sens.* **2015**. [CrossRef]
14. Doerffer, R.; Schiller, H. The MERIS case 2 water algorithm. *Int. J. Remote Sens.* **2007**, *28*, 517–535. [CrossRef]

15. Laliberté, J.; Larouche, P.; Devred, E.; Craig, S. Chlorophyll-a Concentration Retrieval in the Optically Complex Waters of the St. Lawrence Estuary and Gulf Using Principal Component Analysis. *Remote Sens.* **2018**, *10*, 265. [CrossRef]

16. Hamed, G.; Scott, R. Revisiting empirical ocean-colour algorithms for remote estimation of chlorophyll-a content on a global scale. *Int. J. Remote Sens.* **2016**, *37*, 2682–2705. [CrossRef]

17. Ben Mustapha, S.; Bélanger, S.; Larouche, P. Evaluation of ocean color algorithms in the southeastern Beaufort Sea, Canadian Arctic: New parameterization using SeaWiFS, MODIS, and MERIS spectral bands. *Can. J. Remote Sens.* **2012**, *38*, 535–556. [CrossRef]

18. Peña, A.; Nemcek, N. Phytoplankton in Surface Waters along Line P and off the West Coast of Vancouver Island. In *State of the Physical, Biological and Selected Fishery Resources of Pacific Canadian Marine Ecosystems in 2017*; Chandler, P.C., King, S.A., Boldt, J., Eds.; Can. Tech. Rep. Fish. Aquat. Sci.; 2018; Volume 3266, pp. 55–59. Available online: http://waves-vagues.dfo-mpo.gc.ca/Library/40717914.pdf (accessed on 5 November 2019).

19. Head, E.J.H.; Horne, E.P.W. Pigment transformation and vertical flux in an area of convergence in the North Atlantic. *Deep-Sea Res. II* **1993**, *40*, 329–346. [CrossRef]

20. Stuart, V.; Head, E.J.H. The BIO method. In *The Second SeaWiFS HPLC Analysis Round-Robin Experiment (SeaHARRE-2)*; NASA/TM 2005-212785; Hooker, S.B., Ed.; NASA, Goddard Space Flight Center: Greenbelt, MD, USA, 2005; p. 112.

21. Zapata, M.; Rodríguez, F.; Garrido, J. Separation of chlorophylls and carotenoids from marine phytoplankton: A new HPLC method using a reversed phase C8 column and pyridine-containing mobile phases. *Mar. Ecol. Prog. Ser.* **2000**, *195*, 29–45. [CrossRef]

22. Hijmans, R.J. Geosphere: Spherical Trigonometry (R Package Version 1.5-7). 2017. Available online: http://cran.nexr.com/web/packages/geosphere/index.html (accessed on 5 November 2019).

23. Maritorena, S.; Siegel, D.A.; Peterson, A.R. Optimization of a semi-analytical ocean color model for global-scale applications. *Appl. Opt.* **2002**, *41*, 2705–2713. [CrossRef]

24. R Core Team. *R: A Language and Environment for Statistical Computing*; R Foundation for Statistical Computing: Vienna, Austria, 2016.

25. Davison, A.C.; Hinkley, D.V. *Bootstrap Methods and Their Applications*; Cambridge University Press: Cambridge, UK, 1997; ISBN 0-521-57391-2.

26. Canty, A.; Ripley, B.D. Boot: Bootstrap R (S-Plus) Functions (R Package Version 1.3-22). 2019. Available online: https://cran.rapporter.net/bin/linux/ubuntu/disco-cran35/Packages (accessed on 5 November 2019).

27. Garver, S.A.; Siegel, D.A. Inherent optical property inversion of ocean color spectra and its biogeochemical interpretation: 1. Time series from the Sargasso Sea. *J. Geophys. Res. Oceans* **1997**, *102*, 18607–18625. [CrossRef]

28. Gordon, H.R.; Brown, J.W.; Evans, R.H.; Brown, J.W.; Smith, R.C.; Baker, K.S.; Clark, D.K. A semianalytic radiance model of ocean color. *J. Geophys. Res.* **1988**, *93*, 10909–10924. [CrossRef]

29. Pope, R.M.; Fry, E.S. Absorption spectrum (380–700 nm) of pure water. II. Integrating measurements. *Appl. Opt.* **1997**, *36*, 8710–8723. [CrossRef]

30. Smith, R.; Baker, K. Optical properties of the clearest natural waters (200–800 nm). *Appl. Opt.* **1981**, *20*, 177–184. [CrossRef] [PubMed]

31. IOCCG. *Remote Sensing of Inherent Optical Properties: Fundamentals, Tests of Algorithms, and Applications*; Reports of the International Oean-Colour Coordinating Group, No. 5; Lee, Z.P., Ed.; IOCCG: Dartmouth, NS, Canada, 2006.

32. Lee, Z.P.; Carder, K.L.; Arnone, R. Deriving inherent optical properties from water color: A multi-band quasi-analytical algorithm for optically deep waters. *Appl. Opt.* **2002**, *41*, 5755–5772. [CrossRef] [PubMed]

33. Bricaud, A.; Claustre, H.; Oubelkheir, K. Natural variability of phytoplankton absorption in oceanic waters: influence of the size structure of algal populations. *J. Geophys. Res.* **2004**, *110*. [CrossRef]

34. Legendre, P. Lmodel2: Model II Regression (R Package Version 1.7-3). 2018. Available online: https://cran.r-project.org/web/packages/lmodel2/index.html (accessed on 5 November 2019).

35. Seegers, B.; Stumpf, R.; Schaeffer, B.; Loftin, K.; Werdell, J. Performance metrics for the assessment of satellite data products: An ocean color case study. *Opt. Express* **2018**, *26*, 7404. [CrossRef] [PubMed]

36. Brewin, B.; Sathyendranath, S.; Müller, D.; Brockmann, C.; Deschamps, P.Y.; Devred, E.; Doerffer, R.; Fomferra, N.; Franz, B.; Grant, M.; et al. The Ocean Colour Climate Change Initiative: III. A round-robin comparison on in-water bio-optical algorithms. *Remote Sens. Environ.* **2015**, *162*. [CrossRef]

37. Wickham, H. *ggplot2: Elegant Graphics for Data Analysis*; Springer: New York, NY, USA, 2009.

38. Claustre, H. The trophic status of various oceanic provinces as revealed by phytoplankton pigment signatures. *Limnol. Oceanogr.* **1994**, *39*, 1206–1210. [CrossRef]

39. Jeffrey, S.W.; Vesk, M. Introduction to marine phytoplankton and their pigment signature. In *Phytoplankton Pigments in Oceanography: Guidelines to Modern Methods*; Jeffrey, S.W., Mantoura, R.F.C., Wright, S.W., Eds.; UNESCO Publishing: Paris, France, 1997; pp. 37–84.

40. Garcia, V.M.T.; Signorini, S.; Garcia, C.A.E.; McClain, C.R. Empirical and semi-analytical chlorophyll algorithms in the south-western Atlantic coastal region (25–40°S and 60–45°W). *Int. J. Remote Sens.* **2006**, *27*, 1539–1562. [CrossRef]

41. Sun, L.; Guo, M.; Wang, X. Ocean color products retrieval and validation around China coast with MODIS. *Acta Oceanol. Sin.* **2010**, *29*, 21–27. [CrossRef]

42. Jiang, W.; Knight, B.R.; Cornelisen, C.; Barter, P.; Kudela, R. Simplifying Regional Tuning of MODIS Algorithms for Monitoring Chlorophyll-a in Coastal Waters. *Front. Mar. Sci.* **2017**, *4*, 151. [CrossRef]

43. Hooker, S.B.; Esaias, W.E.; Feldman, G.C.; Gregg, W.W.; McClain, C.R. *An Overview of SeaWiFS and Ocean Color*; Tech. Memo. 104566; Hooker, S., Firestone, E., Eds.; NASA Goddard Space Flight Center: Greenbelt, MD, USA, 1992; Volume 1, 24p.

44. Eplee, R.E.; Robinson, W.D.; Bailey, S.W.; Clark, D.K.; Werdell, P.J.; Wang, M.; Barnes, R.A.; McClain, C.R. Calibration of SeaWiFS. II. Vicarious techniques. *Appl. Opt.* **2001**, *40*, 6701–6718. [CrossRef]

45. Stuart, V.; Sathyendranath, S.; Platt, T.; Maass, H.; Irwin, B.D. Pigments and species compositon of natural phytoplankton populations: effect on the absorption spectra. *J. Plankton Res.* **1998**, *20*, 187–217. [CrossRef]

46. Sathyendranath, S.; Cota, G.; Stuart, V.; Maass, H.; Platt, T. Remote sensing of phytoplankton pigments: A comparison of empirical and theoretical approaches. *Int. J. Remote Sens.* **2001**, *22*, 249–273. [CrossRef]

47. Devred, E.; Sathyendranath, S.; Stuart, V.; Maass, H.; Ulloa, O.; Platt, T. A two-component model of phytoplankton absorption in the open ocean: theory and applications. *J. Geophys. Res.* **2006**, *111*. [CrossRef]

48. Sathyendranath, S.; Watts, L.; Devred, E.; Platt, T.; Caverhill, C.; Maass, H. Discrimination of diatoms from other phytoplankton using ocean-colour data. *Mar. Ecol. Prog. Ser.* **2004**, *272*, 59–68. [CrossRef]

49. IOCCG. *Atmospheric Correction for Remotely-Sensed Ocean-Colour Products*; Reports of the International Ocean Colour Coordinating Group; IOCCG: Dartmouth, NS, Canada, 2010; Volume 10. [CrossRef]

50. Morel, A.; Prieur, L. Analysis of variation in ocean color. *Limnol. Oceanogr.* **1977**, *22*, 709–722. [CrossRef]

51. Carswell, T.; Costa, M.; Young, E.; Komick, N.; Gower, J.; Sweeting, R. Evaluation of MODIS-Aqua Atmospheric Correction and Chlorophyll Products of Western North American Coastal Waters Based on 13 Years of Data. *Remote Sens.* **2017**, *9*, 1063. [CrossRef]

Letter

Spatio-Temporal Variability of the Habitat Suitability Index for the *Todarodes pacificus* (Japanese Common Squid) around South Korea

Dabin Lee [1], Seung Hyun Son [2], Chung-Il Lee [3], Chang-Keun Kang [4] and Sang Heon Lee [1,*]

[1] Department of Oceanography, Pusan National University, Geumjeong-gu, Busan, 46241, Korea; ldb1370@pusan.ac.kr

[2] CIRA, Colorado State University, Fort Collins, CO 80523, USA; ssnocean@gmail.com

[3] Department of Marine Bioscience, Gangneung-Wonju National University, Gangneung, 25457, Korea; leeci@gwnu.ac.kr

[4] School of Earth Sciences & Environmental Engineering, Gwangju Institute of Science and Technology, Gwangju, 61005, Korea; ckkang@gist.ac.kr

* Correspondence: sanglee@pusan.ac.kr; Tel.: +82-51-510-2256

Received: 21 October 2019; Accepted: 18 November 2019; Published: 20 November 2019

Abstract: The climate-induced changes in marine fishery resources in South Korea have been a big concern over the last decades. The climate regime shift has led to not only a change in the dominant fishery resources, but also a decline in fishery landings in several species. The habitat suitability index (HSI) has been widely used to detect and forecast fishing ground formation. In this study, the catch data of the *Todarodes pacificus* (Japanese Common Squid) and satellite-derived environmental parameters were used to estimate the HSI for the *T. pacificus* around South Korea. More than 80% of the total catch was found in regions with a sea surface temperature (SST) of 14.91–27.26 °C, sea surface height anomaly (SSHA) of 0.05–0.20 m, chlorophyll-a of 0.32–1.35 mg m^{-3}, and primary production of 480.41–850.18 mg C m^{-2} d^{-1}. Based on these results, the HSI model for *T. pacificus* was derived. A strong positive relationship (R^2 = 0.9260) was found between the HSI and the fishery landings. The climatological monthly mean HSI from 2002 to 2016 showed several hotspots, coinciding with the spawning and feeding grounds of *T. pacificus*. This outcome implies that our estimated HSI can yield a reliable prediction of the fishing ground for *T. pacificus* around South Korea. Furthermore, the approach with the simple HSI model used in this study can be applied elsewhere, and will help us to understand the spatial and temporal distribution of fishery resources.

Keywords: Japanese common squid; *Todarodes pacificus*; habitat suitability index (HSI); the Yellow Sea; the South Sea of South Korea

---

1. Introduction

Todarodes pacificus (Japanese Common Squid) is a commercially important fish species in the South Korea, and is widely distributed around the Korean Peninsula. The fishery production value of *T. pacificus* accounts for about 10% of total fishery production value in South Korea [1]. The commercial catch of *T. pacificus* is mainly performed by jigging and trawling [1]. *T. pacificus* is known as a migratory species which usually migrates to the Yellow Sea and the East/Japan Sea for feeding and moves back to the East China Sea for spawning [2–5]. The community of *T. pacificus* around the Korean Peninsula can be divided into two groups based on the spawning seasons [2–5]. The groups can be expressed as the 'winter spawners' and 'autumn spawners'. Although the spawning ground of both groups are distributed around the South Sea and the East China Sea, their feeding grounds are distributed differently [2–5]. Many previous studies reported that the feeding grounds of *T. pacificus* in Korean waters are located mainly in the Yellow Sea and the East/Japan Sea for 'winter spawners' and 'autumn spawners', respectively [2–5]. The main prey of the juvenile *T. pacificus* is the zooplankton. The main food source of adult squid is small pelagic fishes. Moreover, the zooplankton is an important food source for the small pelagic fishes Consequently, chlorophyll-*a* (Chl-*a*) concentration and primary production will have an indirect relationship with the distribution of *T. pacificus*, thus they will be useful indicator of habitat distribution of the *T. pacificus*.

In the 1980s, the dominant fishery resources were saury (*Cololabis saira*), cod (*Gadus macrocephalus*), and walleye pollock (*Gadus chalcogramma*) [6,7]. However, in the 1990s, the dominant fish catch was changed to squid (*Todarodes pacificus*) [6,7]. There could be many reasons for the recent changes in fishery resource such as climate change and overfishing [8–11]. However, the main causes for the recent changes still remain unclear. To examine the recent changes in fishery resources in South Korea, we need to understand the relationship between major environmental parameters and the species' habitat formation.

The habitat suitability index (HSI) has been widely used to investigate the distribution of marine fish population [12–17]. The development of HSI would allow us to obtain a useful database to establish a resource management strategy [12,18,19]. Generally, the species abundance has a strong dependency on environmental conditions, and each species has a different environmental preference [20–22]. The HSI model can estimate the relative abundance of a species through this approach. A general form of the HSI model is composited of a number of suitability indices based on relationships between environmental factors and species' abundance. The composited HSI index is a non-dimensional value ranging from 0 to 1. Since the population dynamics of marine fishes are very complex and the HSI model is defined by a few key factors such as sea surface temperature (SST), Chl-*a*, and salinity [14], selecting the main parameters is a significant process in the HSI model derivation. However, several studies already have reported that the spatial distribution of the marine fish population is largely influenced by several environmental variables such as temperature, salinity, chlorophyll-*a* (Chl-*a*) concentration, and sea surface height anomaly in other regions [14,23–25]. Therefore, the derivation of the HSI model will be possible by considering several environmental conditions including these environmental variables. Moreover, the simplified HSI model with a few environmental variables will allow us to understand the spatial and temporal variations of the habitat distribution of the *T. pacificus*.

The objectives of this study can be divided mainly into three goals: (1) to investigate the preferred environmental conditions of the *T. pacificus* using the satellite dataset, (2) to develop the HSI model for the *T. pacificus* around South Korea, and (3) to investigate seasonal and spatial variations of the HSI in the Korean waters.

2. Materials and Methods

2.1. Fishery Data

The commercial fishing records for the *T. pacificus* around the South Sea from 2010 to 2016 were obtained from the Large Purse Seine Fishery Cooperatives of South Korea. The total number of reported catches was 9401, and the dataset contains locations and dates, as well as the amount of catch (metric ton, M/T) (Figure 1). The fishing locations were collected at a spatial resolution of 0.17 × 0.17 degrees (latitude × longitude).

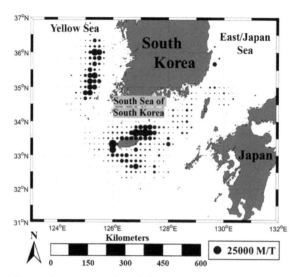

Figure 1. Summarized commercial catch data for the *T. pacificus* during 2010 to 2016.

2.2. Satellite Dataset

We obtained the satellite ocean color data from the Moderate Resolution Imaging Spectroradiometer (MODIS) onboard the satellite Aqua platform provided by the Ocean Biology Processing Group at NASA Goddard Space Flight Center (https://oceandata.sci.gsfc.nasa.gov/MODIS-Aqua/Mapped/8-Day/4km/). We used the MODIS Level-3 8-day composite data from July 2002 to December 2016 at 4 km of spatial resolution covering the South Korea and nearby oceans. SST, Chl-*a*, photosynthetically available radiation (PAR), and diffuse attenuation coefficient at 490 nm (Kd(490)) datasets were used in this study [26–29].

The primary production was derived using a regional algorithm based on the Vertically Generalized Productivity Model (VGPM) [30] with satellite ocean color data described as:

$$PP_{eu} = 0.66125 \times P^B_{opt} \times [E_0/E_0 + 4.1] \times Z_{eu} \times Chl\text{-}a \times DL \tag{1}$$

where PP_{eu} is the daily primary production integrated from euphotic depth (mg C m^{-2} d^{-1}), P^B_{opt} is the optimal carbon fixation rate (mg C (mg Chl)$^{-1}$ h^{-1}), E_0 is the amount of incident photosynthetically available radiation (PAR) during the day (E m^{-2} d^{-1}), Z_{eu} is the euphotic depth (m) which is derived by the equation 4.6/K$_d$(490) that represent 1% penetration depth of 490 nm radiation [31], Chl-*a* is the concentration of chlorophyll-a (mg Chl m^{-2}), and DL is the photoperiod (h) which is computed with latitude and date mathematically.

The P^B_{opt} is derived from a multiple regression equation with SST and Chl-*a* [32] as follows:

$$P^B_{opt} = \frac{0.071 \times SST - 3.2 \times 10^{-3} \times SST^2 + 3.0 \times 10^{-5} \times SST^3}{Chl\text{-}a} + [1.0 + 0.17 \times SST - 2.5 \times 10^{-3} \times SST^2 - 8.0 \times 10^{-5} \times SST^3] \tag{2}$$

In addition to the ocean color data, the sea surface height anomaly (SSHA) datasets were also used for the physical environmental variables in this study. The satellite altimetry products have been produced by SSALTO/DUACS and distributed by AVISO with support from Centre National d'Etudes Spatiales (CNES). We obtained the SSALTO/DUACS SSHA datasets on a Cartesian grid with spatial resolution of 1/4 × 1/4 degrees.

To investigate the environmental conditions when fishing occurred, we extracted mean values in 3 × 3 pixels of each environmental parameter on every catch point from the satellite dataset. The number of catch data matched with the satellite dataset was 5960 (Table 1). Many parts of the satellite data were unavailable due to cloudiness, and thus, only approximately 63% of the catch data was matched with the satellite data.

Table 1. The number of commercial catch data matched with satellite datasets.

Year	Reported Catches	No. of Matchable Catches
2010	797	486
2011	1593	956
2012	1479	918
2013	1372	834
2014	1219	775
2015	1640	1099
2016	1301	892
Total	9401	5960

2.3. HSI Model

Many previous studies derived HSI models with several environmental variables for a single species [14,16,17,33]. In this study, we chose SST, SSHA, Chl-*a*, and primary production (PP) to calculate HSI. Primary production is a very important factor for biological productivity, so it can be an important factor in determining habitats of marine fishes.

The SIs for four parameters were calculated based on Chen et al. [14]. Each SI was estimated as a value between 0 and 1. The SIs were estimated by using an equation as follows:

$$SI = \exp(A \times (X + B)^2) \tag{3}$$

where A and B are constants, and X is the values for the parameters. After deriving the SIs for four parameters, the indices were combined into a single index.

When combining the SIs for the four variables, several types of models, such as the Continued Product Model (CPM), Minimum Model (MINM), Arithmetic Mean Model (AMM), or Geometric Mean Model (GMM), could be used [14,30–38]. However, several previous studies already reported that the AMM is the most appropriate model for the HSI of several fish species [14,16,17,33]. Thus, we used the AMM as described below:

$$HSI = (SI_1 \times SI_2 \times SI_3 \times SI_4)/4 \tag{4}$$

3. Results

3.1. Preferred Environmental Conditions of the Todarodes Pacificus

Monthly distributions of the number of fishing records and the amount of fishery landings showed a high seasonal dependency. Most of the catches were occurred on January, August, and December (Figure 2).

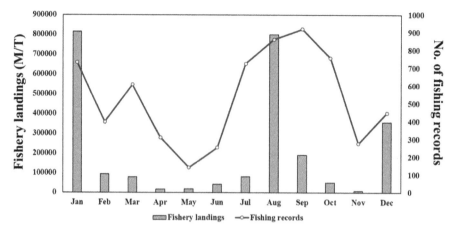

Figure 2. Monthly distribution of total fishery landings (M/T) and the number of fishing records.

The *T. pacificus* around South Korea were distributed in a wide range of environmental conditions. Our match-up results showed that the squid catches were distributed in SST ranging from 5.11 to 31.94 °C, SSHA from –0.15 to 0.46 m, Chl-*a* from 0.15 to 25.43 mg m^{-3}, and PP from 258.53 to 1209.75 mg C m^{-2} d^{-1} (Figure 3). Based on Kaschner et al. [39], we defined the optimum ranges of the three parameters as the 10th percentile and the 90th percentile of each parameter. The optimum ranges were 14.91–27.26 °C, 0.05–0.20 m, 0.32–1.35 mg m^{-3}, and 480.41–850.18 mg C m^{-2} d^{-1}, for SST, SSHA, Chl-*a*, and PP, respectively. In the optimum conditions of SST, SSHA, Chl-*a*, and PP, the accounted amount of total catches was approximately 80% for each factor.

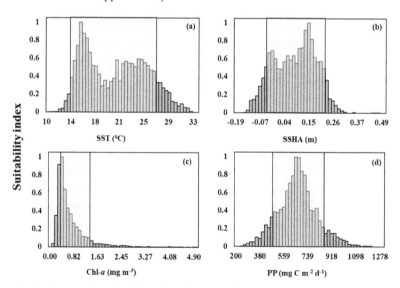

Figure 3. Distribution of the suitability index of (**a**) sea surface temperature (SST) (°C), (**b**) sea surface height anomaly (SSHA) (m), (**c**) chlorophyll-*a* (Chl-*a*) (mg m^{-3}), and (**d**) primary production (PP) (mg C m^{-2} d^{-1}) on the fishing locations for the *T. pacificus*. Gray square represents an optimum range for each parameter.

Among these results, double peaks are found in the frequency distributions of SST and SSHA. To derive an appropriate suitability index model, the seasonal frequency distribution was derived from the results (Figure 4). The distribution of suitability index during winter to spring, and summer to autumn showed a similar SST and SSHA range. Based on these results, the suitability index distributions were analyzed on two periods (winter-spring and summer-autumn; Figure 5). Each period showed remarkably different peaks of SST and SSHA. Accordingly, the suitability index models of SST and SSHA for winter to spring and summer to autumn were derived separately.

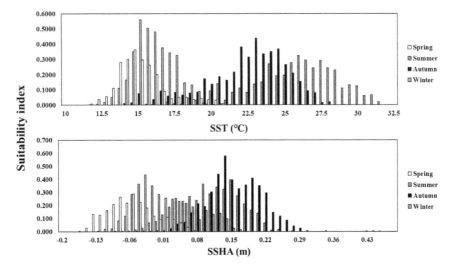

Figure 4. Seasonal distributions of suitability index for the SST (°C) and SSHA (m) on the fishing locations for the *T. pacificus*.

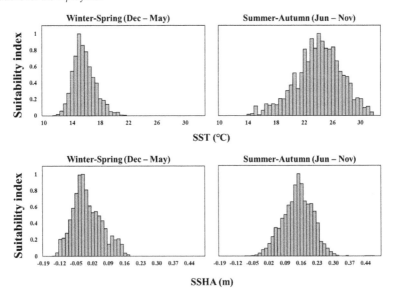

Figure 5. Seasonality of frequency distributions of SST (°C) and SSHA (m) on the fishing locations for the *T. pacificus*.

3.2. HSI Model Derivation

To derive the HSI values for the *T. pacificus*, six empirical models were used (Table 2). The SI models for the environmental parameters were adapted from Chen et al. [14]. As mentioned above, the suitability index models for the SST and the SSHA were divided into the two parts; winter to spring and summer to autumn. The constants of the models were obtained from the least squares fitting to derive optimized models for the study area (Figure 6). In the case of Chl-*a*, we used a natural logarithmic form to fit to the asymmetric distribution of the Chl-*a* concentration.

Table 2. SI models derived from three environmental parameters (SST, SSHA, Chl-*a*, and PP).

Variables	SI models	RMSE	R^2
SST_{WS}	$\exp(-0.2929(X_{SST} - 15.44)^2)$	0.0500	0.9654
SST_{SA}	$\exp(-0.0657(X_{SST} - 24.22)^2)$	0.0745	0.9461
$SSHA_{WS}$	$\exp(-184.8(X_{SST} + 0.01325)^2)$	0.0871	0.9063
$SSHA_{SA}$	$\exp(-157(X_{SST} - 0.1514)^2)$	0.0667	0.9468
Chl-*a*	$\exp(-16.98(\ln(X_{Chl}) + 0.5997)^2)$	0.0640	0.9165
PP	$\exp(-0.00004637(X_{PP} - 658.7)^2)$	0.0876	0.9072

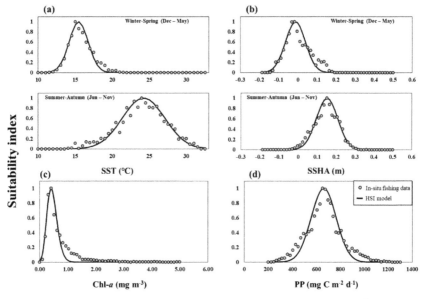

Figure 6. Least squares fitting results of (a) SST (°C), (b) SSHA (m), (c) Chl-*a* (mg m^{-3}), and (d) PP (mg C m^{-2} d^{-1}) with the number of fishing sets (solid line: habitat suitability index (HSI) model, black dot: in situ fishing data).

The RMSE of the SI models for SST_{WS}, SST_{SA}, $SSHA_{WS}$, $SSHA_{SA}$, Chl-*a*, and PP were 0.0500, 0.0745, 0.0871, 0.0667, 0.0640, and 0.0876, respectively. The coefficients of determination (R^2) for the SI models were 0.9654, 0.9461, 0.9063, 0.9468, 0.9165, and 0.9072 for SST_{WS}, SST_{SA}, $SSHA_{WS}$, $SSHA_{SA}$, Chl-*a*, and PP, respectively.

The climatological monthly distribution (July 2002–December 2016) of the HSI by the AMM model was derived by averaging datasets for each month. It showed that a relatively high HSI in the South Sea of South Korea. In addition, the high HSI regions were observed in the Yellow Sea and the East/Japan Sea from early summer to autumn (Figure 7).

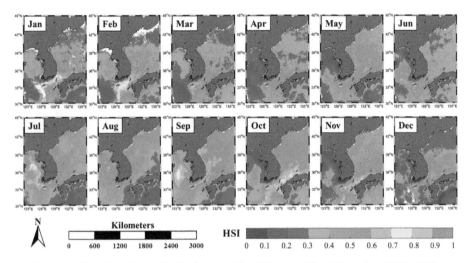

Figure 7. Climatological monthly distribution of the (HSI) around South Korea from 2002 to 2016.

3.3. HSI Model Evaluation

To validate our HSI model, we compared the HSI values from our model with fishery catches. The HSI was divided into 10 classes with a step of 0.1. Then, we summed the landed fisheries from the region with the HSI values corresponding to each class. A strong positive linear relationship ($R^2 = 0.9260$) was observed between the HSI and the fishery landings (Figure 8). In addition, spatial distributions between the HSI and the fish catches for the *T. pacificus* appeared well matched throughout 2010 to 2016 (Figure 9).

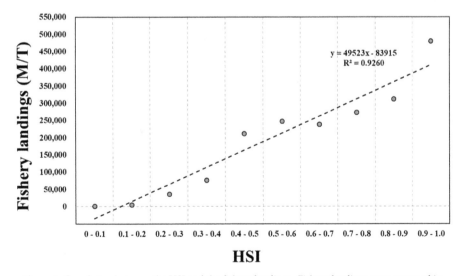

Figure 8. Correlation between the HSI and the fishery landings. Fishery landings were summed in each range of the HSI value.

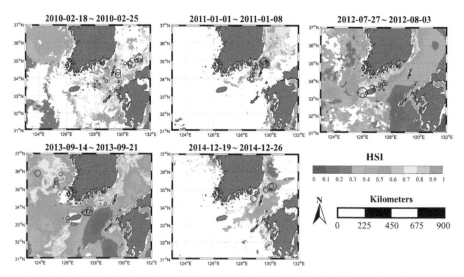

Figure 9. Spatial distribution of the eight-day composited HSI and the *T. pacificus* catches (black open circle) at corresponding periods in the South Sea.

4. Discussion

4.1. Environmental Conditions

Among the several environmental variables, temperature and salinity are well known for the most important factors affect spatial and temporal distribution of marine fish species [40–42]. Several previous studies revealed the habitat formation is closely related with distribution of the SST, SSHA, Chl-*a*, and sea surface salinity (SSS) [14,16,39]. However, salinity was not considered in this study, since its variation in the East/Japan Sea and the South Sea mostly fell within the reported optimal ranges (from 31.93 to 35.7) [39] for the *T. pacificus*. Also, mixed layer depth (MLD) which can be a good indicator for the habitat formation of several species also not considered. The MLD dataset is hard to use in this study due to its relatively poor spatial. We need spatial resolution of at least 0.17×0.17 degrees due to the resolution of the catch data. Thus, the MLD is not considered in this study. Instead, we tried to examine the optimal conditions for the *T. pacificus* with other environmental variables.

The optimal environmental conditions for *T. pacificus* in this study showed similar results with several previous studies [5,39]. Alabia et al. [5] already reported that the environmental ranges of the potential squid habitat in the East/Japan Sea. According to them, the SST and the SSHA ranged from 10 to 30 °C and –0.16 to 0.36 cm, respectively, which correspond well with the results from our study. The Species Environmental Envelope (HSPEN) in AquaMaps global distribution model [39] also showed similar environmental range with this study. The species' preferred environmental conditions for the *T. pacificus* in AquaMaps are 9.24–23.54 °C and 364–939 mg C m^{-2} d^{-1} for temperature and PP, respectively [39].

In this study, we tried to derive the HSI for the *T. pacificus* with SST, SSHA, Chl-*a*, and primary production around the South Korea. The habitat depth of *T. pacificus* is known to be at least 30 m and up to 100 m. It seems to be nonsense that HSI derivation for the *T. pacificus* with SST data, but the seasonal habitat distribution would be reflected in the range of SST. On the other hand, phytoplankton is not a direct prey of *T. pacificus*. However, the intermediate consumers in the food web can largely influenced by the distribution of phytoplankton [43]. Indeed, Lee et al. [44] already reported that the spatial distribution of the common minke whale (*Balaenoptera acutorostrata*) is closely related to the Chl-*a* distribution in the East/Japan Sea [44], although there is not any direct linkage in the food web. Moreover, the HSI derivation for the chub mackerel (*Scomber japonicus*) was also successfully conducted in the South Sea and the East/Japan Sea using primary production dataset [17]. As mentioned before, the zooplankton is an important food source for the juvenile *T. pacificus* and the small pelagic fishes which are main prey for the adult *T. pacificus*. Although there are only a few studies that have used primary production as an environmental variable for the HSI model, not only Chl-*a* but also primary production can be useful indicator of spatio-temporal distribution of the *T. pacificus*.

However, *T. pacificus* distributed mostly around South Korea and Japan (or northwestern pacific). Thus, the HSI model derived from regional dataset should be applied to global ocean or other regions carefully with understanding of the native distribution range of the target species.

4.2. Seasonal Variation of the HSI

In this study, seasonal variations of the HSI for the *T. pacificus* were observed which are not seen in map of the native habitat ranges. The optimal environmental conditions for the *T. pacificus* in this study showed high seasonal dependences. The optimum ranges of the SST and the SSHA for Summer–Autumn were significantly higher than those of Winter–Spring. The monthly distribution of the catches suggested that January and August are representative fishing time for each group (Figure 2). The catches occurred in January were mainly distributed in the southern part of the East/Japan Sea and the eastern part of the South Sea, while catches occurred in August mostly distributed in the Yellow Sea and the western part of the South Sea. As mentioned above, *T. pacificus* is well known for their characteristic migration routes around the Korean Peninsula. Also, they consist of two groups, the 'winter spawners' and 'autumn spawners'. Their characteristic migration routes suggest that the caught population in January and August might be composed with different group each other. Consequently, the differences in the optimal environmental conditions between Summer–Autumn and Winter–Spring might be caused by these seasonal population distributions of *T. pacificus* in Korean waters.

4.3. Regions with High Catch Probability

For the investigation of the hotspots, which are the regions with high catch probability, we defined the hotspot as regions with an HSI over 0.7, because more than half of the total catches occurred in areas when the HSI ranged to 0.7 or more. The results of our hotspot analysis showed a seasonal distribution (Figure 10). One of the major hotspots was observed around the South Sea in winter (January–Febuary) and autumn (October–November). As mentioned above, not only the South Sea is well known for the spawning ground of the *T. pacificus*, they spawn during winter and autumn [2–5]. These characteristic spawning ground distributions of the *T.pacificus* might be reflected in the hotspot located in the South Sea. Another major hotspot located in the Yellow Sea was observed on July and September, while the hotspot located in the East/Japan Sea was observed on September and October. Generally, from late spring to early summer, the 'winter spawners' group begins to migrate mainly to the Yellow Sea which is well known for their feeding grounds, and their feeding activity continues until Autumn [3]. In case of the 'autumn spawners', they begin to migrate to the East/Japan Sea from spring to summer, and their feeding activity continues until late summer to autumn [3]. Consequently, the seasonal distribution of the hotspots was in agreement with the general migration patterns of the *T. pacificus* reported previously around the Korean Peninsula.

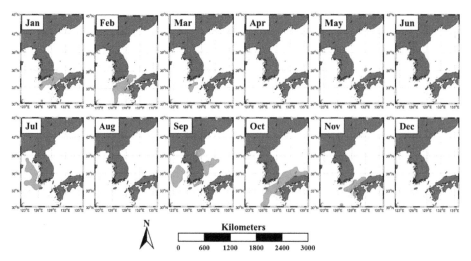

Figure 10. Climatological monthly distribution of the HSI hotspots (HSI > 0.7) for the *T. pacificus* around South Korea from 2002 to 2016.

5. Summary and Conclusions

In this study, the HSI of the *T. pacificus* was derived by using commercial catch data and satellite datasets between 2010 and 2016. Four environmental parameters (SST, SSHA, Chl-*a*, and PP) were selected as key variables in the habitat formation of the *T. pacificus*.

The optimum environmental conditions for SST, SSHA, Chl-*a*, and PP were 14.91–27.26 °C, 0.05–0.20 m, 0.32–1.35 mg m^{-3}, and 480.41–850.18 mg C m^{-2} d^{-1}, respectively. More than 80% of the total catch obtained from the region within the optimum ranges. SST and SSHA showed high seasonality in frequency distributions (Figure 4). Based on these results, suitability index models for SST and SSHA were derived separately for winter to spring and summer to autumn (Figure 6). In the derivation of the HSI model, the AMM was used to combine the four SIs (SST, SSHA, Chl-*a*, and PP) since many previous studies reported that the AMM is the most appropriate model for marine fish species.

Based on the results from the HSI model, we found a strong positive relationship (R^2 = 0.9260) between the HSI and the fishery landings (Figure 8) and good match for the spatial distributions of the *T. pacificus* (Figure 9). The seasonal variations of the spatial distribution of the HSI for the *T. pacificus* were observed in climatological monthly distribution of the HSI (Figure 7). In addition, the hotspot analysis the hotspot analysis revealed several major hotspots around the South Korea (Figure 10). The hotspot observed in the South Sea is consistent with the reported spawning grounds for *T. pacificus*. The hotspots observed in the Yellow Sea and the East/Japan Sea appear to be similar to the feeding grounds of the 'winter spawners' and 'autumn spawners', respectively. Consequently, the seasonal and spatial variations of the HSI match well with the migration patterns of the *T. pacificus* reported previously in the Korean waters. The HSI model derived from this study could help us to predict fishing grounds with high catch probability of the *T. pacificus* around South Korea. The monitoring of the HSI for the *T. pacificus* at the Yellow Sea and the East/Japan Sea can help us to successful fishery management. Ultimately, long-term analysis in spatial and temporal distributions of the HSI for the *T. pacificus* will allow us to understand the recent changes in fishery resources around South Korea.

Author Contributions: Conceptualization, D.L. and S.H.L.; data curation, D.L.; funding acquisition, C.-K.K. and S.H.L.; methodology, D.L.; project administration, C.-K.K. and S.H.L.; supervision, S.H.S. and S.H.L.; validation, D.L.; visualization, D.L.; writing—original draft, D.L.; writing—review and editing, S.H.S., C.-I.L., and S.H.L.

Funding: This research was supported by "Development of the integrated data processing system for GOCI-II" and and "Long-term change of structure and function in marine ecosystems of Korea" funded by the Ministry of Ocean and Fisheries, Korea.

Acknowledgments: The authors would like to thank the anonymous reviewers and the handling editors who dedicated their time for providing the authors with constructive and valuable recommendations.

Conflicts of Interest: The authors declare no conflict of interest.

References

1. Korea Statistics; Korean Statistical Information Service (KOSIS). *Fishery Production Survey*; Korean Statistical Information Service (KOSIS): Daejeon, Korea, 2019.
2. Kiyofuji, H.; Saitoh, S. Use of Nighttime Visible Images to Detect Japanese Common Squid Todarodes Pacificus Fishing Areas and Potential Migration Routes in the Sea of Japan. *Mar. Ecol. Prog. Ser.* **2004**, *276*, 173–186. [CrossRef]
3. Kim, G.B.; Stapleton, H.M. PBDEs, Methoxylated PBDEs and HBCDs in Japanese Common Squid (Todarodes Pacificus) from Korean Offshore Waters. *Mar. Pollut. Bull.* **2010**, *60*, 935–940. [CrossRef] [PubMed]
4. Waska, H.; Kim, G.; Kim, G.B. Comparison of S, Se, and 210 Po Accumulation Patterns in Common Squid Todarodes Pacificus from the Yellow Sea and East/Japan Sea. *Ocean Sci. J.* **2013**, *48*, 215–224. [CrossRef]
5. Alabia, I.; Dehara, M.; Saitoh, S.; Hirawake, T. Seasonal Habitat Patterns of Japanese Common Squid (Todarodes Pacificus) Inferred from Satellite-Based Species Distribution Models. *Remote Sens.* **2016**, *8*, 921. [CrossRef]
6. Park, C.; Kim, Y.; Park, J.; Kim, Z.; Choi, Y.; Lee, D.; Choi, K.; Kim, S.; Hwang, K. *Ecology and Fishing Grounds of Major Commercial Fish Species in the Coastal and Offshore Korean Waters*; National Fisheries Research and Development Institute Busan: Busan, Korea, 1998. (In Korean)
7. Kang, Y.S.; Kim, J.Y.; Kim, H.G.; Park, J.H. Long-term Changes in Zooplankton and its Relationship with Squid, Todarodes Pacificus, Catch in Japan/East Sea. *Fish. Oceanogr.* **2002**, *11*, 337–346. [CrossRef]
8. Jones, D.D.; Walters, C.J. Catastrophe Theory and Fisheries Regulation. *J. Fish. Board Can.* **1976**, *33*, 2829–2833. [CrossRef]
9. Parmesan, C.; Yohe, G. A Globally Coherent Fingerprint of Climate Change Impacts Across Natural Systems. *Nature* **2003**, *421*, 37. [CrossRef]
10. Perry, A.L.; Low, P.J.; Ellis, J.R.; Reynolds, J.D. Climate Change and Distribution Shifts in Marine Fishes. *Science* **2005**, *308*, 1912–1915. [CrossRef]
11. Daskalov, G.M.; Grishin, A.N.; Rodionov, S.; Mihneva, V. Trophic Cascades Triggered by Overfishing Reveal Possible Mechanisms of Ecosystem Regime Shifts. *Proc. Natl. Acad. Sci. USA* **2007**, *104*, 10518–10523. [CrossRef]
12. Brooks, R.P. Improving Habitat Suitability Index Models. *Wildl. Soc. Bull.* **1997**, *25*, 163–167.
13. Morris, L.; Ball, D. Habitat Suitability Modelling of Economically Important Fish Species with Commercial Fisheries Data. *ICES J. Mar. Sci.* **2006**, *63*, 1590–1603. [CrossRef]
14. Chen, X.; Li, G.; Feng, B.; Tian, S. Habitat Suitability Index of Chub Mackerel (Scomber Japonicus) from July to September in the East China Sea. *J. Oceanogr.* **2009**, *65*, 93–102. [CrossRef]
15. Galparsoro, I.; Borja, Á.; Bald, J.; Liria, P.; Chust, G. Predicting Suitable Habitat for the European Lobster (Homarus Gammarus), on the Basque Continental Shelf (Bay of Biscay), using Ecological-Niche Factor Analysis. *Ecol. Model.* **2009**, *220*, 556–567. [CrossRef]
16. Li, G.; Chen, X.; Lei, L.; Guan, W. Distribution of Hotspots of Chub Mackerel Based on Remote-Sensing Data in Coastal Waters of China. *Int. J. Remote Sens.* **2014**, *35*, 4399–4421. [CrossRef]
17. Lee, D.; Son, S.; Kim, W.; Park, J.; Joo, H.; Lee, S. Spatio-Temporal Variability of the Habitat Suitability Index for Chub Mackerel (Scomber Japonicus) in the East/Japan Sea and the South Sea of South Korea. *Remote Sens.* **2018**, *10*, 938. [CrossRef]
18. Rubec, P.J.; Bexley, J.C.; Norris, H.; Coyne, M.; Monaco, M.; Smith, S.; Ault, J. Suitability Modeling to Delineate Habitat Essential. *Am. Fish. Soc. Symp.* **1999**, *22*, 108–133.
19. Vinagre, C.; Fonseca, V.; Cabral, H.; Costa, M.J. Habitat Suitability Index Models for the Juvenile Soles, Solea Solea and Solea Senegalensis, in the Tagus Estuary: Defining Variables for Species Management. *Fish. Res.* **2006**, *82*, 140–149. [CrossRef]
20. Attrill, M.J.; Power, M. Climatic Influence on a Marine Fish Assemblage. *Nature* **2002**, *417*, 275. [CrossRef]

21. Boeuf, G.; Payan, P. How should Salinity Influence Fish Growth? *Com. Biochem. Physiol. C Toxicol. Pharmacol.* **2001**, *130*, 411–423. [CrossRef]

22. Pörtner, H.O.; Knust, R. Climate Change Affects Marine Fishes through the Oxygen Limitation of Thermal Tolerance. *Science* **2007**, *315*, 95–97. [CrossRef]

23. Zainuddin, M.; Kiyofuji, H.; Saitoh, K.; Saitoh, S. Using Multi-Sensor Satellite Remote Sensing and Catch Data to Detect Ocean Hot Spots for Albacore (Thunnus Alalunga) in the Northwestern North Pacific. *Deep Sea Res. II* **2006**, *53*, 419–431. [CrossRef]

24. Robinson, C.J.; Gómez-Gutiérrez, J.; de León, D.A.S. Jumbo Squid (Dosidicus Gigas) Landings in the Gulf of California Related to Remotely Sensed SST and Concentrations of Chlorophyll a (1998–2012). *Fish. Res.* **2013**, *137*, 97–103. [CrossRef]

25. Tian, Y.; Akamine, T.; Suda, M. Variations in the Abundance of Pacific Saury (Cololabis Saira) from the Northwestern Pacific in Relation to Oceanic-Climate Changes. *Fish. Res.* **2003**, *60*, 439–454. [CrossRef]

26. NASA Goddard Space Flight Center; Ocean Ecology Laboratory; Ocean Biology Processing Group. *Moderate-Resolution Imaging Spectroradiometer (MODIS) Aqua 11μm Day/Night Sea Surface Temperature Data*; 2018 Reprocessing; NASA OB.DAAC: Greenbelt, MD, USA, 2018. Available online: https://oceandata.sci.gsfc.nasa.gov/MODIS-Aqua/Mapped/Monthly/4km/sst/ (accessed on 11 September 2018). [CrossRef]

27. NASA Goddard Space Flight Center; Ocean Ecology Laboratory; Ocean Biology Processing Group. *Moderate-Resolution Imaging Spectroradiometer (MODIS) Aqua Chlorophyll Data*; 2018 Reprocessing; NASA OB.DAAC: Greenbelt, MD, USA, 2018. Available online: https://oceandata.sci.gsfc.nasa.gov/MODIS-Aqua/Mapped/Monthly/4km/chlor_a/ (accessed on 11 September 2018). [CrossRef]

28. NASA Goddard Space Flight Center; Ocean Ecology Laboratory; Ocean Biology Processing Group. *Moderate-Resolution Imaging Spectroradiometer (MODIS) Aqua Photosynthetically Available Radiation Data*; 2018 Reprocessing; NASA OB.DAAC: Greenbelt, MD, USA, 2018. Available online: https://oceandata.sci.gsfc.nasa.gov/MODIS-Aqua/Mapped/Monthly/4km/par/ (accessed on 11 September 2018). [CrossRef]

29. NASA Goddard Space Flight Center; Ocean Ecology Laboratory; Ocean Biology Processing Group. *Moderate-Resolution Imaging Spectroradiometer (MODIS) Aqua Downwelling Diffuse Attenuation Coefficient Data*; 2018 Reprocessing; NASA OB.DAAC: Greenbelt, MD, USA, 2018. Available online: https://oceandata.sci.gsfc.nasa.gov/MODIS-Aqua/Mapped/Monthly/4km/Kd_490/ (accessed on 11 September 2018). [CrossRef]

30. Behrenfeld, M.J.; Falkowski, P.G. Photosynthetic Rates Derived from satellite-based Chlorophyll Concentration. *Limnol. Oceanogr.* **1997**, *42*, 1–20. [CrossRef]

31. Kirk, J.T. *Light and Photosynthesis in Aquatic Ecosystems*, 2nd ed.; Cambridge University Press: Cambridge, UK, 1994; pp. 129–169.

32. Yamada, K.; Ishizaka, J.; Nagata, H. Spatial and Temporal Variability of Satellite Primary Production in the Japan Sea from 1998 to 2002. *J. Oceanogr.* **2005**, *61*, 857–869. [CrossRef]

33. Yen, K.; Lu, H.; Chang, Y.; Lee, M. Using Remote-Sensing Data to Detect Habitat Suitability for Yellowfin Tuna in the Western and Central Pacific Ocean. *Int. J. Remote Sens.* **2012**, *33*, 7507–7522. [CrossRef]

34. Chen, X.; Feng, B.; Xu, L. A Comparative Study on Habitat Suitability Index of Bigeye Tuna, Thunnus obesus in the Indian Ocean. *J. Fish. Sci. China* **2008**, *15*, 269–278.

35. Grebenkov, A.; Lukashevich, A.; Linkov, I.; Kapustka, L.A. A habitat suitability evaluation technique and its application to environmental risk assessment. In *Ecotoxicology, Ecological Risk Assessment and Multiple Stressors*; Springer: Berlin/Heidelberg, Germany, 2006; pp. 191–201.

36. Hess, G.R.; Bay, J.M. A Regional Assessment of Windbreak Habitat Suitability. *Environ. Monit. Assess.* **2000**, *61*, 239–256. [CrossRef]

37. Lauver, C.L.; Busby, W.H.; Whistler, J.L. Testing a GIS Model of Habitat Suitability for a Declining Grassland Bird. *Environ. Manag.* **2002**, *30*, 88–97. [CrossRef]

38. Van der Lee, G.E.M.; Van der Molen, D.T.; Van den Boogaard, H.F.P.; Van der Klis, H. Uncertainty Analysis of a Spatial Habitat Suitability Model and Implications for Ecological Management of Water Bodies. *Landsc. Ecol.* **2006**, *21*, 1019–1032. [CrossRef]

39. Kaschner, K.; Kesner-Reyes, K.; Garilao, C.; Rius-Barile, J.; Rees, T.; Froese, R. *AquaMaps: Predicted Range Maps for Aquatic Species*; Version August 2016; World Wide Web Electronic Publication; Available online: www.aquamaps.org (accessed on 2 January 2019).

40. Castillo, J.; Barbieri, M.; Gonzalez, A. Relationships between Sea Surface Temperature, Salinity, and Pelagic Fish Distribution off Northern Chile. *ICES J. Mar. Sci.* **1996**, *53*, 139–146. [CrossRef]
41. Jaureguizar, A.J.; Menni, R.; Guerrero, R.; Lasta, C. Environmental Factors Structuring Fish Communities of the Río de la Plata Estuary. *Fish. Res.* **2004**, *66*, 195–211. [CrossRef]
42. Sutcliffe, W., Jr.; Drinkwater, K.; Muir, B. Correlations of Fish Catch and Environmental Factors in the Gulf of Maine. *J. Fish. Board Can.* **1977**, *34*, 19–30. [CrossRef]
43. Smith, R.; Dustan, P.; Au, D.; Baker, K.; Dunlap, E. Distribution of Cetaceans and Sea-Surface Chlorophyll Concentrations in the California Current. *Mar. Biol.* **1986**, *91*, 385–402. [CrossRef]
44. Lee, D.; An, Y.R.; Park, K.J.; Kim, H.W.; Lee, D.; Joo, H.T.; Oh, Y.G.; Kim, S.M.; Kang, C.K.; Lee, S.H. Spatial Distribution of Common Minke Whale (Balaenoptera acutorostrata) as an Indication of a Biological Hotspot in the East Sea. *Deep Sea Res. II* **2017**, *143*, 91–99. [CrossRef]

 *remote sensing*

Article

Evaluation of Spaceborne GNSS-R Retrieved Ocean Surface Wind Speed with Multiple Datasets

Zhounan Dong [1,2] and Shuanggen Jin [1,3,*]

1   Shanghai Astronomical Observatory, Chinese Academy of Sciences, Shanghai 200030, China;
     zndong@shao.ac.cn
2   University of Chinese Academy of Sciences, Beijing 100049, China
3   School of Remote Sensing and Geomatics Engineering, Nanjing University of Information Science and
     Technology, Nanjing 210044, China
*   Correspondence: sgjin@shao.ac.cn; Tel.: +86-21-34775292

Received: 26 September 2019; Accepted: 20 November 2019; Published: 22 November 2019

Abstract: Spaceborne Global Navigation Satellite Systems-Reflectometry (GNSS-R) can estimate the geophysical parameters by receiving Earth's surface reflected signals. The CYclone Global Navigation Satellite System (CYGNSS) mission with eight microsatellites launched by NASA in December 2016, which provides an unprecedented opportunity to rapidly acquire ocean surface wind speed globally. In this paper, a refined spaceborne GNSS-R sea surface wind speed retrieval algorithm is presented and validated with the ground surface reference wind speed from numerical weather prediction (NWP) and cross-calibrated multi-platform ocean surface wind vector analysis product (CCMP), respectively. The results show that when the wind speed was less than 20 m/s, the RMS of the GNSS-R retrieved wind could achieve 1.84 m/s in the case where the NWP winds were used as the ground truth winds, while the result was better than the NWP-based retrieved wind speed with an RMS of 1.68 m/s when the CCMP winds were used. The two sets of inversion results were further evaluated by the buoy winds, and the uncertainties from the NWP-derived and CCMP-derived model prediction wind speed were 1.91 m/s and 1.87 m/s, respectively. The accuracy of inversed wind speeds for different GNSS pseudo-random noise (PRN) satellites and types was also analyzed and presented, which showed similar for different PRN satellites and different types of satellites.

Keywords: spaceborne GNSS-R; DDM; ocean surface wind speed; GMF; CYGNSS

---

1. Introduction

Rapidly acquiring global high temporal and spatial resolution ocean surface wind field has extremely significant in many fields. Spaceborne GNSS-R is a relatively new remote sensing technique with promising prospects, which receives reflected GNSS signals from the Earth's surface. With the GNSS-R receiver mounted on a low Earth orbit (LEO) microsatellite, it can form a spaceborne bistatic radar scatterometer to sense wind speed near the sea surface. The hardware instrument of the GNSS-R payload is lightweight and low in power, which can greatly reduce the deployment cost of this remote sensing technique. Through reasonable satellite constellation designing, continuous and rapid measurement of the global sea surface wind speed can be reached, which will effectively compensate for the shortcomings of the traditional monostatic scatterometer and radiometer.

The idea of using the signals of opportunity from GNSS for geophysical parameter detection has been through nearly 30 years of development. The passive reflectometry and interferometry system (PARIS) mesoscale ocean altimetry concept was first proposed by Martin-Neira in 1993 [1]. In 1996, Katzberg conceived of receiving GNSS signals from ocean surface reflection using receivers mounted on LEO satellites to remotely sense ocean states and ocean surface physical parameters [2]. In 2000, Zavorotny and Voronovich presented the GNSS-R bistatic radar scattering model [3], which is based on the geometric optical approximation of the Kirchhoff (KA-GO) model that describes the average scattering power of the GNSS signal reflected by the sea surface to the receiver direction over different time delays and Doppler frequency shifts. Subsequently, simulation based on the theoretical model [4–6] and the feasibility test of the real satellite-based GNSS-R mission have been gradually carried out.

In 1998, Garrison and Katzberg conducted the airborne GNSS-R experiment that preliminarily demonstrated the potential of GNSS-R to sense the ocean state [7,8]. The first spaceborne GNSS-R experiment was implemented on Shuttle Imaging Radar with Payload C (SIR-C) carried on the space shuttle to verify the feasibility and determine the expected signal-to-noise (SNR) ratio on LEO [9]. The UK Disaster Monitoring Constellation (UK-DMC) was launched in 2003, and successfully verified that GPS signals could be received from the ice-ocean, snow, ocean water, and even in the case of land on the LEO [10]. On July 8, 2014, the UK Technology Demonstration Satellite-1 (TDS-1) satellite was successfully launched with a specific Spaceborne GNSS Receiver REmote Sensing Instrument (SGR-ReSI) payload specifically designed for generating Delay/Doppler map (DDM) in real-time [11]. On 15 December 2016, NASA launched the CYGNSS satellite constellation consisting of eight small satellites, which became the first microsatellite constellation dedicated to GNSS-R ocean wind remote sensing [12]. In addition, GNSS-R experiments on-board Soil Moisture Active and Passive (SMAP) [13] and GNSS Reflectometry, Radio Occultation, and Scatterometry Onboard the International Space Station (GEROS-ISS) [14] missions both aim to exploit the GNSS-R technique for geophysical parameters remote sensing.

The onboard Delay-Doppler mapping instrument (DDMI) outputs DDM after a series of digital signal processing and precise calibration [15]. The scattering power in different DDM pixels is the average cross-correlation power of the sea surface reflected signal received by the nadir left-hand circular polarization (LHCP) antenna and a locally clear replica GNSS navigation code in the receiver. In most missions, this mode is called traditional satellite-based GNSS-R, abbreviated as cGNSS-R. There is another GNSS-R operation mode that directly uses the reflected and direct signals for correlation processing that is called iGNSS-R [16]. In the early stage of airborne GNSS-R experiments, only a small number of observations was obtained, and the wind speed was inferred mainly through optimal correlation fitting between the observed time delay waveform and the simulation waveform from the Z-V model [3]. However, the current on-orbit UK Technology Demonstration Satellite-1 (TDS-1) satellite and CYGNSS constellation provide massive observation data. The sea surface wind speed retrieval method, similar to the backward microwave scatterometer [17], is commonly employed, which regresses so-called DDM observables against the collocated wind from other observing data sources, and quantifies the mapping relationship to form the empirical geophysical model function (GMF) for future wind predicting.

Previous studies have demonstrated that compared with the DDM model-fitting method, the formed empirical GMF can obtain better performance [18,19]. Before establishing a retrieval model through empirical regression, it is necessary to further calibrate DDM to remove the influence of non-geophysical effects. The UK TDS-1 mission provided DDM calibrated to SNR [20], while for the CYGNSS mission, the DDM can be calibrated to the bistatic radar cross-section (BRCS) with more comprehensive ancillary parameters [15,21]. However, the calibrated DDM still cannot be used directly for wind speed inversion but needs to further extract the feature quantity, it should sensitive to wind speed in a specific size of the DDM window determined in terms of the requirement of the inversion spatial resolution. Clarizia and Ruf studied a fitting method by using CYGNSS DDM observables to build empirical GMF, adaptively performed minimum variance (MV) estimates based on DDM range-corrected gain (RCG) [22], and the Bayesian estimate was also adopted to calculate the weighting coefficient for each observable [23]. The results of these studies show that when the real wind speed over sea surface is less than 20 m/s, the wind speed retrieval error is less than 2 m/s, but a larger error appears at a higher wind speed range [24]. The latest studies have carried out some new attempts: [25] proposed sequential processing based on the extended Kalman filter, which is evaluated by the simulated wind field data; and [26] presented a method to train the neural network for wind speed prediction by using the entire DDM and different feature information. Both showed promising results, but need to further validate with a large amount of measured data.

Although scatterometers, radiometers, buoys, the NWP product, and even ship observations can be used to construct a robust empirical GMF model, the accuracy and spatial-temporal resolution of the reference wind speed provided by different data sources are different. Therefore, it is important to evaluate the effect of the reference wind and the reliability of the inversion model by the retrieval results obtained from different ground truth winds. In this paper, a refined wind speed retrieval algorithm was used to establish the GMF from three different wind data sources, and the inversion results were analyzed and evaluated by multi-dataset. The rest of this manuscript is organized as follows. Section 2 introduces the GNSS-R remote sensing theory and the improved wind speed inversion algorithm. Section 3 depicts the dataset and presents the performance of the wind speed retrieval algorithm compared to different ground truth wind speeds. The discussion is given in Section 4, and Section 5 summarizes the main conclusions.

2. Theory and Methods

2.1. Bistatic Radar Equation

The spaceborne GNSS-R remote sensing sea surface wind speed based on the onboard DDMI, which is capable of cross-correlation reflected signals with the local replica code in the receiver and mapping the scattering power over a range of time delay and Doppler frequency bins, is known as DDM. To generate the DDM, the coherent integration time commonly takes 1 ms during signal processing in the receiver to avoid the influence of strong speckle noise in short-time correlation, and 1 s non-coherent integration is performed to obtain higher SNR DDM. Both the current TDS-1 and CYGNSS projects belong to this traditional cGNSS-R. The bistatic radar equation (BRE) theoretically explains the physical meaning of DDM [3].

$$\left\langle \left| Y_s\left(\hat{t}, f\right) \right|^2 \right\rangle = \frac{P_T G_T \lambda^2 T_I^2}{(4\pi)^3} \iint_A \frac{G_R \sigma^0}{R_T^2 R_R^2} \Lambda^2(\hat{t} - \tau) sinc^2\left(\hat{f} - f\right) dA \tag{1}$$

where P_T is the transmit power of GNSS satellite; G_T is the transmit antenna gain; G_R is the antenna gain of the receiver; λ is the wavelength of the signal carrier; T_I is the relevant integration time; R_T, R_R represent the distance from the transmitter to the sea surface and surface to the receiver, respectively; $\Lambda(\hat{t} - \tau)$ is the GNSS signal spreading function in delay; and $\hat{t}$ and τ are the replica signal and incoming signal time delays, respectively. $sinc^2\left(\hat{f} - f\right)$ is the frequency response of the GNSS signal; and $\hat{f}$ and f are the replica signal and incoming signal frequencies, respectively. A is the effective scattering area of DDM and dA is a differential area within A. σ^0 is the normalized bistatic radar scattering cross-section, which can be expressed as:

$$\sigma^0 = \frac{\pi |\mathfrak{R}|^2 \vec{q}^4}{q_z^4} P\left(-\frac{\vec{q}_\perp}{q_z}\right) \tag{2}$$

where $\mathfrak{R}$ is the Fresnel reflection coefficient associated with the signal polarization characteristics; $\vec{q}$ is the scattering unit vector; $\vec{q}_\perp$ is the horizontal component of the scattering unit vector; q_z is the vertical component (surface normal direction); P is the probability density function of the rough sea surface slope. The scattering coefficient is related to geophysical parameters in the bistatic radar equation. The scattering intensity from the ocean surface is mainly affected by the sea surface roughness. For the wind speed retrieval, the sea surface roughness is affected due to the influence of local winds, and the change in sea surface roughness will be reflected in the variation of the scattering power.

2.2. Delay-Doppler Map Observables

So far, two types of operational wind speed retrieval algorithm have been used in airborne or spaceborne GNSS-R: (1) through the optimal fitting between the simulated DDM by the Z-V theoretical model and actual observed DDM; and (2) using the DDM derived observables to matchup the collocated wind from another observing system to model an empirical GMF, then using GMF to map the new observable to predict wind speed. The method of using the DDM deconvolution to restore the scattering coefficient, then computing the probability density function (PDF) of the surface slope to relate the wind state, is still undergoing further theoretical research [27]. Usually, multiple DDM observables can be used to establish the separated GMF and predict corresponding wind speed, therefore, an optimal weighting estimator is needed to estimate the weighting coefficients to calculate the final weighted wind speed. Due to the influence of parameter uncertainty such as DDM observation noise, receiver antenna gain, and the effective isotropic radiated power (EIRP) of different GNSS satellites, it is difficult to generate very accurate reference DDM from the DDM simulator [6]. At the same time, the whole DDM from a spaceborne platform corresponds to a larger sea surface area, so the glistening area can be more than 400 km in diameter [28,29]. To meet the specific spatial resolution of inversed wind speed, the latter approach is commonly used in the current GNSS-R field. Before modeling the empirical GMF, the original DDM needs to be calibrated and wrapped according to the bistatic radar forward equations to remove non-geophysical effects, which are usually calibrated as bistatic radar cross-sections [15,21]. Based on the requirement of retrieved spatial resolution, the wind speed inversion model has been established to extract observables in a specific delay Doppler window from DDM [24], which is less affected by observed noise, and sensitive to wind speed around the specular point. Furthermore, the processing of time average is carried out to improve the SNR of observables [24].

Clarizia et al. proposed five DDM observables for the UK-DMC project to exploit the characteristics of DDM that sensitively respond to variations in wind speed, including the Delay-Doppler Map Average (DDMA), Leading Edge Slope (LES), Trailing Edge Slope, Delay-Doppler Map Variance (DDMV), and Allan Delay-Doppler Map Variance (ADDMV) [22]. However, the current UK TDS-1 and CYGNSS mission can only extract DDMA and LES, as both projects directly provide DDM after non-coherent processing in DDMI without raw intermediate frequency (IF) signals published. Since the limitation of spatial resolution, TES observables also cannot be adopted [24]. DDMA represents the average value of the BRCS near the specular point. The LES is the leading edge slope of the integral delay waveform (IDW), and the IDW is obtained from scattering power DDM by summing the columns along the Doppler axis in the specific Delay/Doppler window. Readers can refer to the method in [24] to calculate the inversion observations.

2.3. Wind Speed Retrieval Algorithm

The fundamental process of the spaceborne GNSS-R wind speed retrieval algorithm is shown in Figure 1 as follows. (a) calibrate the DDM and compute DDM observables, we directly used the DDM of BRCS in the CYGNSS Level 1 dataset, where the calibration can be found in [15,21]; (b) improving the SNR by time-averaging while satisfying the requirement of spatial resolution; (c) DDM observables matchup the ground surface truth wind speed from other observing techniques to establish training samples; (d) considering the geometry configuration of the bistatic radar system to form the mapping relation between DDM observables and referenced wind speed; (e) using the algebraic parametric model to smooth the 2D GMF to remove the influence of insufficient training dataset and observation noise; (f) estimating the weighting coefficients of different observables by the minimize variance (MV) estimator; and (g) using the model for wind speed predictions.

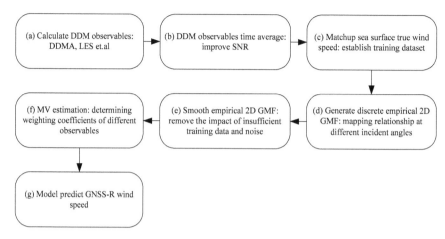

Figure 1. Flow chart of spaceborne GNSS-R wind speed retrieval

In this paper, a refined wind speed algorithm is presented based on [19], the NWP and CCMP wind products are used as ocean surface truth wind speed. Matching the observables with reference wind to form the inversion model, counterparts were obtained by bilinear interpolation in space and linear interpolation in time. In fact, the modeled GMF is a series of mapping relations between the DDM observables and referenced wind speed at different incidence angles but presented as a discrete point. The dynamic range of derived wind speed is limited below 35 m/s, and the incidence angle ranges from 0° to 70°. This study does not distinguish the effects of different sea surface states on GNSS-R wind speed retrieval.

In the incident angle dimension, GMF is modeled under different incident angles starting from 0.5° in a step length of 1°. At a certain incidence angle, wind speeds starting from 0.05 m/s increments to 35 m/s in the step length of 0.1 m/s, and the weighted average observables at different wind speed bins are calculated to form discrete empirical mapping relationships. In order to expand the training samples, the training data can be overlapped in both dimensions. At a certain incident angle, all the matched data pairs falling within the left and right two step length intervals (steps by 1°) are taken as the model training samples. At a certain wind speed range, the weighted DDM observables are calculated by taking the data in the left and right two step length intervals as well, but it should be noted that the step length of the wind speed interval needs to be re-determined according to the wind speed probability density. The strategy in [19] was directly used in this paper. In both dimensions, the training data within different intervals use different weights, and samples within the first step length interval from certain incidence angle/wind speed take twice the weight. The weighting strategy can be explained in Figure 2. There are four different cases for the sample points to construct a specific discrete point for GMF to map between the observable and wind speed at a certain incidence angle: (1) the sample located in the first step length interval around the incident angle and the first step length around specific wind speed, as shown in the blue region in Figure 2, with a scale factor is 4; (2) the sample located within the first step length range of the incident angle and the second step length size around the wind speed, as shown in the purple area in Figure 2 with a scale factor is 2; (3) the sample point is located in the second step length interval around the current incident angle and in the interval of the first step length around the wind speed, as shown in the green area in Figure 2 with a scale factor is 2; and (4) the sample point is located in the second step length interval around the current incident angle and in the second step length interval around the wind speed, as shown in the gray area in Figure 2 with a scale factor is 1. The number of samples in the different intervals is multiplied by the corresponding scale factor as the numerator of the sample weight coefficients in a certain interval, and the sum of the four intervals is used as the denominator of the weight coefficient to compute the weighted value of the discrete points.

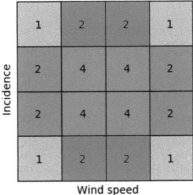

Figure 2. Weighting strategy of observables to form empirical geophysical model function.

To eliminate the fluctuation of the empirical model caused by insufficient training samples and system noise and ensure the accuracy of the empirical GMF, first, the maximum probability density bin of the wind speed in the training dataset is calculated at each incident angle because the probability density distribution of the wind speed slightly varies at different incident angle bins, as shown in Figure 3, it presents the PDF of the matched NWP wind for the CYGNSS DDM observables in each incidence angle bin. Then, the corresponding weighted DDM observables were calculated with the wind speed close to its maximum probability density, the observables above or below the wind speed were sequentially computed with the step size of 0.1 m/s, and the discrete GMF was also forced to be a monotone function with wind speed, which means that the GMF values could be same if monotonicity is violated during calculation. Since a single function form was not found to fit the discrete GMF well, we directly chose a piecewise function to obtain the final GMF model as the CYGNSS science operations center. The smoothing function smaller or larger than the piecewise point is shown in Equations (3) and (4), respectively:

$$\text{obs} = a_0 + a_1 u^{-1} + a_2 u^{-2} \tag{3}$$

$$\text{obs} = b_0 + b_1 u + b_2 u^2 \tag{4}$$

where *obs* means the DDM observables and u represents the ground truth wind speed U_{10}.

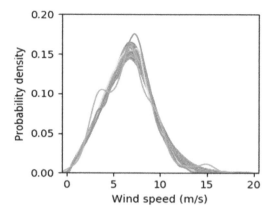

Figure 3. Distribution of P_{wind} as a function of incidence angle (unlabeled colored lines indicate the probability density distribution of wind speed at each incident angle bins).

The nonlinear least-squares fitting was performed for the GMF with wind speeds smaller than the segmented point, a new constraint was added, which requires all coefficients of formula (3) are limited to be non-negative. For discrete points of GMF where the wind speed is greater than the segmented point, the parabolic function fitting means that the values of two functions are equal at the piecewise point, and the first derivatives are equal. Furthermore, new constraints have been added such as limiting the opening of the parabola downward and its axis of symmetry on the left side of the piecewise point to force the established mapping relationship to be smoother and more consistent with the distribution of training data. The smoothing procedure of the discrete GMF is transformed into a nonlinear least square fitting and convex quadratic programming problem. The standard form of the convex quadratic programming is:

$$\min \ \tfrac{1}{2}x^T P x + q^T$$
$$s.t. \ Gx \le h \tag{5}$$
$$Ax = b$$

One of the important things to apply a piecewise function is to determine the segmentation point. In order to determine the optimal piecewise point, first, the smooth processing is performed at each GMF discrete point to find one with the smallest fitting residual. Then, do the smoothing again in the interval of a step length (0.1 m/s) around the discrete point with a smaller step. Finally, find the best location with the smallest fitting residual as the final transition point to re-smooth the final model. After obtaining the GMF function $u(\theta, obs_i)$ (where θ denotes the incident angle and i denotes DDMA or LES) of the spaceborne GNSS-R, more precisely, it is a lookup table. When using the GMF model for the wind speed prediction. The model $u(\theta, obs_i)$ is linearly interpolated to the incident angle θ corresponding to the observable getting $u_\theta(obs_i)$ at first, then $u_\theta(obs_i)$ linearly interpolates again to obtain the inversed wind speed u_{θ,obs_i} corresponding to the observable.

Establishing the separated empirical GMF for both DDMA and LES can benefit quality control. When the retrieved wind speed between the two types of observables is greater than 3 m/s, the inversion results are considered unreliable. Finally, the MV estimation is used to dynamically adjust the DDM SNR variation caused by GNSS-R geometry changes. The optimal wind speed estimator is obtained by weighting the wind speeds from DDMA and LES [22].

3. Results and Validation

The CYGNSS dataset became available in March 2017, and the experiments in this work used the V2.1 version of CYGNSS level 1 data downloaded from the Physical Oceanography Distributed Active Archive Center (PO.DAAC). In the process of forming GMF, the RCG of samples is required to be greater than 10 to ensure the quality of the GNSS-R observations. The training dataset used was from 1 July, 2017 to 30 November, 2017. Data collected in December 2017 were used as a test dataset. The inversed results were compared with the NWP-based and CCMP-based wind, also evaluated by the buoy data. Figure 4 shows the distribution of specular point tracks of the 8 CYGNSS satellite on 1st December 2017 and the distribution of the used moored buoys, which are indicated by the red dot.

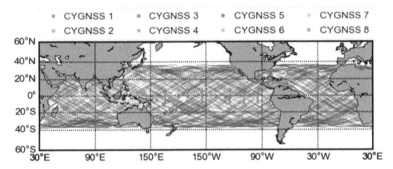

Figure 4. The distribution of daily specular point tracks of 8 CYGNSS satellites and buoys.

3.1. Wind Speed Retrieval Based on ERA5 and GDAS

The atmospheric reanalysis numerical weather prediction (NWP) products from ECMWF ERA5 provided by the Climate Data Store (ADS) and the Global Data Assimilation System (GDAS) products provided by the Research Data Archive (RDA) were used as the ground truth wind speed in the GNSS-R wind speed retrieval. ERA5 is the fifth generation of NWP products provided by the European Center for Medium-range Weather Forecast (ECMWF), the spatial resolution of the grid wind field product is $0.25° \times 0.25°$ and the time resolution reaches per hour [30]. National Centers for Environmental Prediction (NCEP) operates a global data assimilation system and the surface flux grid for the NCEP GDAS/FNL global surface flux products uses the T574 Gaussian global grid with a time resolution of 6 hours for wind speed products [31]. Both NWP products provide the U10 wind field. The wind speed data are interpolated bilinearly in space and linearly in time to matchup the DDM observables. In order to improve the accuracy of the reference wind speed in the training samples, samples with a deviation larger than 3 m/s from the two reference winds were removed. When the wind speed was less than 20 m/s, only the matched reference wind from ERA5 was used. When the wind speed was greater than 20 m/s, but less than 25 m/s, the average of two matching wind speeds was used. While the wind speed was greater than 25 m/s, only the GDAS wind was adopted.

When DDMA and LES take the log scale, the fitting wind speed inversion model for DDMA and LES observables as shown in Figure 5, it is important to note that we have removed the parts of GMF that exceeded the range of the coordinate axis. Different levels of fold appeared where the wind speed was greater than 20 m/s, especially for the LES, which was obviously due to insufficient training samples. The correlation between the DDM observables and wind speed decreased as the wind speed increased, and when the wind speed was greater than 10 m/s, the first-order derivative change rate of the GMF function was small. This can be seen more clearly in Figure 6, which presents a group of DDMA and LES GMF at specific incidence angles. However, there is an apparent difference in the sensitivity between the two types of DDM wind speed observables, the LES could not respond to the variation of wind speed when it reached 25 m/s, while the DDMA was still sensitive to the change in the observables under the condition of strong wind speed. If the training sample is small or the DDM noise is large, the modeling error is easily shifted to GMF, which will definitely amplify the observation error in the final wind retrieval. However, the incident angle has little effect on the inversion model at low wind speed range, but as the wind speed increases, the influence becomes distinct.

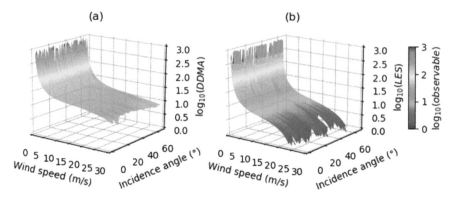

Figure 5. Delay-Doppler Map Average (**a**) and Leading Edge Slope (**b**) geophysical model function using Numerical Weather Prediction winds as the ground truth wind speed.

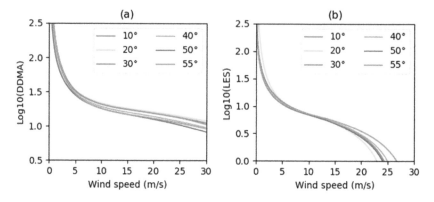

Figure 6. Delay-Doppler Map Average (**a**) and Leading Edge Slope (**b**) geophysical model function at specific incidence angles using Numerical Weather Prediction winds as the ground truth wind speed.

When the NWP product was used as the surface truth wind, the density scatterplot of the retrieved wind speed against reference winds, residual versus reference winds, incident angle, and RCG is shown in Figure 7, where the blue dash line in Figure 7a represents 1:1, the wind speed deviation refers to the inverted wind speed subtracts the ground truth wind speed. The total number of matched test samples was 12,356,042 with bias and the Root Mean Square (RMS) of inferred winds are 0.14 m/s and 1.84 m/s, respectively. The inversed uncertainty of different wind intervals are also counted, and a wind speed greater than 10 m/s accounts for 6.71% with an uncertainty of 2.76 m/s; when the wind speed is larger than 15 m/s, the test samples account for 0.19% of the total training dataset with an uncertainty of 3.24 m/s, and a larger retrieval error appeared at higher wind speed range. The dependence of retrieval error on the NWP-derived ground truth wind is shown in Figure 7b, where positive biases appeared at reference wind speeds of 5–12 m/s, while negative biases can be seen at ground truth winds above 12 m/s. The dependence of retrieval error on the incidence angle is shown in Figure 7c, where the highest density of retrieval errors is generally distributed near zero error. There is a pyramid distribution between the retrieval error and RCG as shown in Figure 7d; since RCG represents the received signal strength, larger RCGs can mean better-received signal quality. Figure 8 shows the average deviation and RMS of the inversed wind speed at different wind speed bins. The maximum matching reference wind speed of the test samples was 23 m/s, and the bias and RMS had large fluctuations around 20 m/s. The main reason is that there are fewer test samples under this wind speed range, which caused large errors.

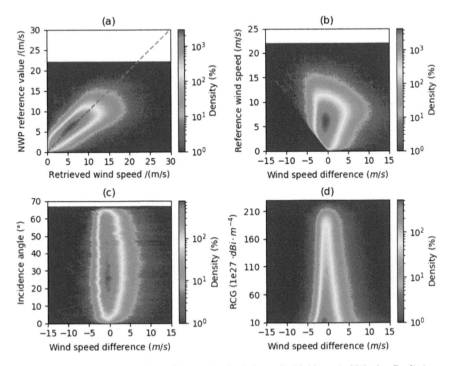

Figure 7. Log(density) scatterplots of the retrieved wind speed with Numerical Weather Prediction ground truth wind speed (**a**), residual versus truth wind speed (**b**), incident angle (**c**), and range-corrected gain (**d**).

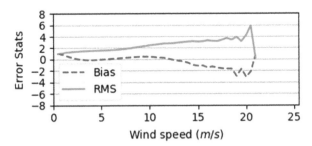

Figure 8. Root Mean Square and bias from Numerical Weather Prediction derived sea surface wind speed in different wind bins.

3.2. Wind Speed Retrieval Based on CCMP

The cross-calibrated multi-platform ocean surface wind vector analysis product V2 (CCMP) combines cross-calibrated satellite microwave winds and in-situ wind, using a variational analysis method (VAM) to produce high-resolution (0.25°) gridded product with a time resolution of 6 h [32], it is also provided by RDA. Satellite-based passive and active microwave measurements mainly contribute to this wind product. The main reason for adopting the CCMP wind as ground truth wind speed is not only because it has high accuracy, but also because it is mainly based on a satellite-based microwave scatterometer and radiometer. Using CCMP wind can further compare the accuracy of wind retrieval between GNSS-R and other satellite-based techniques.

The CCMP data are also interpolated bilinearly in space and linearly in time to match the DDM observables for the training dataset. The empirical wind speed retrieval algorithm mentioned in Section 2.3 was used to construct the 2D GMF, Figure 9 shows the GMF established by DDMA and LES at an incident angle of 30°. Magenta dots in the figure represent the empirical GMF directly obtained by weighting the training samples, and the black line represents the GMF obtained after parametric smoothing. It should be noted that the x-axis magnitude of the two subgraphs is different. Compared with DDMA, the distribution of the LES samples is closer to the y-axis. The comparison of the empirical GMF and parametric smoothed GMF shows that the parametric model of DDMA GMF was better than LES, and the latter had a slightly bigger gap at a lower wind speed range between the smoothed parametric GMF and original GMF, where the phenomenon indeed appeared at the full incidence angle dimension. We infer that the reason for this is that the number of samples used to construct the GMF was still quite small.

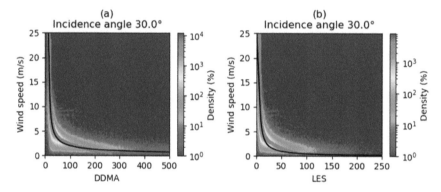

Figure 9. Delay-Doppler Map Average (**a**) and Leading Edge Slope (**b**) geophysical model function at an incident angle of 30° with the CCMP wind.

Since the wind speed products provided by the CCMP dataset are only available until 30 December, 2017, the CCMP testing dataset only included 30 days of samples, and the matchup wind speed pairs reach a total of 10,272,765 groups. The density scatterplot of retrieved wind speed and reference winds, residual versus reference wind speed, incident angle, and RCG are shown in Figure 10 with the logarithmic in the number density of samples. Figure 10a shows that the highest density of samples occurred along the 1:1 blue dash line, and the statistic indicates that the bias and RMS of the CCMP-based inversed winds are 0.05 m/s and 1.68 m/s, respectively. In order to further clarify the inversion accuracy at the different wind speed ranges, the percent and uncertainty of the test samples were also presented, and the wind speed values in training dataset greater than 10 m/s account for 8.48% of the training set with an RMS 2.56 m/s, and wind speeds greater than 15 m/s account for 0.26% with 3.13 m/s. The behavior of the inversed residual with the CCMP wind, incidence angle, and RCG was similar to the NWP-based results, but the residual was closer to the zero error.

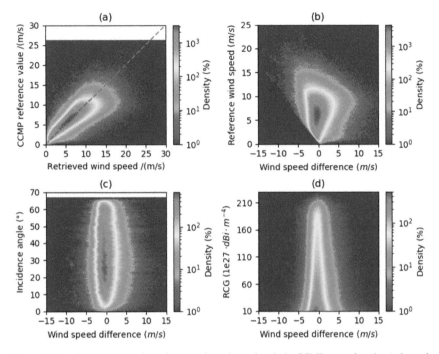

Figure 10. Log(density) scatterplots of retrieved wind speed with the CCMP ground truth wind speed (**a**), residual versus truth wind speed (**b**), incident angle (**c**), and RCG (**d**).

Figure 11 shows the mean bias and RMS of the CCMP-derived wind speed at different wind speed bins. The RMS uncertainty rose gradually with the wind speed below approximately 10 m/s, with bias around 0 m/s and RMS below 2 m/s. However, the inversion uncertainty increased sharply and is accompanied by fluctuations after the wind speed is above 15 m/s, the bias increasing is even surprising. However, the behavior of the RMS and bias is related to many factors. The primary contribution is from the insufficient testing samples at winds above 15 m/s, which also occurred on the training samples, where it appears as a large modeling error, and the observation is very sensitive to the variant of wind speed.

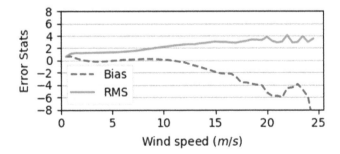

Figure 11. Root Mean Square and bias from CCMP-derived sea surface wind speed in different wind bins.

3.3. GNSS-R Wind Speed Validation with Buoy Observation

Continuous high temporal-resolution wind speed data were obtained from moored buoy measurements provided by the National Data Buoy Center (NDBC). The time resolution of the wind speed product is 10 minutes, the dynamic range of wind speed is 0.0–35 m/s, and the measurement uncertainty is 0.3 m/s or 3%. Figure 4 shows the distribution of the 93 moored buoys used in this study, buoy wind is employed as an additional data source to validate NWP-derived and CCMP-derived wind speed. It should be noted that the wind of the Tropical Atmosphere Ocean (TAO) program buoy array provided by the NDBC is not calibrated to the standard wind speed reference height U10, and is directly corrected by the following formula [33]:

$$U_{10} = 8.87403 \times U_Z / ln(z/0.0016) \tag{6}$$

where U_Z is the measured wind speed at the anemometer height of z above the sea surface in meters. When matching the buoy wind speed with GNSS-R retrieved wind speed, the distance between the specular point of observables and the buoy is limited to less than 50 km, and the reference wind speed was obtained by linear interpolation in the time domain.

Buoy wind speed is obtained by in situ observations, so it has the highest accuracy. Since most of the buoys are located on the offshore coast, the number of training samples collected under the matching condition is very small, so it is difficult to model a 2D GMF. Through analyzing the training dataset, it is found that the buoy reference wind speed is almost below 15 m/s in the matched dataset, even if we tried to neglect the influence of incident angle, the 1D GMF still lacks the reference value for application, so we only used the buoy wind speed as external data source to further evaluate the wind speed inversion results of the other two datasets.

Figure 12 compares the NWP-derived and CCMP-derived winds using the buoy wind as a reference value, where RLM represents a robust regression line, and the histogram shows the distribution of winds. It can be seen from the wind distribution histogram that the wind speed range of the matched samples was basically in the range of 0–15 m/s. The test wind speed pairs from NWP-derived winds and buoy winds only had 23,812 pairs in December 2017, only the sample less than four times of the standard deviation was selected for accuracy statistics, which are contained in the patch of the figure. The bias and RMS are −0.18 m/s and 1.91 m/s, respectively, and the Pearson correlation coefficient between the two sets of wind is 0.78. The CCMP-derived inversed winds demonstrated a slightly better performance compared to NWP, where there are 20,604 groups of opportunity matched wind speed pairs obtained, with the bias and RMS of 0.11 m/s and 1.87 m/s, respectively, the Pearson correlation coefficient between the estimated winds and buoy winds is 0.78, at the same time, the linear regression line had a larger slope. Generally, the accuracy of the retrieved wind speed from two sets of reference wind data sources is considerable. Both the NWP and CCMP wind speed products can be employed as the truth reference wind speed for spaceborne GNSS-R ocean wind speed remote sensing.

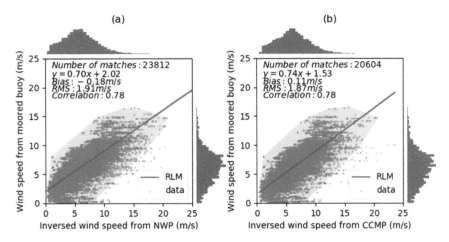

Figure 12. Retrieved wind speed from NWP-based (**a**) and CCMP-based surface truth wind (**b**) versus buoy wind speed. The gray histogram indicates the distribution of wind speed in testing samples, and statistics only compute samples less than four times of the standard deviation within the blue patch.

4. Discussion

Comparing the GMF model of DDMA and LES in Section 3, it can be seen that the sensibility of both observables to the variety of wind speed decreases with the increase of wind speed, and the LES completely loses its response to wind speed changes when the real ocean surface wind speed is close to 25 m/s. On one hand, it is due to the influence of the characteristics of the two wind speed indicators [34], which also means that the accurate calibration of the DDM is very important in improving the inversion accuracy of the retrieval model under high wind speed. On the other hand, because the spaceborne GNSS-R bistatic radar wind speed retrieval algorithm strongly relies on the size of the training dataset and its reliability. Figure 13 shows the probability density distribution of wind speed in the training samples establishing GMF with CCMP wind products as the sea surface truth wind speed. The wind speed in the training dataset is mainly concentrated below 10 m/s, which can also be confirmed in Figure 3 with the NWP reference wind. The most probable wind speed in the dataset is 7 m/s, and the statistics show that wind speeds greater than 10 m/s only accounted for 11.71% in the training samples, and a wind speed greater than 15 m/s only accounted for 0.52%. Therefore, to further improve the inversion accuracy in the medium-strong wind speed range, it is necessary to expand the training data volume in this wind speed range. It should be noted that when CCMP was used as the surface truth wind to model GMF, the retrieved results are better than the NWP. However, when the wind speed was larger than 15 m/s, the bias of the CCMP-based retrieved wind speed was larger than the NWP-based retrievals.

Furthermore, the accuracy of the inversed wind for different GNSS types and PRN satellites is performed. Figure 14 depicts the histogram of the bias, RMS, and the testing dataset size of the inferred wind speed corresponding to different GPS PRN satellites with NWP-derived and CCMP-derived surface truth winds. Currently, the latest published V2.1 version of the CYGNSS data has removed the newly launched GPS block II-F satellite data because their transmit power monitoring is still inaccurate, so only GPS block II-R and GPS block IIR-M related DDMs are available. It can be seen that the accuracy of the inversed wind speed with different PRN satellites is approximately similar to the two different surface reference wind speed sources and there are also no significant differences between the different types of satellites.

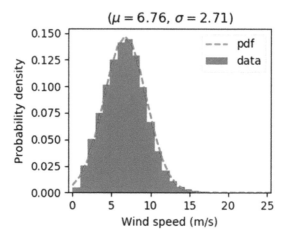

Figure 13. CCMP reference wind speed probability density distribution in the training samples.

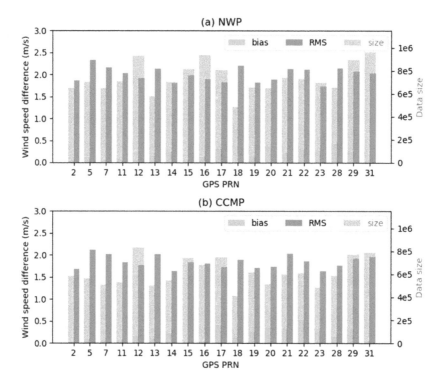

Figure 14. Wind speed inversion accuracy from the NWP-based reference wind (**a**) and CCMP-based reference wind (**b**) corresponding to different GPS PRN satellite types.

5. Conclusions

In this paper, the NWP and CCMP wind products are used as the ground truth wind to establish 2D GMF for spaceborne GNSS-R ocean surface wind speed remote sensing, and the buoy data are included to validate the results of the two groups' inversion results. In order to improve the accuracy of the retrieval algorithm, new constraints are added in the process of smoothing the discrete empirical GMF to eliminate fluctuations caused by the measurement noise and insufficient training samples to ensure that the model is consistent with the trend of the actual physical process. The established 2D GMF is sensitive to the variation of DDM observables under medium-strong wind speed conditions. The results show that the inversion accuracy could reach 1.84 m/s if the surface reference wind speed is given by the NWP when the wind speed was less than 20 m/s, while the inversion accuracy of the CCMP-based retrievals was 1.68 m/s. There are no large deviations between the derived wind based on different reference wind sources, and this result further proves the reliability of GNSS-R derived wind speed. From the distribution of the winds in the training dataset, it shows that the training samples are mainly concentrated below 15 m/s, and therefore, further expanding the size of the training dataset at a high wind range can improve the accuracy of the wind speed retrieval algorithm.

Author Contributions: S.J. and Z.D. conceived and designed the experiments and Z.D. performed the experiments and analyzed the data. Both authors contributed to the writing of the paper.

Funding: This work was supported by the National Key Research and Development Program of China Sub-Project (grant no. 2017YFB0502802), the Strategic Priority Research Program Project of the Chinese Academy of Sciences (grant no. XDA23040100), the Jiangsu Province Distinguished Professor Project (grant no. R2018T20), and the Startup Foundation for Introducing Talent of NUIST (grant no. 2243141801036).

Acknowledgments: The authors would like to thank ECMWF and GDAS for providing the reanalysis data, and the NASA CYGNSS team. The CCMP wind products were provided by the National Center for Atmospheric Research's Research Data Archive, and we also appreciate the work for this dataset.

Conflicts of Interest: The authors declare no conflicts of interest.

References

1. Martineira, M. A Passive Reflectometry and Interferometry System (Paris)—Application to Ocean Altimetry. *Esa J. Eur. Space Agency* **1993**, *17*, 331–355.

2. Katzberg, S.; Garrison, J. *Utilizing GPS To Determine Ionospheric Delay over the Ocean*; NASA Technical Memorandum 4750; NASA: Greenbelt, MD, USA, 1996.

3. Zavorotny, V.; Voronovich, A. Scattering of GPS signals from the ocean with wind remote sensing application. *IEEE Trans. Geosci. Remote Sens.* **2000**, *38*, 951–964. [CrossRef]

4. Saïd, F.; Soisuvarn, S.; Jelenak, Z.; Chang, P.S. Performance Assessment of Simulated CYGNSS Measurements in the Tropical Cyclone Environment. *IEEE J. Sel. Top. Appl. Earth Obs. Remote Sens.* **2016**, *9*, 4709–4719. [CrossRef]

5. O'Brien, A.; Johnson, J.T.; IEEE. Comparing the Cygnss Simulator forward Scattering Model with Tds-1 and Cygnss on-Orbit Ddms. In *2017 IEEE International Geoscience and Remote Sensing Symposium*; IEEE: New York, NY, USA, 2017; pp. 2657–2658.

6. Hoover, K.E.; Mecikalski, J.R.; Lang, T.J.; Li, X.; Castillo, T.J.; Chronis, T. Use of an End-to-End-Simulator to Analyze CYGNSS. *J. Atmos. Ocean. Technol.* **2018**, *35*, 35–55. [CrossRef]

7. Garrison, J.L.; Hill, M.I.; Katzberg, S.J. Effect of sea roughness on bistatically scattered range coded signals from the Global Positioning System. *Geophys. Res. Lett.* **1998**, *25*, 2257–2260. [CrossRef]

8. Lin, B.; Katzberg, S.J.; Garrison, J.L.; Wielicki, B.A. Relationship between GPS signals reflected from sea surfaces and surface winds: Modeling results and comparisons with aircraft measurements. *J. Geophys. Res. Space Phys.* **1999**, *104*, 20713–20727. [CrossRef]

9. Lowe, S.T.; LaBrecque, J.L.; Zuffada, C.; Romans, L.J.; Young, L.E.; Hajj, G.A. First spaceborne observation of an Earth-reflected GPS signal. *Radio Sci.* **2002**, *37*, 28. [CrossRef]

10. Clarizia, M.P.; Gommenginger, C.P.; Gleason, S.T.; Srokosz, M.A.; Galdi, C.; Di Bisceglie, M. Analysis of GNSS-R delay-Doppler maps from the UK-DMC satellite over the ocean. *Geophys. Res. Lett.* **2009**, *36*, 5. [CrossRef]

11. Foti, G.; Gommenginger, C.; Jales, P.; Unwin, M.; Shaw, A.; Robertson, C.; Rosello, J. Spaceborne GNSS reflectometry for ocean winds: First results from the UK TechDemoSat-1 mission. *Geophys. Res. Lett.* **2015**, *42*, 5435–5441. [CrossRef]

12. Ruf, C.S.; Atlas, R.; Chang, P.S.; Clarizia, M.P.; Garrison, J.L.; Gleason, S.; Katzberg, S.J.; Jelenak, Z.; Johnson, J.T.; Majumdar, S.J.; et al. New Ocean Winds Satellite Mission to Probe Hurricanes and Tropical Convection. *Bull. Am. Meteorol. Soc.* **2016**, *97*, 385–395. [CrossRef]

13. Carreno-Luengo, H.; Lowe, S.; Zuffada, C.; Esterhuizen, S.; Oveisgharan, S. Spaceborne GNSS-R from the SMAP Mission: First Assessment of Polarimetric Scatterometry over Land and Cryosphere. *Remote Sens.* **2017**, *9*, 362. [CrossRef]

14. Wickert, J.; Cardellach, E.; Martin-Neira, M.; Bandeiras, J.; Bertino, L.; Andersen, O.B.; Camps, A.; Catarino, N.; Chapron, B.; Fabra, F.; et al. GEROS-ISS: GNSS REflectometry, Radio Occultation, and Scatterometry Onboard the International Space Station. *IEEE J. Sel. Top. Appl. Earth Obs. Remote Sens.* **2016**, *9*, 4552–4581. [CrossRef]

15. Gleason, S.; Ruf, C.S.; O'Brien, A.J.; McKague, D.S. The CYGNSS Level 1 Calibration Algorithm and Error Analysis Based on On-Orbit Measurements. *IEEE J. Sel. Top. Appl. Earth Obs. Remote Sens.* **2019**, *12*, 37–49. [CrossRef]

16. Shah, R.; Garrison, J.L.; Grant, M.S. Demonstration of Bistatic Radar for Ocean Remote Sensing Using Communication Satellite Signals. *IEEE Geosci. Remote Sens. Lett.* **2012**, *9*, 619–623. [CrossRef]

17. Hersbach, H.; Stoffelen, A.; De Haan, S. An improved C-band scatterometer ocean geophysical model function: CMOD5. *J. Geophys. Res. Space Phys.* **2007**, *112*, 18. [CrossRef]

18. Ruf, C.S.; Gleason, S.; McKague, D.S. Assessment of CYGNSS Wind Speed Retrieval Uncertainty. *IEEE J. Sel. Top. Appl. Earth Obs. Remote Sens.* **2019**, *12*, 87–97. [CrossRef]

19. Ruf, C.S.; Balasubramaniam, R. Development of the CYGNSS Geophysical Model Function for Wind Speed. *IEEE J. Sel. Top. Appl. Earth Obs. Remote Sens.* **2019**, *12*, 66–77. [CrossRef]

20. Lin, W.M.; Portabella, M.; Foti, G.; Stoffelen, A.; Gommenginger, C.; He, Y.J. Toward the Generation of a Wind Geophysical Model Function for Spaceborne GNSS-R. *IEEE Trans. Geosci. Remote Sens.* **2019**, *57*, 655–666. [CrossRef]

21. Gleason, S.; Ruf, C.S.; Clarizia, M.P.; O'Brien, A.J. Calibration and Unwrapping of the Normalized Scattering Cross Section for the Cyclone Global Navigation Satellite System. *IEEE Trans. Geosci. Remote Sens.* **2016**, *54*, 2495–2509. [CrossRef]

22. Clarizia, M.P.; Ruf, C.S.; Jales, P.; Gommenginger, C. Spaceborne GNSS-R Minimum Variance Wind Speed Estimator. *IEEE Trans. Geosci. Remote Sens.* **2014**, *52*, 6829–6843. [CrossRef]

23. Clarizia, M.P.; Ruf, C.S. Bayesian Wind Speed Estimation Conditioned on Significant Wave Height for GNSS-R Ocean Observations. *J. Atmos. Ocean. Technol.* **2017**, *34*, 1193–1202. [CrossRef]

24. Clarizia, M.P.; Ruf, C.S. Wind Speed Retrieval Algorithm for the Cyclone Global Navigation Satellite System (CYGNSS) Mission. *IEEE Trans. Geosci. Remote Sens.* **2016**, *54*, 4419–4432. [CrossRef]

25. Huang, F.; Garrison, J.L.; Rodriguez-Alvarez, N.; O'Brien, A.J.; Schoenfeldt, K.M.; Ho, S.C.; Zhang, H. Sequential Processing of GNSS-R Delay-Doppler Maps to Estimate the Ocean Surface Wind Field. Available online: https://ieeexplore.ieee.org/abstract/document/8807371 (accessed on 14 November 2018).

26. Liu, Y.; Collett, I.; Morton, Y.J. Application of Neural Network to GNSS-R Wind Speed Retrieval. Available online: https://ieeexplore.ieee.org/abstract/document/8802279 (accessed on 14 November 2018).

27. Park, H.; Valencia, E.; Rodriguez-Alvarez, N.; Bosch-Lluis, X.; Ramos-Perez, I.; Camps, A.; IEEE. New Approach to Sea Surface Wind Retrieval from Gnss-R Measurements. In *2011 IEEE International Geoscience and Remote Sensing Symposium*; IEEE: New York, NY, USA, 2011; pp. 1469–1472. [CrossRef]

28. Marchan-Hernandez, J.; Camps, A.; Rodriguez-Alvarez, N.; Valencia, E.; Bosch-Lluis, X.; Ramos-Perez, I. An Efficient Algorithm to the Simulation of Delay–Doppler Maps of Reflected Global Navigation Satellite System Signals. *IEEE Trans. Geosci. Remote Sens.* **2009**, *47*, 2733–2740. [CrossRef]

29. Clarizia, M.P.; Ruf, C.S. On the Spatial Resolution of GNSS Reflectometry. *IEEE Geosci. Remote Sens. Lett.* **2016**, *13*, 1064–1068. [CrossRef]

30. Hersbach, H.; de Rosnay, P.; Bell, B.; Schepers, D.; Simmons, A.; Soci, C.; Abdalla, S.; Alonso-Balmaseda, M.; Balsamo, G.; Bechtold, P.; et al. Operational global reanalysis: Progress, future directions and synergies with NWP. *ERA Rep. Ser.* **2018**, *27*, 65.

31. National Centers for Environmental Prediction/National Weather Service/NOAA/U.S. Department of Commerce. 2015, Updated Daily. NCEP GDAS/FNL Global Surface Flux Grids. Research Data Archive at the National Center for Atmospheric Research, Computational and Information Systems Laboratory. Available online: https://rda.ucar.edu/datasets/ds084.4/ (accessed on 14 November 2018).

Remote Sens. **2019**, *11*, 2747

32. Wentz, F.J.; Scott, J.; Hoffman, R.; Leidner, M.; Atlas, R.; Ardizzone, J. Cross-Calibrated Multi-Platform Ocean Surface Wind Vector Analysis Product V2, 1987—ongoing. Research Data Archive at the National Center for Atmospheric Research, Computational and Information Systems Laboratory. 2016. Available online: https://rda.ucar.edu/datasets/ds745.1/ (accessed on 14 November 2018).

33. Thomas, B.R.; Kent, E.C.; Swail, V.R. Methods to homogenize wind speeds from ships and buoys. *Int. J. Climatol.* **2005**, *25*, 979–995. [CrossRef]

34. Jin, S.G.; Feng, G.; Gleason, S. Remote sensing using GNSS signals: current status and future directions. *Adv. Space Res.* **2011**, *47*, 1645–1653. [CrossRef]

Article

Evaluation of HY-2A Scatterometer Ocean Surface Wind Data during 2012–2018

Ke Zhao [1] and Chaofang Zhao [1,2,*]

[1] Department of Marine Technology, Ocean University of China, Qingdao 266100, China; zhaoke@stu.ouc.edu.cn

[2] Laboratory for Regional Oceanography and Numerical Modeling, Pilot National Laboratory for Marine Science and Technology(Qingdao), Qingdao 266237, China

* Correspondence: zhaocf@ouc.edu.cn

Received: 29 September 2019; Accepted: 9 December 2019; Published: 11 December 2019

Abstract: This study focuses on the evaluation of global Haiyang-2A satellite scatterometer (HSCAT) operational wind products from 2012 to 2018. In order to evaluate HSCAT winds, HSCAT operational wind products were collocated with buoy measurements and rainfall data. Error varieties under different atmospheric stratification and rainfall conditions were taken into consideration. After data quality control, the average bias and root mean square error (RMSE) between buoys and HSCAT data were 0.1 m/s and 1.3 m/s for wind speed, and 1° and 27° for wind direction, respectively. Especially, the varieties of the wind direction difference change a lot under non-neutral atmospheric conditions. HSCAT wind speeds are overestimated with an increasing rainfall rate while wind directions tend to be perpendicular to buoys'. In brief, the HSCAT wind product qualities are not stable during 2012 to 2018, especially for the data in 2015 and 2016. Atmospheric stratification and rain effects should be considered in wind retrieval and marine application.

Keywords: HY-2A; scatterometer; sea surface wind field; evaluation

---

1. Introduction

The sea surface wind field is one of the direct factors affecting small-scale and large-scale seawater movement and plays an important role in marine environmental monitoring, weather forecasting, as well as other marine and atmospheric sciences studies. Owing to the spatial and temporal constraints, traditional wind measurement methods, such as ships, buoys, weather stations, and other in situ observations, are not satisfied with the requirements of high-quality and continuous wind observations for ocean research. Since the 1970s, satellite observations have become an indispensable approach to collect sea surface wind field data [1].

There are four main satellite microwave sensors for observing the global sea surface winds: Microwave scatterometer, microwave radiometer, altimeter, and synthetic aperture radar [1]. The microwave scatterometer is the main sensor for providing wind data with its all-weather and all-time capacity [2]. The earliest satellite scatterometer is the Ku-band fan beam Seasat-A Satellite Scatterometer (SASS), launched by National Aeronautics and Space Administration (NASA) in 1978. SASS opened a new era of satellite-observing sea surface wind fields [3]. Afterwards, series C-band and Ku-band scatterometers were successfully launched [4–9]. The Ku-band scatterometer of the first Chinese microwave ocean environment satellite Haiyang-2A (HSCAT), launched on 16 August 2011, is a pencil beam scatterometer with a 1700-km swath and 25-km spatial resolution [10].

Evaluating the quality of scatterometer data is indispensable for sea surface wind field study. There are abundant evaluation investigations of different sensors studied by worldwide researchers. Masuko et al. (2000) evaluated NSCAT wind vectors with Japan Meteorological Agency (JMA) buoys, National Space Development Agency of Japan (NASDA) buoys, and six JMA research vessels in the seas around Japan. They found that no significant systematic errors with wind speeds and incidence angles exist for NSCAT wind products [11]. Ebuchi et al. (2002) compared QuikSCAT/SeaWinds winds with buoy data from National Data Buoy Center (NDBC), Tropical Atmosphere Ocean (TAO), Pilot Research Moored Array in the Tropical Atlantic (PIRATA) projects, and the Japan Meteorological Agency (JMA). They found that a weak positive correlation exists between wind speed bias and significant wave height but no significant correlation with sea surface temperature or atmospheric stability [12]. Pickett et al. (2003) compared QuikSCAT with 12 nearshore and 3 offshore buoys along U.S. West Coast, and found that root mean square errors (RMSEs) of the QuikSCAT wind speed and direction were 1.3 m/s and 26°. QuikSCAT data could satisfy coastal studies although the nearshore data were less accurate than offshore data [13]. Tang et al. (2004) reproduced high-resolution QuikSCAT wind data and compared them with NDBC buoy data in coastal regions. The results showed that the modified algorithm could improve the accuracy of nearshore data, especially the accuracy of wind direction [14]. Ebuchi (2005) compared ADEOS-II SeaWinds with AMSR wind speeds and found that SeaWinds data could better match aircraft flights and other platforms data [15]. Satheesan et al. (2007) compared QuikSCAT with buoy data in the Indian Ocean. Their results showed that the consistencies in the North Indian Ocean were better than in the Equatorial Indian Ocean. Under different sea surface temperatures, QuikSCAT wind speed was generally higher, which might be due to the low accuracy at low speeds and rain contamination [16]. Yang et al. (2011) intercompared wind data from ENVISAT Advanced Synthetic Aperture Radar (ASAR), ASCAT, NDBC buoys, and the U.S. Navy Operational Global Atmospheric Prediction System (NOGAPS) model. They found ASAR data had a similar accuracy with ASCAT and NOGAPS model data, and the differences between ASCAT and ASAR-averaged wind data at different spatial resolutions changed little [17]. Sudha et al. (2013) compared OSCAT winds with buoy data and found that the OSCAT wind direction accuracy could not satisfy the mission requirements (20°) but wind direction biases decreased as speeds increased [18]. Wang et al. (2013) validated the first six months of HSCAT wind products by using NDBC buoys, R/V Polarstern, Aurora Australis, Roger Revelle (R/Vs), and PY30-1 oil platforms measurements. The RMSEs were 1.3 m/s and 19.19° compared with NDBC buoys. Similar results were also shown in R/Vs and oil platforms' comparisons. These results verified the practicability of HSCAT wind products in marine research [19]. Xing et al. (2016) also validated HSCAT wind products by using NDBC buoys, TAO, the European Centre for Medium Range Weather Forecasting (ECMWF) reanalysis data (ERA-Interim), and ASCAT data during 2012–2014. They found that rain has a significant effect on overestimating HSCAT wind speed at low and moderate wind speeds [20]. Bhaskar et al. (2016) generated and evaluated two-day OSCAT wind fields with tropical Indian Ocean buoys and ASCAT data [21]. Lindsley et al. (2016) applied an ultrahigh resolution (UHR) to the ASCAT wind product and compared ASCAT UHR winds with standard 25-km Level-2 ASCAT winds, near-coastal buoy winds, and open ocean SAR winds. The results showed that UHR product agrees well with these products and indicated that ASCAT UHR products may be used in near coastal and storm research [22]. Verhoef et al. (2017) evaluated long-term SeaWinds and ASCAT wind datasets with buoy and numerical weather prediction (NWP) winds. They found that SeaWinds and ASCAT wind had strong stability over time [23]. Wentz et al. (2017) introduced ocean wind consistent climate data records (OW-CDRs) and mentioned four evaluation methods, including a comparison with buoy winds, comparison with numerical model winds, comparison between two different platform sensors, and comparison between different data providers. They also planned some methods of calibrating other instruments, which would be merged into OW-CDRs in the future [24]. Yang et al. (2018) evaluated ISS-RapidScat wind vectors by using various buoys and ASCAT data. The average biases of ISS-RapidScat winds were 1.42 m/s and 19.5° compared with buoys while the biases were 1.15 m/s and 15.21° compared with ASCAT [9]. Hutchings et al. (2019) applied direction interval retrieval techniques and other processing improvements to RapidScat 2.5-km ultrahigh-resolution (UHR) winds. They also validated the new products with Level 2B (L2B) winds, numerical weather prediction winds, and buoy winds. The results showed that the new retrieval algorithm can improve the spatial consistency of UHR winds [25].

At present, studies of satellite scatterometer wind field evaluation are mostly regional and in a short-time period. Our research focused on a global and long-term evaluation to study the stability of the HSCAT-operated wind product during 2012–2018. Additionally, atmospheric stratification and rain effects on scatterometer wind retrieval were taken into account.

In this paper, we used HSCAT operational Level 2B wind products from January 2012 to December 2018 and buoy data provided by the European Centre for Medium-Range Weather Forecasts (ECMWF) MARS archive to evaluate and analyze HSCAT wind products. In addition, we also combined HSCAT data with the Special Sensor Microwave Imager Sounder (SSMIS) rainfall data to evaluate the rain effects on HSCAT winds. Scatterometer winds, buoy winds, and SSMIS rainfall data as well as their collocation methods are introduced in Section 2; the main product analysis, atmospheric stratification, and rain effect results are discussed in Section 3; and summaries of this study are provided in Section 4.

2. Data and Data Collection

2.1. HSCAT Data

Haiyang-2A (HY-2A) satellite is the first Chinese microwave ocean environment satellite launched by National Satellite Ocean Application Service (NSOAS) on 16 August 2011. HY-2A is sun-synchronous with a 971-km orbital altitude. Detailed information of the HY-2A satellite is shown in Table 1 [10]. The Ku-band scatterometer on HY-2A (HSCAT) operates at 13.3 GHz with the pencil beam conical scanning mode. The inner antenna works in horizontal polarization (HH) at a 41° incidence angel over a 1350-km swath, and the outer antenna works in vertical polarization (VV) at a 48° incidence angel over a 1700-km swath. HSCAT Level 2B (L2B) wind vectors are retrieved based on NSCAT-2 Geophysical Model Function (GMF) by maximum likelihood estimation (MLE) and the circle median filter is used to remove wind direction ambiguities [26]. The L2B products provide daily sea surface information in orbits with a 25-km resolution, and every single data file consists of 1624 rows and 76 columns. The wind direction in L2B products is defined as 0 degrees when wind is blowing toward the north and increases in the clockwise [27]. The HSCAT L2B wind data from 2012 to 2018 were used in this paper.

Table 1. Haiyang-2A (HY-2A) satellite parameters.

HY-2A Satellite Parameters	Description
Orbit	Sun-synchronous Orbit
Orbital Altitude	971 km
Orbital Inclination	99.34°
Local Time of Descending Node	6:00 a.m.
Repetition (Earlier Stage)	104.46 min
Laps (Earlier Stage)	13+11/14 times
Repetition (Later Stage)	104.50 min
Laps (Later Stage)	13+131/168 times
Instruments	Radar Altimeter; Microwave Scatterometer; Microwave Radiometer; DORIS; GPS; Laser Range Finder

2.2. ECMWF MARS Buoy Data

In situ wind observations are used to evaluate HSCAT winds. Buoy wind vectors are distributed by the Global Telecommunication System (GTS), which have been retrieved and quality controlled by the ECMWF MARS archive. The temporal resolution of buoy winds is one hour by averaging the wind data over 10 min. The wind direction is defined as 0 degrees when wind is blowing from north and increases in the clockwise direction [23]. The data of 248 buoy stations consisting of 30 nearshore buoy stations (within 50 km from the coast) and 218 offshore stations (more than 50 km away from the coast) were collected in this study and their locations are shown in Figure 1.

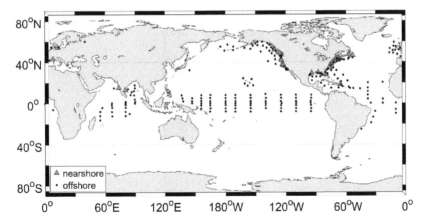

Figure 1. The location of buoys used in this study. The black points are offshore stations (more than 50 km away from the coast), and the red triangles are nearshore stations (within 50 km from the coast).

As scatterometer wind data are regarded as 10-m neutrally stable wind, all buoy winds were converted to 10-m neutral winds (marked as $\overrightarrow{V_{10n}}$) by using the LKB model proposed by Liu, Katsaros and Businger in 1979 [28,29]. However, the scatterometer measures the sea surface normalized radar cross-section (NRCS), which is affected by wind friction directly. Therefore, we converted $\overrightarrow{V_{10n}}$ to stress-equivalent wind (marked as $\overrightarrow{V_{10s}}$). The relation between the stress-equivalent wind and neutral winds is [30]:

$$\overrightarrow{V_{10s}} = \overrightarrow{V_{10n}} \sqrt{\frac{\rho}{\langle \rho \rangle}}, \tag{1}$$

where ρ is the air density (kg/m^3); and $\langle \rho \rangle$ is the global average air density and equals to 1.225 kg/m^3 here.

2.3. SSMIS Data

To study rain effects on HSCAT wind retrieval, we used the Special Sensor Microwave Imager Sounder (SSMIS) rainfall data to extract HSCAT wind data contaminated by rain. The Special Sensor Microwave Imager (SSM/I) and SSMIS are series microwave radiometers payloaded on the near-polar orbiting Defense Meteorological Satellite Program (DMSP) satellites since 1987. SSMIS works at 19.35, 22.235, 37, and 85.5 GHz. The rainfall data are provided with a gridded dataset of a 0.25° spatial resolution generated by Remote Sensing System (REMSS) [31]. To match with the HSCAT descending time, we chose the DMSP F17 platform, of which the descending time (6:37 a.m.) is the closest to HSCAT's (6:00 a.m.) and quite stable during the chosen time period. The ascending and descending equatorial crossing time for different instruments can be inquired from REMSS [32].

2.4. Data Collocation Method

Before collocation, data selection and quality control need to be done firstly. Because of the dependences between the azimuth dispersion of rotating scan scatterometer and the accuracies of wind vector retrieval, small azimuth dispersion causes a lower quality in the winds [33–35]. Cross-track wind cell can be divided as the outer swath (OS, or outer region (OR)) and inner swath (IS) and IS consists of a sweet region (SR) and nadir region (NR), as shown in Figure 2. Since only vertical polarization exists in the outside edge region of the swath, HSCAT OS data were not included in the collocation dataset. For buoy data, some of the buoys changed their moored locations during measurement. We selected the buoy data measured at the same time but with different locations, and averaged the buoy data as one measurement. As for SSMIS rainfall data, we removed all the invalid data marked by REMSS.

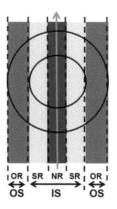

Figure 2. HSCAT cross-track wind cells.

After data quality control, we made the buoy and scatterometer collocation datasets first. HSCAT wind field data and buoy wind data were collocated within $25/\sqrt{2}$ km (about 17.7 km) and 30 min. In this study, wind speeds were used within 0.1 to 25 m/s and HSCAT wind directions were turned 180° to keep with the buoy wind directions before analysis. As for rainfall data, we used the location of HSCAT wind data to find F17 SSMIS rain data within 1 h.

Due to the sea surface status being affected by the terrain in shallow water and the contamination of the island, NRCS is greatly affected near the coast. We classified buoy data as the offshore dataset and nearshore dataset according to the distance between buoy stations and coastline. There are 30 nearshore buoy stations within 50 km from the coastline and others are offshore buoy stations, as shown in Figure 1. Besides, it was found that the amount of nearshore data is about 4% of the offshore data. RMSEs of the nearshore data were about 1.58 m/s and 35.52° while the results of the offshore data were 1.30 m/s and 27.40° during the whole study period. The deviation of the nearshore data is large and therefore only the offshore dataset was used in this study.

3. Results and Discussions

3.1. Statistical Analysis of HSCAT Wind Variations from 2012 to 2018 Year by Year

To evaluate the quality of the HSCAT wind data, HSCAT winds were collocated with buoy data year by year from 2012 to 2018. The spatial-temporal collocation windows are 17.7 km and 30 min as introduced in Section 2.

3.1.1. The Distribution of Collocated Data and Statistical Parameters

The whole-year HSCAT wind data were compared with the contemporaneous buoy wind data and the wind speed and wind direction collocation contours of histograms are shown in Figure 3 from 2012 to 2018. Figure 3a–g represent the comparisons for each year from 2012 to 2018; number 1 and 2 represent the wind speed and wind direction collocation contours of histograms, respectively.

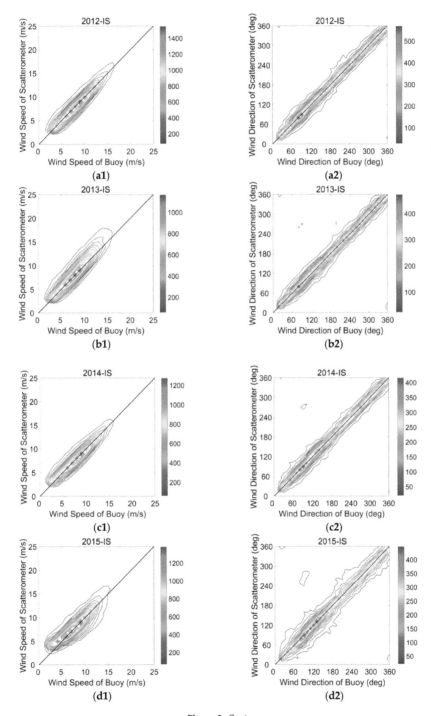

Figure 3. *Cont.*

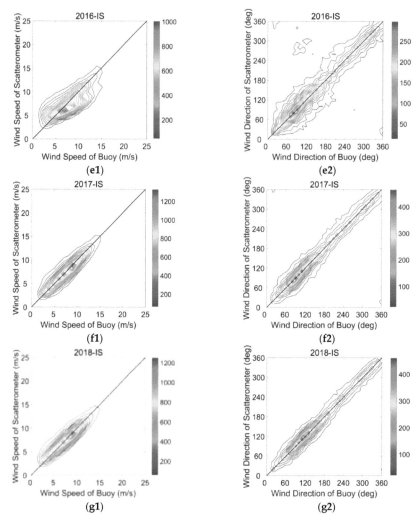

Figure 3. Contours of histograms of comparisons between HSCAT and buoy wind data. (**a**)–(**g**) represent the comparisons for each year from 2012 to 2018; number **1** and **2** represent the wind speed collocation contours of the histograms and the wind direction collocation contours of the histograms, respectively.

Figure 3(a1–g1) show that the HSCAT and buoy wind speeds are mainly distributed in the range from 5 to 12 m/s. The consistencies between HSCAT and buoy wind speed descend from 2012 to 2016 but ascend in 2017 and 2018, which shows the HSCAT wind speed product quality variations. The quantified statistics will be analyzed in this subsection later. Wind speed contours in 2013 have a larger dispersion among the contours from 2012 to 2014. The speed contours in 2015 and 2016 deviate from y = x and HSCAT wind speeds are higher when speeds are less than 5 m/s but lower when speeds are higher than 10 m/s.

Figure 3(a2–g2) show that the HSCAT and buoy wind direction are mainly distributed in the range from 60° to 150°. Similar with the wind speed, consistencies between the HSCAT and buoy wind direction descend from 2012 to 2016 and then ascend in 2017 and 2018.

The quantitative statistics of the wind vector bias in the collocated winds are given in Table 2. Invalid data were removed, the speeds of which are twice as high as the speed's standard deviation(σ) by using the 2σ quality control method proposed by Lin et al. (2013) [26]. The statistical parameters were calculated using the following equation (here, '< >' represents the statistical average) [9,36]:

$$Bias_{spd} = \langle spd_{SCAT} - spd_{BUOY} \rangle, \tag{2}$$

$$RMSE_{spd} = \sqrt{\langle (spd_{SCAT} - spd_{BUOY})^2 \rangle}. \tag{3}$$

Wind direction statistical parameters were calculated by using Equations (4)–(7) to ensure the continuity at 0°and 360° [37]:

$$\overline{U} = \langle \sin(dir_{SCAT} - dir_{BUOY}) \rangle, \tag{4}$$

$$\overline{V} = \langle \cos(dir_{SCAT} - dir_{BUOY}) \rangle, \tag{5}$$

$$Bias_{dir} = \cot(\overline{U}/\overline{V}), \tag{6}$$

$$RMSE_{dir} = \cot \sqrt{(\langle \sin^2(dir_{SCAT} - dir_{BUOY}) \rangle / \langle \cos^2(dir_{SCAT} - dir_{BUOY}) \rangle)}. \tag{7}$$

Table 2. The statistical parameters of the wind vector bias between HSCAT and buoys.

Year	Speed Bias	Speed RMSE	Direction Bias	Direction RMSE	Collocation Number
2012	−0.03	1.08	0.20	21.75	33,462
2013	0.48	1.42	0.60	22.06	31,276
2014	0.06	1.06	1.23	22.15	27,339
2015	0.26	1.37	0.29	26.37	36,211
2016	−0.41	1.87	1.93	29.50	25,898
2017	−0.14	1.20	2.12	24.50	27,511
2018	0.03	1.17	1.32	22.94	24,997

The statistical parameters in Table 2 explain the quantification of the variation trends shown in Figure 3: HSCAT wind vectors agree well with buoy data in 2012 and 2014 but the average speed bias in 2013 is larger by about 0.4 m/s and the root mean square error (RMSE) is larger by about 0.4 to 0.5 m/s than the statistics in 2012 and 2014. However, in 2015 and 2016, both of the wind speed and wind direction statistics have the greatest offset, in that the speed RMSEs are larger by about 0.2 to 0.4 m/s and the direction RMSEs are larger by 4 to 8° compared to other years. This trend also reflects that the data qualities in 2015 and 2016 and wind speed data quality in 2013 are worse than others.

In terms of yearly variations of the statistical parameters, consistencies between HSCAT and buoy winds are good; wind speed consistencies show a descending trend in 2013 but direction consistencies are similar with 2012. From 2014, wind direction consistencies also descend; both the wind speed and direction show great offsets in 2015 and 2016, but the offsets get smaller in 2017 and 2018 and the data quality in 2018 is similar with 2012.

All in all, the HSCAT wind speed accuracy can satisfy the scatterometer mission requirements (2 m/s) but the wind direction accuracy is a little worse. Besides, the wind data in 2015 and 2016 have the greatest offsets. A strict quality control is needed in the application of these datasets.

3.1.2. The Probability Distribution Functions (PDFs) of Wind Speeds and Directions

In order to analyze the entire distributions of the collocated data between HSCAT and buoys, we plotted the probability distribution functions (PDFs) of wind speeds and directions for each year. The PDFs are shown in Figure 4 in which a–g represent the PDF curves for each year from 2012 to 2018; number 1 and 2 represent the wind speed and direction PDFs; and the dotted line is the PDF for the buoy data and the solid line for HSCAT PDF.

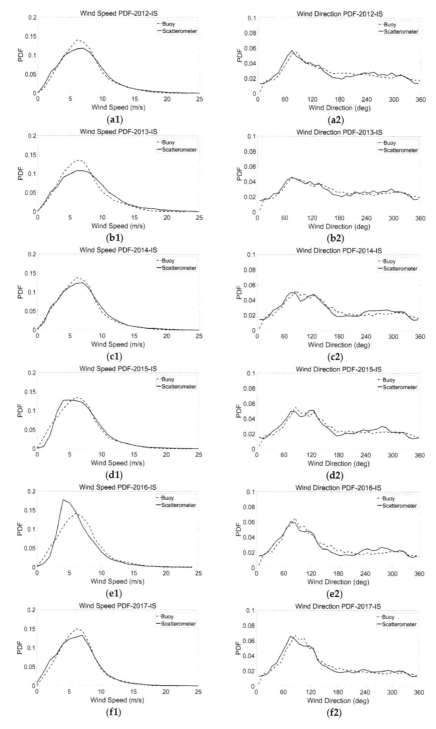

Figure 4. *Cont.*

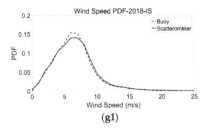
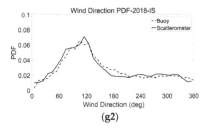

(g1) (g2)

Figure 4. Probability distribution functions (PDFs) between HSCAT and buoys during 2012–2018. (**a**)–(**g**) represent the PDF curves for each year from 2012–2018; number **1** and **2** represent the wind speed and direction PDFs; the dotted line is the PDF for the buoy data and the solid line for HSCAT PDF.

The HSCAT wind vector PDFs have similar distributions to the buoy PDFs, but they do not agree well in detail. Similar with the results in Section 3.1.1, the HSCAT and buoy wind speeds are mainly distributed in the range from 5 to 12 m/s and the wind directions are mainly distributed in the range from 60° to 150°. Until 2014, the distributions are similar and basically agree well at the peaks. However, the HSCAT PDF curves clearly separate from the buoy curves in 2015 and 2016 but converge in 2017 and 2018.

For wind speed PDFs, the peak values of the buoy data are about 6 to 7 m/s and HSCAT data are distributed at 6 to 8 m/s except in 2016, which is less than 5 m/s. Within the wind speed ranges from 0 to 4 m/s, the numbers of the HSCAT wind speed are slightly higher except in 2015 and 2016; within 4 to 8 m/s, the numbers of the HSCAT wind speed are about 2% lower than the buoys; within 8 to 13 m/s, the numbers of the HSCAT wind speed are generally higher; and when speeds are higher than 13 m/s, the HSCAT speed PDFs are basically consistent with the wind speed PDFs of the buoy data.

For wind direction PDFs, both the HSCAT and buoy peak values are distributed stably about 90 to 110° among seven years. The numbers of the HSCAT wind direction are a little higher when the wind directions are less than 90°; in the range 90–100°, the numbers of the HSCAT direction at the peak values are less than the buoys; the numbers of the HSCAT direction are lower in the range 120–220°, and then become higher in 220 to 330° but become lower again when directions are larger than 330°. From 2014, the HSCAT wind direction PDFs show two little peaks (70–80° and 110–120°) on the main central peak region but disappear in 2018.

In a word, for the annual wind vector PDF distributions, the numbers of HSCAT and buoy wind data are basically the same, but the number distributions vary in different wind speed and wind direction bins.

3.2. The Overall Statistical Analysis of HSCAT Wind Variations from 2012 to 2018

3.2.1. The Trends of Residual Variations of Wind Speed and Wind Direction

The dependencies of the wind speed and direction residual ($SCAT - BUOY$) from 2012 to 2018 are shown in Figure 5. In order to study wind speed and direction bias variations in different speed bins, the residuals were calculated in the speed bins of 1 m/s. To avoid a binning effect, we used the average wind speed, ($spd_{SCAT} + spd_{BUOY}$)/2, as the binning wind speed [38]. All collocated data were quality controlled by the 2σ method (Lin et al. 2013). Figure 5a,b represent the wind speed and direction residual distributions; the error bars represent the wind data biases and RMSEs, respectively.

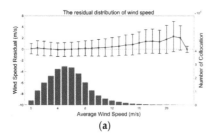
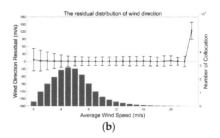

(a)   (b)

Figure 5. Dependencies of the wind speed and direction residual ($SCAT - BUOY$) on the average wind speed. Calculation bins of the average wind speed is 1 m/s. (**a**,**b**) represent the wind speed and direction residual distribution; the error bars represent the wind data biases and root mean square errors (RMSEs).

The collocated wind speed data are mostly distributed in the range 3–8 m/s. Average wind speed biases generally increase as wind speeds increase. When wind speeds are in the range 3–11 m/s, wind speed biases are approximately equal to 0. The wind speed RMSEs are less than 2 m/s and have no obvious changes with the increasing speeds. When wind speeds are higher than 10 m/s, wind speed biases are positive; that is, HSCAT wind speeds are higher than buoy wind speeds. Wind speed biases and RMSEs increase with the increasing wind speeds and the average bias and RMSE are about 2 and 3 m/s when the wind speed is 20 m/s. However, the increasing trend of the speed biases and RMSEs can be related with the decreasing collocation numbers. Generally, less data can cause a large offset.

In terms of wind direction, the whole distributions of wind direction bias remain nearly unchanged around zero as wind speeds increase except the bias at 23 m/s (about 120°), which may be caused by the small collocation numbers at the high wind speed region. When wind speeds are less than 5 m/s, the wind direction RMSEs decrease with the increasing speeds and drop from 60° to 20° rapidly. When speeds are higher than 5 m/s, the wind direction RMSEs remain nearly unchanged around 18°.

Both the wind speed and wind direction average biases are near 0 and the RMSEs are less than 2 m/s and 20° at moderate speeds, which means HSCAT wind products basically meet the scatterometer mission requirement. However, larger offsets are found in high speeds, which may be caused by the small numbers of collocated wind data.

3.2.2. HSCAT Wind Field Data Monthly Variations during 2012 to 2018

To further study HSCAT wind field data variations over time, wind speed and direction residual monthly variations are shown in Figure 6. Figure 6a,b represent the wind speed and direction error bars and collocated data histograms during 2012–2018, respectively. The error bars indicate the wind data biases and RMSEs.

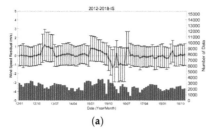
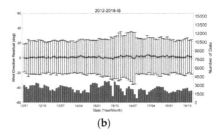

(a)   (b)

Figure 6. Monthly variations of the wind speed and wind direction residual. (**a**) shows the variations of the wind speed residual (above) and the collocated data histograms (below) during 2012–2018; (**b**) represents the variations of the wind direction residual (above) and the collocated data histograms (below) during 2012–2018.

As seen from the collocated data histograms, collocation numbers remain stable at around 2000 to 3000 pairs except some months like September in 2015; March, August, and September in 2016; and February in 2017, which is related with the missing records of the HSCAT L2B data.

For wind speed, wind speed data are stable and have just slight offsets compared with buoy data in 2012. The biases are around 0 and the RMSEs are about 1 m/s. In the first half of 2013, the wind speed biases abruptly increase to 1 m/s and the RMSEs also increase to about 1.5 m/s, but the biases then turn to around 0 and the RMSEs decrease to more than 1 m/s in the second half of 2013 to 2015. From 2015 to 2016, the wind speed biases are quite unstable and the RMSEs are obviously larger than other years. The maximum bias is −1.5 m/s and the maximum RMSE is about 2 m/s. From September 2016 to the end of 2017, wind speed biases and RMSEs start decreasing. Those variations show that the HSCAT wind speed products have large offsets and are unstable in 2013, 2015, and 2016, which is in accordance with the results in Section 3.1.1.

Wind direction biases are distributed stably around zero and the RMSEs of the wind direction are about 25°. However, from the second half of 2015, the wind direction biases tend to a 2° positive deviation and RMSEs also increase and increase to 30° in the end of 2016, which are far away from the mission requirement (20°). In 2017, the direction biases and RMSEs both decrease to about 2° and 20° in 2018. The variations of the wind direction are also similar with the statistical parameters mentioned above.

3.2.3. HSCAT Wind Field Data Variations in 24 Hours

The wind speed and direction residual variations during 24 h are shown in Figure 7. These statistics were gathered by the buoy time. Figure 7a,b represent the wind speed and direction error bar graphs and collocated data histograms during 2012–2018, respectively. The error bars indicate the wind data biases and RMSEs.

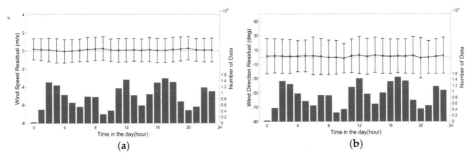

Figure 7. Daily variations of the wind speed and wind direction residual. (**a**,**b**) represent wind speed and direction error bars and collocated data histograms, respectively.

Figure 7 shows that there is a drop of the collocation numbers at 3 to 4 h later than the HY-2A satellite descending time and have peaks from 2 a.m. to 3 a.m., 11 a.m. to 1 p.m., and 4 p.m. to 6 p.m. This correlation is caused by the sun-synchronous orbit of the HY-2A satellite and the positions of buoys. Wind speed bias variations are stably distributed around 0 in 24 h except in the periods that have less collocation pairs and wind speed RMSEs are generally less than 2 m/s. Wind direction biases are also distributed around 0 and fluctuate within ±2°. However, the RMSEs of the direction are stably distributed around 20 to 30°, which does not satisfy the mission requirement. This suggests that the wind speeds of HSCAT can provide stable speed products in one day with accurate qualities. Unlike the stable daily wind speed residual variations, the wind direction products of HSCAT in one day are not stable or accurate enough to be applied in marine research and these are also similar to the statistical parameters mentioned above.

3.3. *The Sea Surface Temperature and Air Temperature Impact on HSCAT Wind Products*

Although scatterometers can measure sea surface parameters with all-weather, all-time capabilities, the actual measurements are the sea surface normalized radar cross-section (NRCS), which is affected by both sea and atmospheric conditions. To study the sea and atmospheric parameter impacts on HSCAT wind field retrieval, we classified the collocated data into two groups according to the difference between the sea surface temperature and air temperature measured by buoys: The neutral group (the absolute values of temperature difference are less than or equal to 1 °C) and non-neutral group (the absolute values of temperature difference are higher than 1 °C). After data quality control, we obtained 217,359 pairs of collocations from 2012 to 2018 consisting of 129,604 pairs of the neutral group and 87,755 pairs of the non-neutral group.

3.3.1. The Probability Distribution Functions (PDFs) of Wind Speeds and Directions

We plotted the PDFs of the wind speed and direction for different groups. The PDFs are shown in Figure 8 in which (a) and (b) represent wind speed and direction PDFs; number 1, 2, and 3 represent the neutral group, non-neutral group, and all collocated data PDFs; and the dotted line is the PDF for the buoy data and the solid line for HSCAT PDF, respectively.

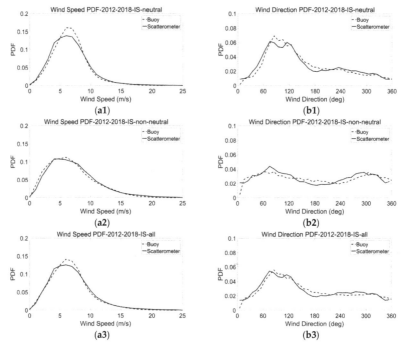

Figure 8. Probability distribution functions (PDFs) between HSCAT and buoys under different stratification conditions. (**a,b**) represent the wind speed and direction PDFs; the number **1**, **2**, and **3** represent the neutral group, non-neutral group, and all collocation data PDFs; the dotted line is the PDF for the buoy data and the solid line is for HSCAT PDF, respectively.

As for wind speed, HSCAT PDFs are consistent with the buoy PDFs except the numbers of HSCAT winds are less at the peak values (6–8 m/s). The HSCAT speed PDF is not in good agreement with the buoy PDF at 6 to 8 m/s for the neutral group while the non-neutral group is more consistent. In the neutral group, HSCAT has more collocation numbers when speeds are less than 5 m/s but less numbers than the buoys at 6 to 8 m/s. However, in the non-neutral group, HSCAT has less numbers when speeds are less than 5 m/s. When speeds are higher than 10 m/s, both the neutral and non-neutral group have consistent collocation results.

As for wind direction, the HSCAT and buoy wind directions are mainly distributed in the range 60–120°. Besides, HSCAT has another peak value distributed in the range 240–300°. In the non-neutral group, there is a double peak characteristic of both the HSCAT and buoy PDFs, and the HSCAT PDFs have larger peak values. In the neutral group, HSCAT has more collocation numbers when the directions are less than 70° but less numbers at 90°, where the buoy direction collocation numbers are the maximum.

Generally, the consistency of speed collocation in the neutral group is worse than the non-neutral group but the direction collocation consistency is better in the neutral group. Since the neutral group is the main part of all the collocated data, the overall PDF distribution during 2012 to 2018 is similar to the neutral group.

3.3.2. The Trends of Residual Variations of Wind Speed and Wind Direction

Figure 9 shows the variations of the wind speed and direction residual on the average wind speed from 2012 to 2018 under different stratification conditions. We also used the average wind speed as the binning wind speed with 1-m/s intervals to avoid the binning effect. In Figure 9, a and b represent the wind speed and direction residual distributions; 1, 2, and 3 represent the neutral group, non-neutral group, and all collocation data residual distributions; and the error bars represent the wind data biases and RMSEs, respectively.

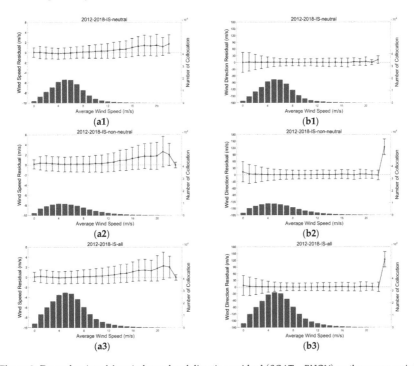

Figure 9. Dependencies of the wind speed and direction residual ($SCAT - BUOY$) on the average wind speed under different stratification conditions. Calculation bins of the average wind speed are 1 m/s. (**a**,**b**) represent the wind speed and direction residual distribution; **1**, **2**, and **3** represent the neutral group, non-neutral group, and all collocation data residual distributions; the error bars represent the wind data biases and RMSEs, respectively.

For both the neutral and non-neutral group, the average speed biases are distributed around 0 in 0 to 12 m/s and the biases increase with increasing speeds. The speed RMSEs of the neutral group are about 1 m/s, but the RMSEs of the non-neutral group are larger (about 2 m/s). When the wind speed is higher than 12 m/s, the average speed biases start increasing in both groups. The speed biases of the neutral group are about 1.5 m/s and the RMSEs increase to about 1.5 m/s in the range 16–20 m/s. Additionally, in this range, the non-neutral group's average wind speed biases and RMSEs are larger. When the speed equals 20 m/s, the bias and RMSE increase to about 2 m/s and 2.5 to 3 m/s, respectively.

For wind direction, the biases remain nearly unchanged around zero. In 0 to 5 m/s, both RMSEs of the neutral and non-neutral groups decrease from 45° to 20°. When the wind speed is higher than 5 m/s, the RMSEs of the neutral group are about 25° while the RMSEs of the non-neutral group are about 30°. Both the wind speed and direction biases of the non-neutral group have large changes at 23 m/s, which may be connected with the small collocation numbers at 23 m/s.

3.3.3. The Statistical Parameters of Collocated Data

For further study of the winds' residual variations under neutral and non-neutral conditions, the quantitative statistics of the wind vector difference between HSCAT and buoys are given in Table 3.

Table 3. The statistical parameters of the wind vector bias between HSCAT and buoys under different stratification conditions.

Year	Wind Speed (m/s)				Wind Direction (°)				Collocation Number	
	Bias-Neutral	Bias-Non-Neutral	RMSE-Neutral	BMSE-Non-Neutral	Bias-Neutral	Bias-Non-Neutral	RMSE-Neutral	BMSE-Non-Neutral	Neutral	Non-Neutral
2012	−0.12	0.14	0.95	1.31	0.15	0.30	19.89	24.47	20,596	13,908
2013	0.37	0.65	1.24	1.66	0.55	0.75	20.28	24.26	17,593	13,691
2014	−0.03	0.19	0.93	1.23	0.93	1.61	19.99	24.81	15,427	11,917
2015	0.18	0.42	1.17	1.67	0.18	0.44	23.84	29.95	21,446	14,793
2016	−0.52	−0.23	1.85	1.92	1.77	2.21	28.97	30.36	15,913	9990
2017	−0.23	0.07	1.05	1.48	1.34	3.61	22.44	27.96	17,474	10,047
2018	−0.03	0.19	1.01	1.48	1.08	1.84	19.99	27.64	15,995	9038
ALL	−0.02	0.25	1.15	1.53	0.82	1.35	24.58	31.68	123,819	83,179

After data quality control, generally, the number of the neutral group is 1.3 to 1.8 times the non-neutral group number. Those statistical parameters also validate the results in Sections 3.3.1 and 3.3.2 that the winds' average biases and RMSEs of the neutral group are less than the non-neutral group; both groups of the HSCAT wind speed can meet the mission requirement but the RMSEs of the wind direction are much higher than 20°, especially in the non-neutral group, which can reach 30°.

For wind speed, the biases of the non-neutral group are about 0.2 m/s higher than the neutral group except in 2016, which is about 0.4 m/s higher, but they all meet the mission requirement. In 2013, 2015, and 2016, both the neutral and non-neutral groups' offsets are much larger than other years. Except in 2015, the average wind speed bias of the neutral group is similar to other years but much larger under non-neutral conditions. Those large offsets decrease in 2017 and 2018.

For wind direction, the average biases are quite small from 2012 to 2015. However, after 2015, the average biases are more than 1° and the RMSEs are quite large whether under neutral conditions or non-neutral conditions, although the RMSEs of the neutral group are less 4 to 6° compared to the non-neutral group. From 2012 to 2014, the direction RMSEs of the neutral group are less than 20° and start increasing from 2015. In 2016, both the neutral and non-neutral groups' RMSEs increase to 30° and then decrease in 2017, but wind direction RMSEs under non-neutral conditions are still about 7° higher than the mission requirement.

HSCAT wind products can meet the mission requirement under neutral conditions. Under non-neutral conditions, wind speeds can also satisfy the requirement while wind direction offsets are large. In terms of the variations of statistics, although wind speed biases in 2013, 2015, and 2016 are larger than the other years, the data can still meet the requirement. Wind direction biases and RMSEs increase from 2015 and RMSEs of the non-neutral group in 2016 can get to 30°. Under non-neutral conditions, variations of the wind direction are larger than the wind speed, which shows that HSCAT wind direction retrieval is more susceptible than the wind speed to sea and atmospheric parameters. Besides, wind speed and direction offsets are large whether for the neutral group or non-neutral group in 2016 and the offset differences are less than other years, which illustrates the worst quality for the 2016 product in an indirect way. After 2015, HSCAT wind products need strict quality control, especially the wind direction products when HSCAT wind data are used in marine studies.

3.4. The Rain Impact on HSCAT Wind Products

As mentioned above, the sea surface wind field is retrieved from scatterometer NRCS, which is affected by sea surface roughness. At different rainfall rates, the changes of the sea surface roughness are complicated. Besides, the rain drop size and density also have an impact on the return signal scattering. Therefore, dealing with rain contaminated data is one of the difficulties in scatterometer wind retrieval. It is especially important for scatterometer data usage under rainy conditions to study HSCAT wind field variations at different rainfall rates [20].

We collocated the HSCAT and buoy collocation dataset with SSMIS data within 1 h. We divided rain collocation data into four groups according to different rainfall rates: Rain free (rainfall rate equals to 0 mm/h), light rain (rainfall rate is from 0 to 3 mm/h), moderate rain (rainfall rate is from 3 to 8 mm/h), and heavy rain (rainfall rate is higher than 8 mm/h). To avoid the extra offsets caused by small numbers of rain sample data, we studied all the rain collocated data from 2012 to 2018.

3.4.1. The Distribution of Collocated Data and Statistical Parameters

We compared HSCAT wind data with buoy wind data, and the wind speed and wind direction collocation contours of histograms are shown in Figure 10. Figure 10a,b represent the wind speed and direction collocation contours of histograms; number 1 to 4 represent collocation contours of histograms of the rain free group, light rain group, moderate rain group, and heavy rain group, respectively.

It is evident that the collocation numbers decrease with an increasing rainfall rate and both the wind speed and direction collocation contours appear as losing trends. As for wind speed, the speed contours move to the top of the figure with the increasing rainfall rate, which means HSCAT-retrieved wind speeds are increasingly overestimated. HSCAT wind speeds are almost twice the buoy speeds and have large deviation in heavy rain. As for wind direction, HSCAT wind direction numbers increase in the range 60–120° and 240–300°, which means that the HSCAT wind directions tend to be perpendicular to the buoy directions. The two-peak characteristic is evidently shown in Figure 10, which increases the wind direction estimation errors.

We also calculated the statistics of the wind vector biases in different rainfall rates, which are given in Table 4.

Table 4. The statistical parameters of the wind vector bias between HSCAT and buoys in different rainfall rates.

Rainfall Rate	Speed Bias	Speed RMSE	Direction Bias	Direction RMSE	Collocation Number
Rain free	0.05	1.73	1.00	24.10	121,080
0–3 mm/h	1.14	2.86	1.36	29.09	11,021
3–8 mm/h	3.73	5.38	−0.04	35.57	1683
>8 mm/h	6.16	7.94	3.42	38.44	291

The statistical parameters also illustrate the decreasing collocation numbers and increase of the wind speed and direction offsets with an increasing rainfall rate. The average biases of the wind speed increased and RMSEs also multiplied with the increasing rainfall rate, i.e., the higher the rainfall rate, the worse the HSCAT-retrieved wind speed product. Besides, the positive biases of the wind speed also increase, which reflects that the overestimation of the retrieved speeds is also increasing with the rainfall rate. In the rain-free group, the wind speed and direction can satisfy the mission requirement.

In the light rain group, the wind speed RMSEs are over 2 m/s, which reflects that the HSCAT-retrieved speed data need quality control even under light rain conditions. Similar to the wind speed, the wind direction data quality greatly decreases with the increasing rainfall rate. The wind direction RMSEs are almost 40° under heavy rain conditions, i.e., twice the mission requirement. The wind direction is also a little larger than the requirement even in the rain-free group.

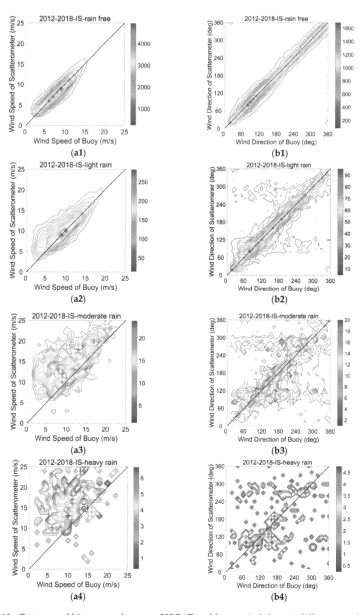

Figure 10. Contours of histograms between HSCAT and buoy wind data in different rainfall rate. (**a**,**b**) represent the wind speed and direction collocation contours of histograms; number **1** to **4** represent the collocation contours of histograms of the rain free group, light rain group, moderate rain group, and heavy rain group, respectively.

Therefore, HSCAT can basically satisfy the marine study requirement under rain-free conditions, although HSCAT wind directions have larger offsets. In rainfall conditions, both the speed and direction offsets are larger than the requirement. When the rainfall rate is higher than 3 mm/h, we do not recommend the use of HSCAT wind products due to their large deviations. It is also a key problem to be resolved in order to improve the accuracy of HSCAT wind retrieval in rainfall conditions.

3.4.2. The Probability Distribution Functions (PDFs) of Wind Speeds and Directions

We plotted the PDFs of wind speeds and directions in different rainfall rates from 2012 to 2018 as shown in Figure 11 with a speed of 1 m/s and a 10° direction interval. (a) and (b) represent the wind speed and direction PDFs; number 1 to 4 represent the rain-free group, light rain group, moderate rain group, and heavy rain group PDFs; and the dotted line is the PDF for the buoy data and the solid line is for HSCAT PDF, respectively.

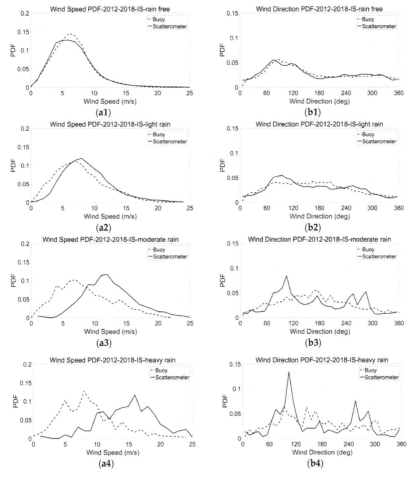

Figure 11. Probability distribution functions (PDFs) between HSCAT and buoys under different rainfall rates. (**a,b**) represent wind speed and direction PDFs; number 1 to 4 represent the rain-free group, light rain group, moderate rain group, and heavy rain group PDFs; dotted line is the PDF for buoy data and solid line for HSCAT PDF, respectively.

The wind speed and direction PDFs reconfirm the results in Section 3.4.1.

For wind speed, the distributions of the buoy speed are mainly distributed in 6 to 8 m/s and are nearly unchanged with the increasing rainfall rate. Therefore, the buoy speed PDFs' fluctuation can be regarded as changes of the ocean environment and the decreasing of collocations with the increasing rainfall rate. However, the entire distribution of the buoy speed is stable so that the buoy speed can be a reliable evaluation standard. However, the peaks of the HSCAT wind speed PDFs move from 5 to 6 m/s under the rain free condition to 16 to 18 m/s under the heavy rain condition. The overestimation of the HSCAT speed increases as the rainfall rate increases, and the HSCAT speeds of the heavy rain group are almost twice the buoy speeds.

For wind direction, the rain free buoy wind directions are mainly distributed in the range 60–120° and then slightly move to 180° with an increasing rainfall rate. However, the movement is quite small so the buoy direction can be seen as a relatively stable evaluation standard. Similar with the results in Section 3.4.1, under the rain-free condition, the HSCAT and buoy direction PDFs are consistent well. However, with the rainfall rate increasing, the HSCAT wind directions have two peaks in the ranges 60–120° and 240–300° with growing peak values, and the extent of the two PDFs' separation increases. It illustrates that HSCAT-retrieved wind directions tend to be perpendicular to the true directions with an increasing rainfall rate, which causes large wind direction offsets under rainy conditions.

4. Conclusions

The stabilities of the HY-2A scatterometer in different times, atmospheric stratification conditions, and rainfall rates were analyzed by comparing HSCAT L2B wind products with buoy wind measurements from 2012 to 2018. In the study, HSCAT and buoy data were selected within the range 0.1–25 m/s and collocated within $25/\sqrt{2}$ km (about 17.7 km) and 30 min. After data quality control, yearly matching data were 34,462, 31,276, 27,339, 36,211, 25,898, 27,511, and 24,997 pairs from 2012 to 2018. According to the collocated data, wind speed biases and RMSEs between HSCAT and buoy measurements were about 0.1 to 0.2 m/s and 1.2 m/s while biases and RMSEs of the wind direction were 1° and 27°, respectively. These suggest that HSCAT wind direction accuracy is slightly worse than the scatterometer mission requirement (20°). The yearly PDF curves also supported these results. In 2015 and 2016, the HSCAT PDFs and buoy PDFs showed clear separation and then tended to agree with each other in 2017. From 2017, HSCAT winds and buoy winds had similar distributions and the offsets tended to decrease.

For wind speed and direction bias variations in different speed bins, the collocated data were mostly distributed in the speed range 6–8 m/s. The minimums of the wind speed bias and RMSE were distributed around 3 to 11 m/s while the minimums of the wind direction biases and RMSEs appeared when speeds are higher than 5 m/s. When speeds are higher than 10 m/s, the average speed biases and RMSEs were increased with increasing binning speeds. The wind direction offsets were barely changed, except the RMSEs of the wind direction increased slightly when speeds were higher than 16 m/s.

According to the HSCAT wind field data variations over time, wind speed and direction average biases were distributed around 0. Wind speed RMSEs were about 2 m/s and wind direction RMSEs were about 20 to 30°. HSCAT wind speed biases were a little larger in 2013, and HSCAT wind products were the most unstable in 2015 and 2016, which had the largest offsets and variations among the seven years.

To study the atmospheric stratification impacts on HSCAT wind products, we classified the collocated data into a neutral group (the absolute values of temperature difference are less than 1 °C) and non-neutral group (the absolute values of temperature difference are higher than 1 °C) according to the differences between the sea surface temperature and air temperature and then analyzed the wind field error features in the different groups. The wind PDFs and error characteristic variations with the wind speeds illustrated that the quality of HSCAT-retrieved wind products are better in neutral conditions. Wind direction has more of an influence from marine and atmosphere environmental

Remote Sens. **2019**, *11*, 2968

parameters and owns large changes in the non-neutral condition. From the yearly statistical parameters of the collocated data, the HSCAT wind product qualities declined from 2015 but were improved from 2017. In 2016, the offset differences between the neutral and non-neutral group were less than other years, which reflects the worst quality in an indirect way and the data in 2016 need to be treated very carefully in applications.

We also studied the rain impacts on HSCAT wind products. We divided rain-collocated data into a rain-free group (0 mm/h), light rain group (0–3 mm/h), moderate rain group (3–8 mm/h), and heavy rain group (>8 mm/h) according to the rainfall rate. From the wind PDFs and collocated contours, we found that the accuracy of the HSCAT-retrieved wind field decreases as the rainfall rate increases: HSCAT-retrieved wind speeds and the RMSEs of collocated speeds increased with an increasing rainfall rate; HSCAT-retrieved wind directions tended to concentrate on the 60–120° and 240–300° ranges, which means HSCAT-retrieved wind directions tend to be perpendicular to the true directions. These results may provide references for the correction of scatterometer rain-contaminated data.

Overall, HSCAT L2B wind products are less accurate and less stable than the mission requirement especially for wind direction products during 2012 to 2018. The qualities of HSCAT operational wind products declined from 2012 to 2016 but were improved from 2017. HSCAT wind products should be treated carefully, especially in 2015 and 2016, or adopt other ways to improve data qualities when HSCAT wind data are used in marine studies.

Long-term evaluation for HSCAT wind products was analyzed in this study, and we also studied the error characteristics in different times, atmospheric stratification conditions, and rainfall rates primarily. However, only a little collocated data was distributed in the high wind speed region due to the lack of high speed data from buoys. The reasons for quality improvement from 2017 were not discussed here. Comparing HSCAT wind data with other satellite scatterometer data, studying the environmental parameter impacts on scatterometers working in different bands, finding improvement methods for HSCAT winds, and other related studies need to be investigated in the future.

Author Contributions: Conceptualization, K.Z. and C.Z.; methodology, K.Z.; software, K.Z.; validation, K.Z.; writing—original draft preparation, K.Z.; writing—review and editing, K.Z. and C.Z.

Funding: This research was funded by the Shandong Joint Fund for Marine Science Research Centers, grant number U1606405, the Marine S&T Fund of Shandong Province for Pilot National Laboratory for Marine Science and Technology (Qingdao), grant number No.2018SDKJ0102-8, and the key techniques for ocean and climate models programs, grant number 2016ASKJ16.

Acknowledgments: The authors would like to thank National Satellite Ocean Application Service (NSOAS) for the HSCAT operational wind data, the European Centre for Medium-Range Weather Forecasts (ECMWF) for buoy data, and Remote Sensing System (REMSS) for SSMIS rainfall data.

Conflicts of Interest: The authors declare no conflict of interest.

References

1. Liu, L. *An Introduction to Satellite Oceanic Remote Sensing*; Wuhan University Press: Wuhan, China, 2005; pp. 245–264.
2. Zhixiong, W. The Improvement of HY2-SCAT Wind Retrieval Algorithm Based on MSS and 2DVAR Method. Master's Thesis, Ocean University of China, Qindao, China, 2014.
3. Wang, X.; Yang, B. Abroad satellite microwave scatterometer application and the trends. *Erospace China* **2006**, 26–29.
4. Yang, J.; Zhang, J. Evaluation of ISS-RapidScat Wind Vectors Using Buoys and ASCAT Data. *Remote Sens.* **2018**, *10*, 648. [CrossRef]
5. Crapolicchio, R.; Lecomte, P. The ERS-2 scatterometer mission: Events and long-loop instrument and data performances assessment. In Proceedings of the 2004 ENVISAT & ERS Symposium, Salzburg, Austria, 6–10 September 2004; pp. 6–10.
6. Spencer, M.W. A Methodology for the Design of Spaceborne Pencil-Beam Scatterometer Systems. Ph.D. Thesis, Brigham YoungUniversity, Provo, UT, USA, 2001.
7. Gelsthorpe, R.; Schied, E.; Wilson, J. ASCAT-Metop's advanced scatterometer. *ESA Bull.* **2000**, *102*, 19–27.

8. Graf, J.; Sasaki, C.; Winn, C.; Liu, W.T.; Tsai, W.; Freilich, M.; Long, D. NASA Scatterometer Experiment1. *Acta Astronaut.* **1998**, *43*, 397–407. [CrossRef]

9. Department of Space, Indian Space Research Organisation. Available online: https://www.isro.gov.in/Spacecraft/oceansat-2 (accessed on 20 September 2019).

10. National Satellite Ocean Application Service. Available online: http://www.nsoas.org.cn/news/content/2018-10/25/44_531.html (accessed on 20 September 2019).

11. Masuko, H.; Arai, K.; Ebuchi, N.; Konda, M.; Kubota, M.; Kutsuwada, K.; Manabe, T.; Mukaida, A.; Nakazawa, T.; Nomura, A. Evaluation of vector winds observed by NSCAT in the seas around Japan. *J. Oceanogr.* **2000**, *56*, 495–505. [CrossRef]

12. Ebuchi, N.; Graber, H.C.; Caruso, M.J. Evaluation of wind vectors observed by QuikSCAT/SeaWinds using ocean buoy data. *J. Atmos. Ocean. Technol.* **2002**, *19*, 2049–2062. [CrossRef]

13. Pickett, M.H.; Tang, W.; Rosenfeld, L.K.; Wash, C.H. QuikSCAT satellite comparisons with nearshore buoy wind data off the US west coast. *J. Atmos. Ocean. Technol.* **2003**, *20*, 1869–1879. [CrossRef]

14. Tang, W.; Liu, W.T.; Stiles, B.W. Evaluation of high-resolution ocean surface vector winds measured by QuikSCAT scatterometer in coastal regions. *IEEE Trans. Geosci. Remote Sens.* **2004**, *42*, 1762–1769. [CrossRef]

15. Ebuchi, N. Intercomparison of Wind speeds observed by AMSR and SeaWinds on ADEOS-II. In Proceedings of the 2005 IEEE International Geoscience and Remote Sensing Symposium (IGARSS'05), Seoul, Korea, 29 July 2005; pp. 3314–3317.

16. Satheesan, K.; Sarkar, A.; Parekh, A.; Kumar, M.R.R.; Kuroda, Y. Comparison of wind data from QuikSCAT and buoys in the Indian Ocean. *Int. J. Remote Sens.* **2007**, *28*, 2375–2382. [CrossRef]

17. Yang, X.; Li, X.; Pichel, W.G.; Li, Z. Comparison of ocean surface winds from ENVISAT ASAR, MetOp ASCAT scatterometer, buoy measurements, and NOGAPS model. *IEEE Trans. Geosci. Remote Sens.* **2011**, *49*, 4743–4750. [CrossRef]

18. Sudha, A.K.; Prasada Rao, C.V.K. Comparison of Oceansat-2 scatterometer winds with buoy observations over the Indian Ocean and the Pacific Ocean. *Remote Sens. Lett.* **2013**, *4*, 171–179. [CrossRef]

19. Wang, H.; Zhu, J.; Lin, M.; Huang, X.; Zhao, Y.; Chen, C.; Zhang, Y.; Peng, H. First six months quality assessment of HY-2A SCAT wind products using in situ measurements. *Acta Oceanol. Sin.* **2013**, *32*, 27–33. [CrossRef]

20. Xing, J.; Shi, J.; Lei, Y.; Huang, X.-Y.; Liu, Z. Evaluation of HY-2A Scatterometer Wind Vectors Using Data from Buoys, ERA-Interim and ASCAT during 2012–2014. *Remote Sens.* **2016**, *8*, 390. [CrossRef]

21. Bhaskar, T.V.S.U.; Jayaram, C.; Bansal, S.; Mohan, K.K.; Swain, D. Generation and Validation of two Day Composite Wind Fields from Oceansat-2 Scatterometer. *J. Indian Soc. Remote Sens.* **2016**, *45*, 113–122. [CrossRef]

22. Lindsley, R.D.; Blodgett, J.R.; Long, D.G. Analysis and validation of high-resolution wind from ASCAT. *IEEE Trans. Geosci. Remote Sens.* **2016**, *54*, 5699–5711. [CrossRef]

23. Verhoef, A.; Vogelzang, J.; Verspeek, J.; Stoffelen, A. Long-Term Scatterometer Wind Climate Data Records. *IEEE J. Sel. Top. Appl. Earth Obs. Remote Sens.* **2017**, *10*, 2186–2194. [CrossRef]

24. Wentz, F.J.; Ricciardulli, L.; Rodriguez, E.; Stiles, B.W.; Bourassa, M.A.; Long, D.G.; Hoffman, R.N.; Stoffelen, A.; Verhoef, A.; O'Neill, L.W. Evaluating and extending the ocean wind climate data record. *IEEE J. Sel. Top. Appl. Earth Obs. Remote Sens.* **2017**, *10*, 2165–2185. [CrossRef]

25. Hutchings, N.; Long, D.G. Improved Ultrahigh-Resolution Wind Retrieval for RapidScat. *IEEE Trans. Geosci. Remote Sens.* **2018**, *57*, 3370–3379. [CrossRef]

26. Lin, M.; Zou, J.; Xie, X.; Zhang, Y. HY-2A Microwave Scatterometer Wind Retrieval Algorithm. *Eng. Sci.* **2013**, *15*, 68–74.

27. HY-2A Microwave Scatterometer Data Format User's Guide, National Satellite Ocean Application Service. May 2012. Available online: http://ftp2.nsoas.org.cn/Data_Format_User\T1\textquoterights_Guide/HY-2AMicrowaveScatterometerDataFormatUser\T1\textquoterightsGuide.pdf (accessed on 20 September 2019).

28. Liu, W.T.; Katsaros, K.B.; Businger, J.A. Bulk parameterization of air-sea exchanges of heat and water vapor including the molecular constraints at the interface. *J. Atmos. Sci.* **1979**, *36*, 1722–1735. [CrossRef]

29. Liu, W.T.; Tang, W.Q. *Equivalent Neutral Wind*; Jet Propulsion Laboratory Publication 96-17; 1996. Available online: https://ntrs.nasa.gov/archive/nasa/casi.ntrs.nasa.gov/19970010322.pdf (accessed on 4 April 2019).

30. de Kloe, J.; Stoffelen, A.; Verhoef, A. Improved use of scatterometer measurements by using stress-equivalent reference winds. *IEEE J. Sel. Top. Appl. Earth Obs. Remote Sens.* **2017**, *10*, 2340–2347. [CrossRef]

31. Remote Sensing Systems. Available online: http://www.remss.com/missions/ssmi (accessed on 20 September 2019).
32. Remote Sensing Systems. Available online: http://www.remss.com/support/crossing-times/ (accessed on 20 September 2019).
33. Portabella, M.; Stoffelen, A. *Qual. Control. Wind Retrieval for Sea Winds*; Ministerie van Verkeer en Waterstaat, Koninklijk Nederlands Meteorologisch: De bilt, The Nederlands, 2002; pp. 7–9.
34. Stiles, B.W.; Pollard, B.D.; Dunbar, R.S. Direction interval retrieval with thresholded nudging: A method for improving the accuracy of QuikSCAT winds. *IEEE Trans. Geosci. Remote Sens.* **2002**, *40*, 79–89. [CrossRef]
35. Portabella, M.; Stoffelen, A. A probabilistic approach for SeaWinds data assimilation. *Q. J. R. Meteorol. Soc.* **2004**, *130*, 127–152. [CrossRef]
36. QuikSCAT Scatterometer Mean Wind Field Products User Manual, The CERSAT (Centre ERS d'Archivage et de Traitement)/Laboratory of Oceanography From Space is part of IFREMER (French Research Institute for Exploitation of the Sea), Version 1.0. May 2002. Available online: http://apdrc.soest.hawaii.edu/doc/qscat_mwf.pdf (accessed on 14 September 2018).
37. Qiu, C.; Li, D. The Calculation Algorithms for Average Wind Direction and Their Comparison. *Plateau Meteorol.* **1997**, 94–98.
38. Stoffelen, A. Toward the true near-surface wind speed: Error modeling and calibration using triple collocation. *J. Geophys. Res. Ocean.* **1998**, *103*, 7755–7766. [CrossRef]

 *remote sensing*

Article

Evaluation of Sentinel-3A Wave Height Observations Near the Coast of Southwest England

Francesco Nencioli * and Graham D. Quartly

EOSA, Plymouth Marine Laboratory, Plymouth PL1 3DH, UK; gqu@pml.ac.uk
* Correspondence: fne@pml.ac.uk

Received: 30 October 2019; Accepted: 10 December 2019; Published: 13 December 2019

Abstract: Due to the smaller ground footprint and higher spatial resolution of the Synthetic Aperture Radar (SAR) mode, altimeter observations from the Sentinel-3 satellites are expected to be overall more accurate in coastal areas than conventional nadir altimetry. The performance of Sentinel-3A in the coastal region of southwest England was assessed by comparing SAR mode observations of significant wave height against those of Pseudo Low Resolution Mode (PLRM). Sentinel-3A observations were evaluated against in-situ observations from a network of 17 coastal wave buoys, which provided continuous time-series of hourly values of significant wave height, period and direction. As the buoys are evenly distributed along the coast of southwest England, they are representative of a broad range of morphological configurations and swell conditions against which to assess Sentinel-3 SAR observations. The analysis indicates that SAR observations outperform PLRM within 15 km from the coast. Within that region, regression slopes between SAR and buoy observations are close to the 1:1 relation, and the average root mean square error between the two is 0.46 ± 0.14 m. On the other hand, regression slopes for PLRM observations rapidly deviate from the 1:1 relation, while the average root mean square error increases to 0.84 ± 0.45 m. The analysis did not identify any dependence of the bias between SAR and in-situ observation on the swell period or direction. The validation is based on a synergistic approach which combines satellite and in-situ observations with innovative use of numerical wave model output to help inform the choice of comparison regions. Such an approach could be successfully applied in future studies to assess the performance of SAR observations over other combinations of coastal regions and altimeters.

Keywords: satellite altimetry; significant wave height; SAR; wave buoy observations; validation; southwest England; coastal altimetry; Sentinel-3A; SRAL

---

1. Introduction

Global consistent wave height data are needed for planning ship routing, designing offshore engineering structures, alerting coastal authorities and predicting transport and dispersal of floating objects [1]. Furthermore, waves also contribute to air-sea transfer of gases [2], and wave-breaking to the mixing of gases, nutrients and plankton in the surface layer of the ocean [3]. Long-term records are also important for studying potential effects of climate change [4]. Traditionally, in-situ observations of significant wave height (SWH) and wave direction are obtained from moored buoys. Although of high quality and high temporal resolution, such fixed-point measurements have sporadic spatial coverage [5]. Thus, they are of limited use even for basin-scale applications [6]. With the advent of satellite-borne radar altimeters, repeated global observations of SWH can also be obtained from remote sensing. The first global mapping of SWH was provided by Seasat, launched in 1978 and operative for only three months. Seasat was followed by Geosat in 1985. Since then, there have been a large number of instruments operating in this conventional mode (so-called "Low Resolution Mode" (LRM)) (see [7] for a review).

Although primarily designed to observe sea level elevation, satellite-borne radar altimetry can also be used to retrieve other physical variables at the sea surface, such as SWH and wind speed. Radar altimeters emit a radio pulse towards the Earth's surface and record the shape and magnitude of its reflection (the so-called waveform). The recorded signals are processed by fitting each waveform with a physical or empirical model (a process called retracking) [8]. The resulting fitting parameters are then used to derive sea level elevation, SWH and surface winds. In particular, SWH is obtained from the slope of the leading edge of the return pulse [9]. For wave height records, the effective instrument footprint of LRM altimetry is a disc from 2 to 20 km in diameter, depending upon wave conditions and satellite altitude [10,11]. Several studies have shown that in the open ocean the accuracy of altimeter SWH is similar to that of in-situ wave buoy observations [12–14]. However, such a large footprint can give particular problems near the coast, where stray reflections from land surfaces more than 10 km away can contaminate the received signal. In addition, under very calm conditions, sheltered bays may provide an almost mirror-like surface, generating specular returns in the waveforms [15,16]. To overcome such limitations, several studies have developed and assessed dedicated coastal retracking algorithms [6,17,18]. These have shown an overall improved accuracy of SWH from LRM altimetry in coastal regions, although performances on a case-by-case basis remain dependent on the specific coastal morphology and the angle between the satellite track and the coastline.

Recent advances in altimetry technology have included Synthetic Aperture Radar (SAR) or Delay-Doppler processing [19]. SAR processing focuses on a smaller strip of the antenna footprint compared with conventional LRM (see Figure 1 in Moreau et al. [20]), enabling observations at higher along-track resolution. Furthermore, because of SAR-specific data processing, SAR observations are also characterised by higher signal to noise ratio. Thus, SAR altimeters have the potential to resolve sea surface signals down to the sub-mesoscale (around 10 km) [21]. Near the coast, the smaller footprint of SAR observations is expected to reduce the signal contamination by spurious returns from nearby land, resulting in an increased number of accurate observations closer to the land–sea interface. The first instrument operating with SAR mode was Cryosat-2 [22]. However, the mode was only used for specific periods and limited regions over the oceans. Despite that, several studies have shown excellent performances over both open ocean and coastal regions [21,23,24]. Sentinel-3A (hereafter S3A, https://sentinel.esa.int/web/sentinel/missions/sentinel-3), launched in February 2016, was the first altimeter to use SAR operations over the global ocean. Direct comparison with in-situ observations have already shown overall improved performance of S3A with respect to previous altimeter missions in the European northwest shelf seas [25]. Along with SAR mode, S3A waveforms are also processed to generate pseudo-low-resolution mode (so-called PLRM) measurements similar to those produced by conventional altimeters. This way, SAR and PLRM data can be directly compared to evaluate the expected advantages of SAR altimetry over conventional in coastal regions.

In this paper, we provide a critical analysis of the quality of S3A SWH data focussing specifically on the coastal region of southwest England. In particular, S3A SAR performance was evaluated against in-situ observations as the altimeter footprint approaches or recedes from the coast, and it was contrasted with what can be achieved with PLRM processing. This work was developed within the activities of the Sentinel-3 Mission Performance Centre (S3MPC), responsible for cal/val, off-line quality control and algorithm correction activities.

The succeeding text commences with a description of the area under study and of the data used, with Section 3 discussing the methods used to compare the altimeter and buoy data, and of how the model was used to inform the selection of comparison points. The results in Section 4 show how the scatter plots and correlation statistics may be used to clarify which buoy and altimeter track pairings are useful for validation, with this process being informed by the innovative use of high-resolution model data. The agreement between good pairings are then used to examine the difference between processing modes (SAR or PLRM), variation with distance from buoy or coast and also to look for any effects of swell. Discussion and conclusions are provided in Section 5.

2. Study Area and Datasets

2.1. Southwest England

The study focused on the coastal region of southwest England (6.75°W–2°W, 49.5°N–51.75°N; Figure 1). The regions was chosen because of the availability of both in-situ (17 buoys) and altimetry (12 S3A tracks) observations near the coast. The distribution of buoys and S3A tracks provides different morphological and geometrical configurations (i.e., type of coast, wave field properties, buoy distance from the track and track orientations with respect to the coast) under which S3A SAR performance can be assessed. Wave characteristics in the region are a mixture of Atlantic swells and local wind seas generated in the English Channel [26]. The region experiences the most energetic wave conditions of all England: 50% of SWH is larger than 1 m, 10% range from 2 m to 5 m and extreme high wave conditions (SWH > 6 m) associated with severe storm events occur on average five times annually [26]. The Atlantic swells are predominantly from west and southwest. After entering the English Channel, swell wave height and direction are modified due to interactions with coastline and bathymetry, as discussed in subsequent sections.

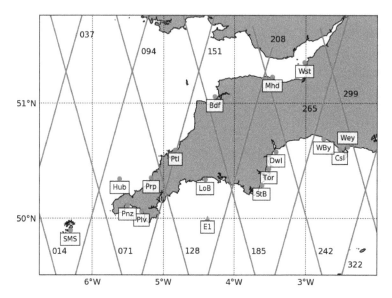

Figure 1. Locations of S3A tracks (blue lines) and wave buoys (red dots) used for this study.

2.2. Sentinel-3A Observations

S3A observations are from the 12 tracks shown in Figure 1. The S3A satellite is in an orbit with a repeat cycle of 27 days, with each track being repeated with this periodicity, and with a deviation of ±1 km in the longitudinal direction. Ascending passes are from south-southeast to north-northwest, whereas the descending ones are from north-northeast to south-southwest. The observations used in this study are from the S3A altimeter SRAL. SRAL data in this study were specifically produced by ACRI (https://www.acri.fr/) for the members of the Sentinel-3 Mission Performance Centre (S3MPC) based on the processing baseline 2.33. Data based on the same processing baseline have also been produced by EUMETSAT (https://www.eumetsat.int) and made available to the general public via the Copernicus Online Data Access (CODA, https://codarep.eumetsat.int/).

The analysis used SAR and PLRM Ku-band observations at 20 Hz (variables swh_ocean_20_ku and swh_ocean_20_plrm_ku) from Cycle 002 in March 2016 to Cycle 041 in February 2019. Cycle 001 had incomplete SAR observations and was not included. Cycles 002 and 003 had incomplete PLRM observations but were included for the SAR analysis. The 20 Hz data have an along-track spatial resolution of ~340 m. To remove bad observations due to land contamination, only data with quality flag = 0 and distance from the coast >0 km were used.

Figure 2 shows an example of SWH observations along track 128 for Cycle 006 (7 July 2016). PLRM and SAR data are both noisy (i.e., have very short wavenumber variability), although SAR is characterised by smaller noise than PLRM. To reduce the noise, all S3A observations were averaged using a moving Gaussian window with a full width at half maximum of 50 samples. This corresponds to ~17 km, the same order of magnitude as the smallest scale that S3A is expected to resolve [21].

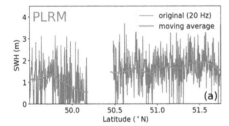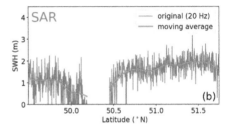

Figure 2. SWH along track 128 for cycle 6: (**a**) PLRM; and (**b**) SAR mode. The gap in both tracks is due to land. In orange are the data averaged with a moving Gaussian window of 50 samples at FWHM (corresponding to σ ~21 samples).

2.3. Buoy Observations

In-situ observations of SWH are from a network of 17 buoys distributed along the SW England coast as shown in Figure 1 (see also Table 1). All buoys except E1 are managed by the National Network of Regional Coastal Monitoring of England funded by the UK Department for Environment, Food and Rural Affairs (DEFRA). The observations are freely distributed through the Channel Coastal Observatory data portal (http://www.channelcoast.org). The buoys are Directional WaveRider buoys produced by Datawell BV (http://www.datawell.nl). All buoys are deployed in ~10–15 m water depth, except for the Hub and SMS buoys which are in ~50 m water depth. Each buoy provides time series of SWH, wave direction and wave period as 30-min averages (Figure 3).

The E1 buoy, located in front of Plymouth Sound (4.375°W 50.043°N; 75 m depth), is part of the Western Channel Observatory (http://westernchannelobservatory.org.uk) [27]. The buoy is financed by the UK Natural Environment Research Council (NERC) and managed by the Plymouth Marine Laboratory (PML). The instrumentation includes a broad array of sensors to measure meteorological and sea surface (physical and biogeochemical) variables. Wave observations are from a Tri-Axys directional wave sensor provided and managed in collaboration with the UK Met Office. Time-series of SWH and wave direction (but not wave period) are provided as 1-h averages.

In this study, we used buoy observations from March 2016 to February 2019, the same time period as the S3A observations. All buoys, except for that at the Hub site, provided almost continuous time series within that period. Occasional gaps in the data are due to buoy maintenance. Hub buoy observations are only up to 6 June 2018, when the buoy was decommissioned. Figure 3 shows examples of SWH time series and directional wave spectra for the Hub, WBy and Pnz buoys. Waves are higher and predominantly from the west at the Hub buoy. The site is the closest to the open Atlantic among the 17 (the SMS buoy is deployed within the Scilly Islands archipelago), thus the swell characteristics are the least impacted by bottom and lateral boundary interactions. On the other hand, the dominant wave characteristics at coastal sites (such as Pnz ad WBy) are markedly different (i.e., smaller SWH and wave direction roughly perpendicular to the coastline) as the waves there are strongly influenced by coastline and bathymetry.

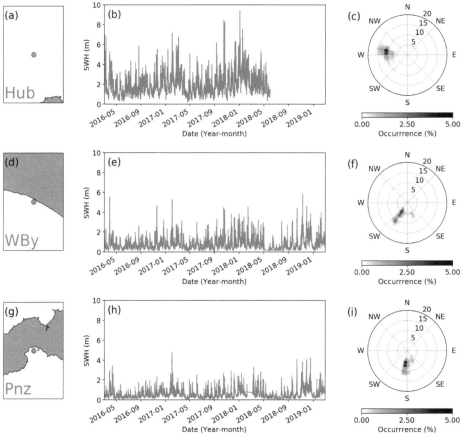

Figure 3. (**a,d,g**) Locations of the Hub, WBy and Pnz buoys (red circles). Each displayed area extends for ~15 km on all sides of the relevant buoy. (**b,e,h**) Time series of SWH for the three buoys. (**c,f,i**) Polar distribution of observed principal wave direction and period for the three buoys.

2.4. Numerical Model Simulations

Our analysis included the use of modelled SWH fields from the North-West European Shelf Wave Analysis and Forecast system distributed by CMEMS (product NORTHWESTSHELF_ANALYSIS_FORECAST_WAV_004_012 at http://marine.copernicus.eu). The fields were generated by the UK Met Office using a WAVEWATCH III (version 4.18) 7 km Atlantic Margin Model (hereafter, denoted as WWIII-AMM7). A full description of the model configuration is available at http://marine.copernicus.eu/documents/PUM/CMEMS-NWS-PUM-004-012.pdf. Modelled SWH fields are provided as instantaneous snapshots at hourly intervals from April 2014 to September 2018. For the study, we focussed exclusively in the area defined in Figure 1. Modelled SWH and wave direction for 15 February 2018 are shown in Figure 4. The figure is an example of how the characteristics of open Atlantic swells vary as the waves approach the SW England coast.

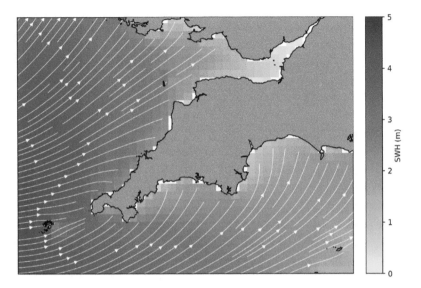

Figure 4. Example of significant wave height (colour) and swell direction (streamlines) from the WWIII-AMM7 model for 15 February 2018 at 01:00.

3. Methods

3.1. S3A and Buoy Correlations

The first step of the analysis was to identify the S3A observations (e.g., track number and locations along that track) to be compared against the in-situ observations from each buoy. To define the appropriate track/buoy combinations, we decided to implement a completely automated approach, so that it can be easily extended in the future to other coastal regions or to Sentinel-3B observations. The approach consisted in first identifying for each buoy the two closest S3A tracks (see Table 1). Along each of those tracks, the S3A observations for the comparison were selected from the closest point to the buoy as well as from a series of other locations north and south of it (see maps in Figures 5–7). Using multiple locations along each track allowed us to assess S3A performance as satellite observations approach or recede from the coast. The distance between each location was set to 50 samples (same size as the moving average window applied to S3A SWH observations) to ensure complete independence between the smoothed observations at each location. Along each track, only locations up to 60 km from the closest point to the buoy were used for the analysis. Furthermore, a latitudinal constraint on the along-track locations was imposed at each buoy to avoid comparing in-situ observations from the northwest shore and satellite observations from the English Channel (and vice versa) for the satellite tracks crossing over UK mainland. Finally, from each buoy time series, we selected the closest observation to the time of each S3A passage over the two nearest tracks to the buoy. As a result, for each combination of wave buoy and S3A track, we obtained a set of time series of concurrent in-situ and satellite SWH observations at varying distance from the buoy.

Table 1. List of the buoys used in our analysis. For each buoy general characteristics and the two closest S3A tracks are provided. Buoy locations and depth are from the latest reports available for each buoy.

Buoy	Full Name	Lat	Lon	Depth (m)	S3A Tracks	
Bdf	Bideford Bay	51°03.48′N	04°16.62′W	11	151	185
Csl	Chesil	50°36.13′N	02°31.37′W	12	265	299
Dwl	Dawlish	50°34.80′N	03°25.04′W	11	242	265
E1	E1 Station	50°01.56′N	04°13.50′W	75	185	208
Hub	Wave Hub	50°20.84′N	05°36.84′W	50	071	094
LoB	Looe Bay	50°20.33′N	04°24.65′W	10	185	208
Mhd	Minehead	51°13.68′N	03°28.15′W	10	208	242
Plv	Porthleven	50°03.76′N	05°18.44′W	15	128	151
Pnz	Penzance	50°06.86′N	05°30.18′W	10	071	094
Prp	Perranporth	50°21.19′N	05°10.48′W	14	128	151
PtI	Port Isaac	50°35.65′N	04°50.07′W	N/A	128	151
SMS	St Mary's Sound	49°53.53′N	06°18.77′W	53	014	094
StB	Start Bay	50°17.53′N	03°36.99′W	10	185	208
Tor	Torbay	50°26.02′N	03°31.08′W	11	242	265
WBy	West Bay	50°41.63′N	02°45.06′W	10	265	299
Wey	Weymouth	50°37.36′N	02°24.85′W	11	208	299
Wst	Weston Bay	51°21.13′N	03°01.23′W	13	299	322

To compare S3A and buoy observations, we computed their linear correlation for each of the time series at a given location. The correlations were obtained using the Theil–Sen estimator, which derives the slope as the median of all slopes between paired values [28]. This method is more robust to outliers than traditional linear regression, thus better suited for our analysis due to the limited numbers of pairs (40 maximum) used to derive the linear correlation. The robustness of the method was confirmed by performing the analysis with a smaller subset of data. Our test showed that 30 pairs (corresponding to S3A Cycles 002–031) were sufficient to return analogous results as those from the full dataset. The slope of the linear fit (regression slope) and the root mean square error (RMSE) from each correlation, as well as the bias between satellite and in-situ observations along each time series, were used in our analysis to assess the S3A SAR performance.

3.2. Areas of Correlation

As shown in Section 4.1, defining track/buoy pairs based on the distance between buoy and satellite tracks is not as reliable in a complex coastal environment as it is in the open ocean. Thus, to properly define the pairs to be used to assess the S3 performance while at the same time maintaining the approach as automated as possible, we implemented a second identification method based on the results from the WWIII-AMM7 model. The SWH fields from the model were used to identify the area around each buoy where wave characteristics remain similar to those observed at the buoy site. Satellite observations within such areas of correlation are expected to match in-situ observations at the corresponding buoy site. Therefore, these areas were used to further refine the optimal S3A track and wave buoy combinations to be included in the analysis.

To identify the area of correlation of each buoy, we first computed the linear correlations between the time series of SWH at each grid point of the model domain and the one at the nearest grid point to the buoy location. For this analysis, the correlations were obtained using traditional linear regression since the time series consisted of a large number of points (38808 hourly values), numerical models are unlikely to produce outliers and the traditional method (as opposed to the Theil–Sen method) can also provide estimates of the correlation coefficient, one of the parameters required for the analysis. The time series comparisons were used to derive the spatial distribution of four parameters all associated with specific characteristics of the linear fit: the correlation coefficient, the RMSE, the regression slope and the regression intercept. The correlation coefficient indicates how strongly the relation between the two time series is linear (the closer it is to 1, the more linear is the relation). The RMSE is associated with the spread of the observations around the linear fit (the larger is the RMSE, the larger is the spread). The slope indicates how close the regression is to the 1:1 relation (occurring when the slope = 1). The intercept indicates (in case of a 1:1 relation) how large the bias between the two time series is (the smaller is the absolute value of the intercept, the smaller is the bias). Because of that, we considered two time series to be similar if the parameters from their linear correlation fell within all the following thresholds: correlation coefficient ≥ 0.95, 0.8 ≤ slope ≤ 1.2, −0.1 m ≤ intercept ≤ 0.1 m, and RMSE ≤ 0.25 m. (Section 4.2 describes how these values were defined.) Thus, the area of correlation of each buoy was defined as the ensemble of the model grid points where these four constraints are satisfied.

4. Results

4.1. S3A and Buoy Correlations

The correlations between S3A and in-situ observations at varying distances from each buoy showed different behaviour depending on the buoy location. As a first example, the Hub buoy shows correlations between satellite and in-situ observations close to the 1:1 relation at all distances from the buoy (Figure 5). As described in Section 2.3, the wave field at the Hub site is the least impacted by bathymetry and coastal effects among all the buoy sites, and thus the most representative of the swell conditions coming from the open Northwest Atlantic. Waves at the various locations along the S3A 094 and 071 tracks also have characteristics similar to those from the open Northwest Atlantic (Figure 4). Thus, results from the Hub buoy indicate good accuracy in the retrieval of SWH from SAR altimetry observations in the open ocean, confirming what has already been reported for conventional altimetry [12–14]. To further support that, results from PLRM observations at the same locations show analogous correlations to those from SAR (not shown). Moreover, the Hub correlations indicate that, as long as waves are not impacted by coastline and bathymetry, validation of satellite SWH observations can be performed using in-situ observations collected within an ample radius from the satellite track (at least 60 km in our case).

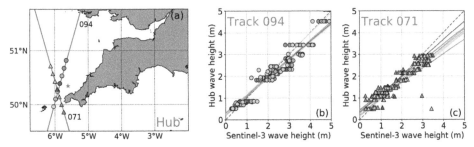

Figure 5. (a) Map showing the Hub buoy position (magenta star), the two nearest S3A tracks to the buoy (tracks 094 and 071, both in blue) and the various locations along each track at which the correlations between satellite and in-situ observations were computed (coloured circles and triangles). (**b**,**c**) Scatter plots between S3A SAR and in-situ observations at each location (symbols and colours are the same as in panel (**a**)) and corresponding regression lines. The gray dashed line indicates the 1:1 relationship.

While the results from the Hub buoy are unique within our dataset, the changes of correlation with distance from the buoy for the rest of the coastal buoys can be grouped into two main types. Results from the Pnz buoy are shown in Figure 6 as an example of the first type of behavior. As shown in the map, the two nearest tracks to the buoy (094 and 071) are the same as for the Hub buoy. However, as opposed to the Hub buoy, the Pnz buoy is located in a sheltered coastal region, where height and direction of open sea waves are strongly refracted and attenuated due to interactions with the bathymetry and the coastline (Figure 4). As a result, the scatter plots show satellite SWH to be larger than the ones observed at the Pnz buoy at all locations along the two tracks. As for the Hub buoy, the corresponding regression lines are similar to each other (since wave characteristics do not vary along the two satellite tracks), but they are consistently below the 1:1 relation.

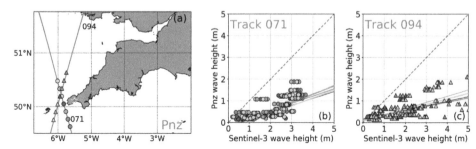

Figure 6. Same as Figure 5 but for the Pnz buoy and S3A tracks 071 and 094.

Examples of the second type of observed cahnges are from the WBy buoy shown in Figure 7. As shown in the map, both closest S3A tracks (265 and 299) cross the coastline in proximity of the buoy (~10 and ~30 km away, respectively). Since the coastline around the WBy buoy is quite uniform, geometrical configurations (south to southwest facing shoreline) and morphological conditions (sandy beaches) are similar at all locations. The scatter plots indicate good agreement between satellite and in-situ observations. Satellite SWH at along-track locations further offshore are higher compared with those observed by the coastal buoy. However, the two are similar at the closest locations to the buoy (and hence to the coastline) for both tracks. This is better evidenced by the regression lines, which approach the 1:1 relationship for locations progressively closer to the buoy.

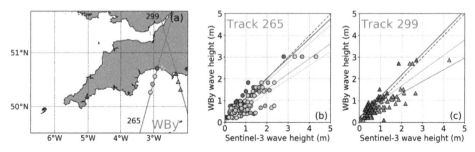

Figure 7. Same as Figure 5 but for the WBy buoy and S3A tracks 265 and 299.

Overall, the three types of behavior observed from our dataset can be visualised by plotting the regression slope as a function of the distance from each buoy (Figure 8). As already mentioned, the Hub buoy represents the only case of open sea buoy in the network analysed in our study. The associated slopes are high at all along-track locations (~0.8 on average) and do not show any variation with respect to the distance from the buoy. The Pnz buoy represents a coastal buoy with bad agreement between in-situ and satellite observations. Similar buoys include Mhd, Plv, PtI, SMS, StB, Tor, Wey and Wst. For such buoys, regression slopes are low (<0.6) at all locations. In some cases, such as the Pnz buoy, the regression slope does not show any variation with respect to the distance from the buoy; in others (as is the case for the Tor and Wey buoys, not shown), the slope shows increasing values with decreasing distance from the buoy. Coastal buoys with good agreement between in-situ and satellite observations (such as the WBy buoy) also show low slope values at the farthest locations from the buoy. However, the values progressively increase as the along-track locations approach the buoy, until reaching values ~1 at the closest locations. This group of buoys include Bdf, Csl, Dwl, E1, LoB and Prp.

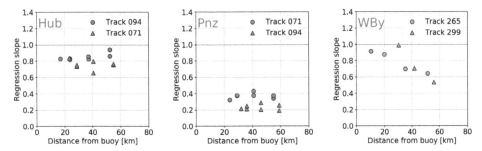

Figure 8. Variation of the regression slope as function of the distance from the buoy for the Hub, Pnz and WBy buoys. The regression slopes were computed at varying locations along the two closest S3A tracks to each buoy.

The good agreement between in-situ and satellite observations obtained for some of the buoys indicate that S3A SAR observations can accurately retrieve SWH near the coast. Moreover, regression slope behavior such as that observed for the WBy buoy suggest that satellite observations can also correctly capture the progressive attenuation of SWH as the open ocean swells propagate towards the coast (see Figures 4 and 7). However, the poor correlations observed at other coastal buoys show that this is not always the case. Those poor correlations can occur because of two reasons: (a) intrinsic inaccuracies of S3A SAR observations in coastal regions; or (b) poor pairing between the in-situ and satellite observations used for the validation. The latter case occurs when in-situ and satellite observations are compared from spatially close locations that are however representative of markedly different SWH conditions. Poor pairing could indeed be a relevant issue in our analysis since most of the SW England coast is characterised by complex morphology and conditions can rapidly vary even between close locations. Because of that, buoy-to-track distance is likely to be inadequate to identify the appropriate track–buoy combinations to be used for validation in such a complex coastal region. Thus, to better assess the nature of the observed poor correlations and identify more accurately the appropriate track–buoy combinations to validate the satellite observations, we decided to integrate in our analysis the results from the wave numerical model described in Section 2.4.

4.2. Areas of Correlation

Figure 9 shows the spatial distribution of the four parameters (correlation coefficient, regression slope, regression intercept and RMSE) derived from the WWIII-AMM7 model simulation for the Hub buoy, as described in Section 3.2. Values of the correlation coefficient (r^2) are larger than 0.8 over most of the model domain, indicating good linear regression between SWH time series at the Hub buoy and those from the rest of the model domain. Although not as large as for the Hub buoy, high values of r^2 extend for good portions of the model domain also for the rest of the coastal buoys (not shown). This is expected, since, although attenuated, SWH at the coast remains related to the SWH in the open sea. This is the reason for the high threshold value defined for r^2 (Section 3.2). For the regression slope parameter, the lower threshold was defined based on the average values observed from the in-situ and satellite correlations obtained for the Hub buoy (Figure 8). The upper one was then defined so that the threshold interval is symmetric around 1. Figure 9 (top) shows that for the Hub buoy a large portion of the model domain is within these thresholds. However, the areas within thresholds are more localised for the coastal buoys, indicating that the chosen values are appropriate for identifying the significant areas of similarity. Among the four parameters, RMSE is the most localised around the Hub buoy (Figure 9d) and, hence, the most restrictive in defining the area of correlation. The same occurs for all coastal buoys. The threshold value of RMSE is in line with the average values observed for traditional altimetry mission in the open ocean [10,29]. In the case of the coastal buoys, it represents roughly 25% of the observed mean SWH (~1 m; see the time series for the WBy and Pnz buoys in Figure 3 as examples). As shown in Figure 9, despite the rapid decrease of RMSE away from the buoy, the area within the threshold still extends for several grid points around the buoy location. Finally, the regression intercept parameter shows a large area of values close to 0 in the case of the Hub buoy, but much reduced areas for the coastal buoys (similarly to what is observed for the distribution of regression slope values near 1). For those buoys, the distribution patterns of the intercept values are analogous to those observed for the RMSE. However, even for an absolute threshold value much smaller than the RMSE, the areas within the intercept thresholds are much larger than those found for the RMSE at all buoys (this also includes the Hub buoy, as shown in Figure 9c). For this reason, the intercept parameter has the least impact among the four in defining the area of correlation around each buoy.

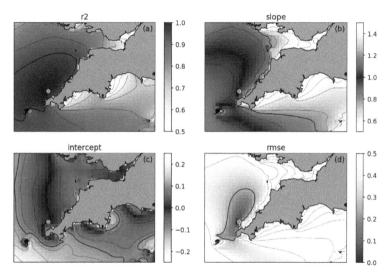

Figure 9. Example of spatial distribution of the correlation coefficient (**a**), regression slope (**b**), regression intercept (**c**) and RMSE (**d**) computed between the time series of the WWIII-AMM7 model simulation nearest to the Hub buoy and those at each of the grid points of the model domain (see Section 3.2 for more details). The red circle marks the position of the Hub buoy. The thicker contours mark the thresholds for each parameter used to define the areas of correlation around each buoy shown in Figure 10: correlation coefficient (r^2) $\geq$ 0.95, 0.8 $\leq$ slope $\leq$ 1.2, -0.1 m $\leq$ intercept $\leq$ 0.1 m and RMSE $\leq$ 0.25 m.

The resulting areas of correlation for the Hub, Pnz and WBy buoys are shown in Figure 10. Comparison with Figure 9 confirms that the areal extent for the Hub buoy is most strongly constrained by the RMSE parameter (Figure 9a). The figure indicates that S3A tracks 071 and 094 can be used for comparison with in-situ observations at the Hub buoy, and that satellite observations are expected to compare well with the in-situ one along a large portion of each track near the buoy (the portion along track 071 being shorter than that along track 094). Thus, it confirms the results discussed in Section 4.1 (Figure 5). Furthermore, although not tested in our analysis, the figure suggests that in-situ observations from the Hub buoy could also be used to validate those from track 128, although the track is much further from the buoy then the two used in our analysis.

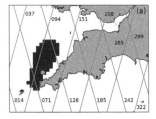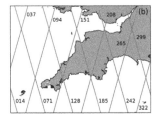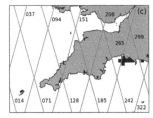

Figure 10. Examples of areas of correlation retrieved from the model output for the Hub (**a**), Pnz (**b**) and WBy (**c**) buoys.

The area of correlation for the Pnz buoy (Figure 9b) is limited to two grid points near the buoy location. The main constraint for such limited extent is due to the RMSE values which rapidly increase to values >0.5 m just two grid points away from the buoy. No satellite tracks intersect this area. This indicates that the poor correlations between in-situ and satellite observations at the Pnz buoy (Figure 6) are due to poor pairing rather than inaccuracies in satellite observations. Analogous results were obtained for some of the other buoys (Mhd, Plv, SMS, StB and Wst) that showed poor correlations between in-situ and satellite observations. For the remaining three buoys (PtI, Tor and Wey), the areas of correlations indicate that satellite observations from at least one of the two closest tracks should compare well against the in-situ observations. Closer inspection of the buoy location showed that all three buoys are located very close to the coast within small bays. Because of the coarse grid resolution, these bays are not represented in the model. As a consequence, all three buoy locations correspond to land points in the model grid. To derive the area of correlation, the analysis retrieves the SWH time-series at the buoy site using SWH from the nearest ocean point in the model. However, it is likely that such points do not correctly represent the SWH conditions observed by the three buoys in much more sheltered locations. As such, we decided to not include results from those buoys in the analysis.

The area of influence for the WBy buoy is shown in Figure 10c. The figure confirms that satellite observations along the portion of track 265 closest to the buoy should be similar to the buoy observations, as shown in Figures 7 and 8. At the same time, despite the good correlations obtained in Figure 7c, it indicates that satellite observations from track 299 should not be validated with the WBy observations. That track intersects the coastline immediately east of Weymouth (Wey). Again, because of the low grid resolution, fine scale coastal features (such as Weymouth Bay and the Isle of Portland immediately south of it) are only coarsely represented in the model. Therefore, it is likely that their sheltering effect on the SWH field is also misrepresented, and that coastal observations along track 299 are indeed analogous to the ones at WBy (as is the case for the model grid point immediately east of the track). For these reasons, we decided to retain track 299 in the analysis (this includes observations from the WBy buoy as well as the Csl one). Similar analysis for the other buoys which showed good correlation with satellite observations confirmed the good pairing between both closest S3A tracks and the Csl, E1 and Prp buoys. For the Bdf and Dwl buoys, the analysis indicated that only one of the two tracks should be used (185 and 242, respectively). Thus, the other two (151 and 265, respectively) were removed from the analysis. For the LoB buoy, the analysis showed that both closest tracks (185 and 208) are not representative of the conditions observed at the buoy location, and thus both pairings were removed from the analysis. Indeed, although characterised by increasing slope values with decreasing distance from the buoys (as seen for the WBy buoy, Figure 8), the four removed satellite track–buoy combinations all showed maximum slope values that remained below 0.8 even at the minimum distance from the buoy. This suggests less accurate correlations than observed for the other pairings, further supporting the results from the model analysis.

Finally, the areas of correlation around buoys LoB and StB indicated that tracks different from the two closest ones should be used. This further confirms that, in a complex coastal environment such as southwest England, the distance from the buoy alone is not a reliable parameter to define the appropriate wave buoys to be used for validating satellite observations along specific tracks. The new tracks correspond to track 128 for the LoB buoy, and to track 242 for the StB one. These two new pairings were included in the analysis.

The final buoy–S3A track pairings are summarised in Table 2 and illustrated in Figure 11. (The Hub buoy is not considered in the subsequent analysis, as it is effectively open ocean rather than coastal.) Sensitivity analysis on the identification of the area of correlation showed that relaxing the thresholds for the four parameters modified the extent of the areas but did not substantially change the identified pairs for each buoy. These pairs are used in the next section to evaluate the performance of S3A SWH observations in our coastal region.

Table 2. List of buoy–track parings used in Section 4.3 to evaluate S3A performance in the coastal zone. In bold are the tracks identified using the area of correlation described in Section 4.2.

Buoy	S3A Tracks	
Bdf	185	–
Csl	265	299
Dwl	242	–
E1	185	208
LoB	**128**	–
Prp	128	151
StB	**242**	–
WBy	265	299

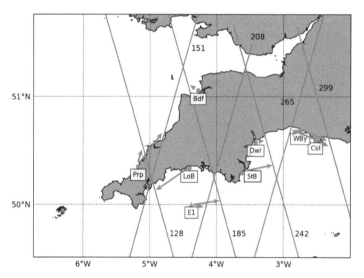

Figure 11. Same as Figure 1, but showing only the wave-buoys and satellite tracks from Table 2 used for the analysis of S3A performance. The green arrows indicate approximately where along each track lie the minima of RMSE with respect to the buoy measurements.

4.3. Evaluation of S3A Performance in the Coastal Region

Figure 12 shows the regression slope and the RMSE between S3A and in-situ SWH observations as a function of the distance from each buoy for the pairings listed on Table 2. Both diagnostics indicate good performance of satellite SAR observations in the coastal zone. The regression slope (Figure 12a) shows an inverse relation with respect to the distance from the buoy, with slope values approaching the 1:1 relation as satellite measurements are collected progressively closer to the buoys. The RMSE decreases from values of about 1 m at 60 km from the buoys to less than 0.5 m at 30 km. For closer distances to the buoys, the RMSE remains bound between values of 0.6 and 0.25 m.

As none of the buoys is positioned directly under a S3A track, the minimum distance between buoy and satellite observations is never below ∼10 km, even for the closest pairings. For some pairings (e.g., Wby–299, Csl–265 and E1–185) the minimum distance is on the order of 30 km, and for others (e.g., LoB–128 and PrP–151) even more. Moreover, under certain geometrical conditions (such as in the case of the Stb–242 pairing), the closest position to the buoy along a satellite track can correspond to a location that is further offshore than the buoy. Because of that, the closest regression slopes to the 1:1 relation can be found at locations more representative of the same coastal conditions found at the buoy location, but further away from the buoy than the closest ones. Thus, as in the case of the pairing selection in Section 4.2, even in this case the distance from the buoy is not the most appropriate variable to be used for the analysis.

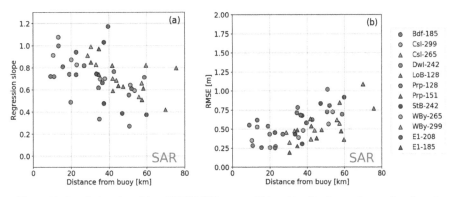

Figure 12. Correlation slope (**a**) and RMSE (**b**) between S3A and in-situ observations as function of distance between along-track location and buoy position. Only the satellite–buoy pairings from Table 2 are included. Both panels are for SAR observations.

Alternatively, the regression slopes can also be plotted as a function of the distance from the coast (Figure 13a). In this case, the slopes still show the same inverse correlation as in Figure 12a. Moreover, the closest values to 1 occur at the minimum distance from the coast for almost all pairings. Exceptions are the E1–208, LoB–128 and Bdf–185 pairings. E1 represents a special case with respect to the other buoys included in the analysis, since it is located in deeper waters and further away from the coast. For this reason, the closest regression slope to the 1:1 relation occurs at about 20 km from the coast, whereas as expected it increases to about 1.2 within 5 km from the coast (i.e., satellite observes smaller SWH values near the coast than those at E1). Regarding the Lob and Bdf buoys, their regression slopes increase with decreasing distance from the coast down to 10 km, but they maintain similar values for the closest points to the coast. Since the satellite tracks associated with both buoys intersect the coast in areas characterised by complex morphology and small bays, it is possible that the coarse resolution of the numerical model leads to inaccurate results in the identification of the area within which satellite observations can be paired with the two buoys. Finally, Dwl–242 is the only other satellite track–buoy pairing that does not show a regression slope above 0.8. As opposed to LoB and Bdf, the buoy is located on a portion of the coast where other buoys show very good correlation with the associated satellite measurements (e.g., StB, Wby and Csl). Thus, coarse model resolution cannot be hypothesised to be the cause for such low regression slopes, and the reasons for the poor performances observed at Dwl remain to be determined.

Regression slopes from PLRM observations (Figure 13b) show worse performance than those from SAR. These regression values are still characterised by a trend towards the 1:1 relation as the distance from the coast decreases down to 15 km. However, they have a broader range of values than SAR at similar distances. Furthermore, the regression slopes quickly decrease to very low values within 15 km from the coast, indicating that in that region PLRM SWH observations are consistently larger than those observed by the buoys. Such high PLRM-based SWH observations are likely resulting from retracking errors due to contaminations in the returned echoes by the presence of land within the satellite footprint near the coast.

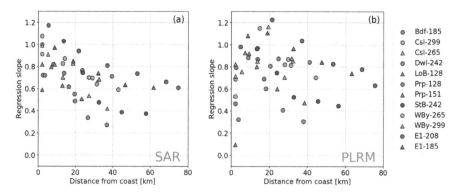

Figure 13. Regression slope between S3A and in-situ observations as a function of the distance from the coast (**a**) SAR observations; and (**b**) PLRM observations. As for Figure 12, only satellite track–buoy pairings from Table 2 are included.

Figure 14 shows the RMSE for SAR and PLRM observations as a function of distance from the coast. While RMSE values of SAR observations (Figure 14a) remain inversely correlated with the distance from the coast, their decrease approaching the coast is less pronounced compared with Figure 12b. The average value within 15 km from the coastline is 0.46 ± 0.14 m, while offshore of 15 km it is 0.60 ± 0.22 m. RMSE values of PLRM observations (Figure 12b) show trend and values similar to SAR offshore of 15 km (average value is 0.63 ± 0.23 m). However, they quickly degrade within 15 km from the coast (average value 0.84 ± 0.45 m), with particularly larger errors when the coast is within 5 km. Thus, as for the regression slope, SAR observations outperform PLRM ones in the coastal region also in terms of RMSE.

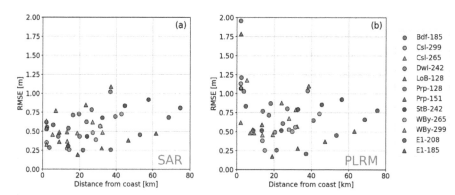

Figure 14. RMSE between S3A and in-situ observations as a function of the distance from the coast: (**a**) SAR observations; and (**b**) PLRM observations. As for Figure 12, only satellite track–buoy pairings from Table 2 are included.

To better understand whether some geophysical factors can potentially contribute to the observed RMSE in SAR mode, we investigated the dependence of the bias between satellite and in-situ observation on the swell characteristics. Figure 15 shows the difference between satellite and buoy observations as function of the swell period for each of the pairings from Table 2. Each point represents the difference of a given satellite observation at the location closest to the coast. The resulting distribution does not evidence any trend or correlation between bias and swell period (an analogous distribution was also obtained for PLRM observations; not shown). Thus, no clear dependence between the two can be identified from our analysis.

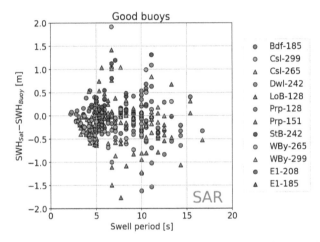

Figure 15. SWH difference between S3A SAR and in-situ observations as a function of swell period. Only the pairing from Table 2 are included. Each point is the bias computed for a given S3A cycle at the location closest to the coast for a specific track–buoy pairing.

The same biases were also analysed as a function of the swell direction to assess their dependence on the relative incident angle between swell and satellite track. The polar plots in Figure 16 show the difference between satellite and buoy observations as a function of swell direction for each satellite track associated with the buoys WBy, E1 and PrP. For each buoy, no clear trends or patterns can be identified in the distributions of the bias either as a function of varying swell directions for a given satellite track, or as a function of different satellite tracks for the same swell direction. An exception is represented by the E1 buoy that shows larger biases along track 208 than along track 185 for swells directions between W and SW. However, this is the only example among all the pairings from Table 2. Thus, as in the case of swell period, no clear dependence between the observed bias and swell direction can be identified from our analysis.

Previous studies based on sea level anomaly has also shown a dependency on altimetry performance based on whether the satellite transition is from sea to land or vice versa. Unfortunately, within our dataset, there are only three buoys paired with satellite tracks with both types of transition: Prp, WBy and Csl (the latter two being very close to each other and paired with the same satellite tracks). With such limited number of observations, it was not possible to identify an analogous dependency of SWH on satellite track transition.

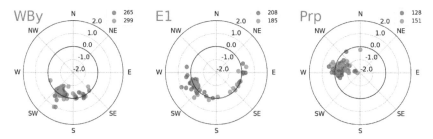

Figure 16. Polar plots with SWH difference between S3A SAR and in-situ observations as a function of swell direction for the WBy, E1, and Prp buoys. The bias varies along the radial direction from −2 m at the center to 2 m along the outermost circle. The thicker circle marks zero bias.

5. Discussion

The first step of our analysis was to implement an automated method to identify which satellite observations could be compared against the observations from each buoy. Our initial approach was to use the distance between buoy locations and satellite tracks as an objective diagnostic to identify the appropriate track–buoy pairings to be used for the validation. While such a diagnostic can be successfully used in the open ocean (where the characteristics of the wave field change gradually and over larger spatial scales), it showed some limitations in our region of study (where wave height and direction can change drastically over short distances along the coast due to wave interactions with land and bathymetry). A second approach involved the use of wave fields from a regional numerical model to identify the areas around each buoy location where wave characteristics are expected to remain similar. In the text, we refer to such areas as "areas of correlation". After those regions were defined, in-situ observations from a given buoy were compared only with satellite observations collected within its corresponding area of correlation. This second approach was better suited to identify the appropriate satellite–buoy parings to be used for the validation.

Comparison with the buoy observations indicated good performance of S3A SAR observations of SWH near the coast. S3A shows decreasing SWH for observations progressively closer to the coast, consistent with what is expected from theory and predicted by numerical models. Thus, correlations with the coastal buoy observations are characterised by an inverse relation with respect to distance from the coast. Regression slopes are between 0.6 and 0.2 at more than 30 km from the coast and progressively increase towards 1 for closer and closer distances. For most of the buoys, the regression slopes closest to the 1:1 relation occur at the minimum distance from the coast. On the contrary, regression slopes based on PLRM observations show a sharp decrease to values between 0.8 and 0.1 within the last 15 km from the coast. Improved performance of SAR observations with respect to PLRM were also observed from the analysis of RMSE as function of distance from the coast. Average value of RMSE within 15 km from the coast was found to be 0.46 ± 0.14 m for SAR observations and 0.84 ± 0.45 m for PLRM. Thus, our analysis confirms the expected advantages of SAR altimetry over traditional nadir altimetry in coastal regions. For conventional LRM mode altimeters, differences in the quality of range data have been noted between tracks approaching and receding from land; it was not possible to identify such distinction in our dataset for SWH data from a SAR-mode instrument.

Dependence of the bias between satellite and in-situ observations on swell characteristics was also explored. However, our analysis could not identify any correlation or trend between the observed biases and swell period or direction. Prior work had shown that the thin rectangular instrument footprint associated with SAR processing can sometimes lead to an underestimate of SWH [20]; we surmise that our results differ because even the data from the buoys well exposed to the North Atlantic did not observe sufficiently long wavelength swell at the time of our S3A overpasses.

Many recent studies have focussed on improving performance of traditional nadir altimetry over coastal regions by developing and assessing dedicated coastal retracking algorithms [6,17,18]. There is

Remote Sens. **2019**, *11*, 2998

the potential that some of these algorithms could be adapted to improve wave height estimates from S3A PLRM in the coastal zone, making use of appropriate quality control of the data. As SAR mode altimetry over the ocean is still new, there is also much scope for innovative algorithms to optimise SWH retrieval in the coastal zone.

6. Conclusions

In this study, we assessed the performance of S3A SAR observations in the coastal region of southwest England using in-situ observations from an array of 17 coastal wave buoys. The analysis indicates that SAR observations outperform PLRM within 15 km from the coast. Within that region, regression slopes between SAR and buoy observations are close to the 1:1 relation, and the average root mean square error between the two is 0.46 ± 0.14 m. On the other hand, regression slopes for PLRM observations rapidly deviate from the 1:1 relation, while the average root mean square error increases to 0.84 ± 0.45 m. The analysis did not identify any dependence of the bias between SAR and in-situ observation on the swell period or direction.

The validation approach outlined in this paper combines satellite and in-situ observations with numerical model results. Its general, principles have been designed with the aim of implementing a validation analysis for complex coastal regions that is as automated as possible. One obvious advantage of such an approach is that it can be easily applied to different regions and/or altimeters. Due mainly to the coarse resolution of the wave numerical model adopted in this study, our analysis still required a certain degree of manual intervention to correctly identify the appropriate satellite–buoy pairings to be used for validation. However, higher resolution models should drastically reduce the need for manual intervention. As such, we are confident that the proposed validation approach could be successfully replicated in future studies to assess the performance of Sentinel-3B in southwest England as well as to extend the assessment of S3A SAR performance over additional coastal regions spanning an even broader range of swell characteristics and conditions.

Author Contributions: G.D.Q. proposed the original idea for the research, which was developed by F.N., who did all the analysis and plotting. Interpretation of the results and writing of the paper was by F.N. and G.D.Q.

Funding: This research was funded by the European Space Agency (ESA) through the Sentinel-3 Mission Performance Centre (Contract No. 4000111836/14/I-LG), for which the overall management is provided by ACRI-ST. The views expressed herein can in no way be taken to reflect the official opinion of either the European Union or the European Space Agency.

Acknowledgments: The authors would like to thank ESA for financing this work. They would also like to thank all the members of the S3MPC for the constructive exchanges and discussions occurred over the many meetings and videoconferences within the last three years.

Conflicts of Interest: The authors declare no conflict of interest.

References

1. Ardhuin, F.; Stopa, J.E.; Chapron, B.; Collard, F.; Husson, R.; Jensen, R.E.; Johannessen, J.; Mouche, A.; Passaro, M.; Quartly, G.D.; et al. Observing Sea States. *Front. Mar. Sci.* **2019**, *6*, 124. [CrossRef]
2. Cronin, M.F.; Gentemann, C.L.; Edson, J.; Ueki, I.; Bourassa, M.; Brown, S.; Clayson, C.A.; Fairall, C.W.; Farrar, J.T.; Gille, S.T.; et al. Air-Sea Fluxes With a Focus on Heat and Momentum. *Front. Mar. Sci.* **2019**, *6*, 430. doi:10.3389/fmars.2019.00430. [CrossRef]
3. D'Asaro, E.A. Turbulence in the Upper-Ocean Mixed Layer. *Annu. Rev. Mar. Sci.* **2014**, *6*, 101–115. doi:10.1146/annurev-marine-010213-135138. [CrossRef] [PubMed]
4. Young, I.R.; Ribal, A. Multiplatform evaluation of global trends in wind speed and wave height. *Science* **2019**, *364*, 548–552. doi:10.1126/science.aav9527. [CrossRef]
5. Bailey, K.; Steinberg, C.; Davies, C.; Galibert, G.; Hidas, M.; McManus, M.A.; Murphy, T.; Newton, J.; Roughan, M.; Schaeffer, A. Coastal Mooring Observing Networks and Their Data Products: Recommendations for the Next Decade. *Front. Mar. Sci.* **2019**, *6*, 180. doi:10.3389/fmars.2019.00180. [CrossRef]

6. Passaro, M.; Fenoglio-Marc, L.; Cipollini, P. Validation of Significant Wave Height From Improved Satellite Altimetry in the German Bight. *IEEE Trans. Geosci. Remote Sens.* **2015**, *53*, 2146–2156. doi:10.1109/TGRS.2014.2356331. [CrossRef]

7. Le Traon, P.Y. From satellite altimetry to Argo and operational oceanography: Three revolutions in oceanography. *Ocean Sci.* **2013**, *9*, 901–915. doi:10.5194/os-9-901-2013. [CrossRef]

8. Gommenginger, C.; Thibaut, P.; Fenoglio-Marc, L.; Quartly, G.; Deng, X.; Gomez-Enri, J.; Challenor, P.; Gao, Y. Retracking altimeter waveforms near the coasts. In *Coastal Altimetry*; Vignudelli, S., Kostianoy, A.G., Cipollini, P., Benveniste, J., Eds.; Springer: Berlin/Heidelberg, Germany, 2011; pp. 61–102.

9. Brown, G. The average impulse response of a rough surface and its applications. *IEEE Trans. Antennas Propag.* **1977**, *25*, 67–74. doi:10.1109/TAP.1977.1141536. [CrossRef]

10. Durrant, T.H.; Greenslade, D.J.M.; Simmonds, I. Validation of Jason-1 and Envisat Remotely Sensed Wave Heights. *J. Atmos. Ocean. Technol.* **2009**, *26*, 123–134. doi:10.1175/2008JTECHO598.1. [CrossRef]

11. Chelton, D.B.; Walsh, E.J.; MacArthur, J.L. Pulse Compression and Sea Level Tracking in Satellite Altimetry. *J. Atmos. Ocean. Technol.* **1989**, *6*, 407–438. [CrossRef]

12. Cotton, P.D.; Carter, D.J.T. Cross calibration of TOPEX, ERS-I, and Geosat wave heights. *J. Geophys. Res. Ocean* **1994**, *99*, 25025–25033. doi:10.1029/94JC02131. [CrossRef]

13. Young, I. An intercomparison of GEOSAT, TOPEX and ERS1 measurements of wind speed and wave height. *Ocean. Eng.* **1998**, *26*, 67–81. doi:10.1016/S0029-8018(97)10016-6. [CrossRef]

14. Zieger, S.; Vinoth, J.; Young, I.R. Joint Calibration of Multiplatform Altimeter Measurements of Wind Speed and Wave Height over the Past 20 Years. *J. Atmos. Ocean. Technol.* **2009**, *26*, 2549–2564. [CrossRef]

15. Gomez-Enri, J.; Vignudelli, S.; Quartly, G.D.; Gommenginger, C.P.; Cipollini, P.; Challenor, P.G.; Benveniste, J. Modeling Envisat RA-2 Waveforms in the Coastal Zone: Case Study of Calm Water Contamination. *IEEE Geosci. Remote. Sens. Lett.* **2010**, *7*, 474–478. doi:10.1109/LGRS.2009.2039193. [CrossRef]

16. Wang, X.; Ichikawa, K. Coastal waveform retracking for Jason-2 altimeter data based on along-track Echograms around the Tsushima Islands in Japan. *Remote Sens.* **2017**, *9*, 762. doi:10.3390/rs9070762. [CrossRef]

17. Hithin, N.K.; Remya, P.G.; Balakrishnan Nair, T.M.; Harikumar, R.; Kumar, R.; Nayak, S. Validation and Intercomparison of SARAL/AltiKa and PISTACH-Derived Coastal Wave Heights Using In-Situ Measurements. *IEEE J. Sel. Top. Appl. Earth Obs. Remote. Sens.* **2015**, *8*, 4120–4129. [CrossRef]

18. Peng, F.; Deng, X. Validation of Improved Significant Wave Heights from the Brown-Peaky (BP) Retracker along the East Coast of Australia. *Remote Sens.* **2018**, *10*, 1072. doi:10.3390/rs10071072. [CrossRef]

19. Raney, R.K. The delay/Doppler radar altimeter. *IEEE Trans. Geosci. Remote. Sens.* **1998**, *36*, 1578–1588. doi:10.1109/36.718861. [CrossRef]

20. Moreau, T.; Tran, N.; Aublanc, J.; Tison, C.; Gac, S.L.; Boy, F. Impact of long ocean waves on wave height retrieval from SAR altimetry data. *Adv. Space Res.* **2018**, *62*, 1434–1444. doi:10.1016/j.asr.2018.06.004. [CrossRef]

21. Raynal, M.; Labroue, S.; Moreau, T.; Boy, F.; Picot, N. From conventional to Delay Doppler altimetry: A demonstration of continuity and improvements with the Cryosat-2 mission. *Adv. Space Res.* **2018**, *62*, 1564–1575. doi:10.1016/j.asr.2018.01.006. [CrossRef]

22. Wingham, D.; Francis, C.; Baker, S.; Bouzinac, C.; Brockley, D.; Cullen, R.; de Chateau-Thierry, P.; Laxon, S.; Mallow, U.; Mavrocordatos, C.; et al. CryoSat: A mission to determine the fluctuations in Earth's land and marine ice fields. *Adv. Space Res.* **2006**, *37*, 841–871. Natural Hazards and Oceanographic Processes from Satellite Data, doi:10.1016/j.asr.2005.07.027. [CrossRef]

23. Fenoglio-Marc, L.; Dinardo, S.; Scharroo, R.; Roland, A.; Sikiric, M.D.; Lucas, B.; Becker, M.; Benveniste, J.; Weiss, R. The German Bight: A validation of CryoSat-2 altimeter data in SAR mode. *Adv. Space Res.* **2015**, *55*, 2641–2656. doi:10.1016/j.asr.2015.02.014. [CrossRef]

24. Boy, F.; Desjonquères, J.; Picot, N.; Moreau, T.; Raynal, M. CryoSat-2 SAR-Mode Over Oceans: Processing Methods, Global Assessment, and Benefits. *IEEE Trans. Geosci. Remote. Sens.* **2017**, *55*, 148–158. doi:10.1109/TGRS.2016.2601958. [CrossRef]

25. Wiese, A.; Staneva, J.; Schulz-Stellenfleth, J.; Behrens, A.; Fenoglio-Marc, L.; Bidlot, J.R. Synergy of wind wave model simulations and satellite observations during extreme events. *Ocean Sci.* **2018**, *14*, 1503–1521. doi:10.5194/os-14-1503-2018. [CrossRef]

Remote Sens. **2019**, *11*, 2998

26. Scott, T.; Masselink, G.; Russell, P. Morphodynamic characteristics and classification of beaches in England and Wales. *Mar. Geol.* **2011**, *286*, 1–20. doi:10.1016/j.margeo.2011.04.004. [CrossRef]

27. Smyth, T.; Atkinson, A.; Widdicombe, S.; Frost, M.; Allen, I.; Fishwick, J.; Queiros, A.; Sims, D.; Barange, M. The Western Channel Observatory. *Prog. Oceanogr.* **2015**, *137*, 335–341. doi:10.1016/j.pocean.2015.05.020. [CrossRef]

28. Sen, P.K. Estimates of the Regression Coefficient Based on Kendall's Tau. *J. Am. Stat. Assoc.* **1968**, *63*, 1379–1389. [CrossRef]

29. Ray, R.D.; Beckley, B.D. Calibration of Ocean Wave Measurements by the TOPEX, Jason-1, and Jason-2 Satellites. *Mar. Geod.* **2012**, *35*, 238–257. doi:10.1080/01490419.2012.718611. [CrossRef]

Article

Retrieval of Particulate Backscattering Using Field and Satellite Radiometry: Assessment of the QAA Algorithm

Jaime Pitarch [1,2], Marco Bellacicco [3,*], Emanuele Organelli [2,4], Gianluca Volpe [2], Simone Colella [2], Vincenzo Vellucci [5] and Salvatore Marullo [2,3]

[1] Department of Coastal Systems, NIOZ Royal Netherlands Institute for Sea Research and Utrecht University, 1790 Texel, The Netherlands; jaime.pitarch@nioz.nl

[2] Istituto di Scienze del Mare (ISMAR)-CNR, Via Fosso del Cavaliere, 100 Rome, Italy; emanuele.organelli@artov.ismar.cnr.it (E.O.); gianluca.volpe@cnr.it (G.V.); simone.colella@cnr.it (S.C.)

[3] Energy and Sustainable Economic Development (ENEA), Italian National Agency for New Technologies, 00044 Frascati, Italy; salvatore.marullo@enea.it

[4] Laboratoire d'Océanographie de Villefranche (LOV), Sorbonne University, CNRS, 06230 Villefranche-sur-Mer, France

[5] Institut de la Mer de Villefranche (IMEV), Sorbonne University, CNRS, F-06230 Villefranche-sur-Mer, France; enzo@obs-vlfr.fr

* Correspondence: marco.bellacicco@enea.it

Received: 29 November 2019; Accepted: 20 December 2019; Published: 24 December 2019

Abstract: Particulate optical backscattering (b_{bp}) is a crucial parameter for the study of ocean biology and oceanic carbon estimations. In this work, b_{bp} retrieval, by the quasi-analytical algorithm (QAA), is assessed using a large in situ database of matched b_{bp} and remote-sensing reflectance (R_{rs}). The QAA is also applied to satellite R_{rs} (ESA OC-CCI project) as well, after their validation against in situ R_{rs}. Additionally, the effect of Raman Scattering on QAA retrievals is studied. Results show negligible biases above random noise when QAA-derived b_{bp} is compared to in situ b_{bp}. In addition, R_{rs} from the CCI archive shows good agreement with in situ data. The QAA's functional form of spectral backscattering slope, as derived from in situ radiometry, is validated. Finally, we show the importance of correcting for Raman Scattering over clear waters prior to semi-analytical retrieval. Overall, this work demonstrates the high efficiency of QAA in the b_{bp} detection in case of both in situ and ocean color data, but it also highlights the necessity to increase the number of observations that are severely under-sampled in respect to others environmental parameters.

Keywords: particulate optical backscattering; Raman scattering; QAA algorithm; ESA OC-CCI

1. Introduction

Retrieval of water inherent optical properties (IOPs) from both field and ocean color radiometry is at the base of several biogeochemical and physical oceanographic studies [1,2]. IOPs of algal and non-algal particles can be derived from remote sensing reflectance spectra (R_{rs}; units of sr^{-1}) by using appropriate algorithms [3–5]. Among IOPs, the particulate optical backscattering coefficient (b_{bp}; in m^{-1}) is related to the particle concentration in seawater, on their size distribution, refractive index, shape and structure [6–8]. Former research suggested that b_{bp} is mostly influenced by submicron non-algal particles [9–11]. However, it has been recently shown that most of b_{bp} is due to particles with equivalent diameters between 1 and 10 μm [8], thus including the contribution of phytoplankton cell and supporting the use of b_{bp} for the retrieval of: (i) particulate organic concentration (POC) [12,13]; (ii) particle size distribution [14,15]; and (iii) phytoplankton carbon biomass concentration (C_{phyto}; mg m^{-3}) [16–18], a key parameter also for phytoplankton physiology studies [2,19,20]. Efficiency in the b_{bp} retrieval is crucial for ocean biology and global ocean carbon estimations.

On one hand, radiative transfer theory provides the link between b_{bp} and optical radiometry [21]. Therefore, inversion algorithms for b_{bp} detection from optical radiometry can be developed. In particular, the quasi-analytical algorithm [3,22] is a multi-level algorithm that concatenates a sequence of empirical, analytical, and semi-analytical steps to retrieve spectral total non-water light absorption and backscattering (a_{nw} and b_{bp}) first and to decompose a_{nw} into its CDOM, algal and non-algal contributions. Specifically, about b_{bp}, some studies suggested some degree of b_{bp} overestimation by the QAA [23,24], but their reference b_{bp} data were sub-products of chlorophyll-a (Chl) measurements. QAA estimations from satellite R_{rs} showed a bias of +16.4% with respect to in situ b_{bp} for the Adriatic Sea [25]. Using the in situ NOMAD dataset [26], a b_{bp} overestimation of +38% by the QAA with respect to the observed value was reported [27]. Other results, based on in situ matchups, showed a bias of +2.5–8.8% for the QAA-derived b_{bp} in Arctic waters, and of +9.5% to +16.4% in low-latitude waters [28]. Pitarch et al. [29] reported a slight underestimation within 10% in the Mediterranean Sea. Most recently, QAA-derived b_{bp} from different satellite sensors (i.e., MODIS, VIIRS, OLCI) showed good performance with respect to a large in situ b_{bp} dataset collected on biogeochemical (BGC)-Argo floats [30].

Currently, in the European Space Agency (ESA) Ocean Colour (OC) Climate Change Initiative (CCI), QAA is the selected algorithm to retrieve b_{bp}. Specifically, the ESA OC-CCI project aims at creating a long-term, consistent, uncertainty-characterized time series of ocean color products, for use in climate-change studies [5,31]. In such a context, while in the case of Chl the uncertainties are fully provided, the b_{bp} satellite products lack such information that is also crucial for POC and C_{phyto} estimations [1,32]. This absence of statistical assessment is influenced by the paucity of a sufficient number of in situ observations for the determination of uncertainties.

Nowadays, the uncertainties associated to QAA-based b_{bp} retrievals globally are not known. In order to provide a best-effort b_{bp} uncertainty assessment, this work aims at evaluating the efficiency of QAA for global b_{bp} retrievals by using a large database of corresponding in situ R_{rs} and b_{bp} data ($N = 2881$). In details, we use the updated version of the recent in situ global bio-optical dataset [33] together with field measurements from the BOUSSOLE buoy [34] and two different oceanographic cruises in the Tyrrhenian and Adriatic Seas. Unlike previous studies [29], here, the QAA performance is considered at multiple bands that further allow the evaluation of the b_{bp} spectral slope retrievals against in situ measurements. The goals of this paper are thus: (i) to define the accuracy of the QAA for b_{bp} retrievals using only in situ R_{rs} data; (ii) to validate the CCI R_{rs} with in situ corresponding data; and (iii) to evaluate the performance of the QAA using satellite CCI R_{rs} as input data.

2. Data and Methods

2.1. Assessment of the Quasi-Analytical Algorithm (QAA)

The original algorithm [3] has undergone many updates and developments by several researchers. The QAA version here used is based on the algorithm for the CCI bands which is currently integrated in the SeaWiFS data analysis system (SeaDAS) [22].

The sub-surface R_{rs} (named r_{rs}) is calculated as $r_{rs} = R_{rs}/(0.52 + 1.7R_{rs})$ and modeled as a function of the IOPs as: $r_{rs} = g_0 u + g_1 u^2$, with $u = b_b/(a + b_b)$, $g_0 = 0.089$ and $g_1 = 0.1245$. This approach provided good results in the Mediterranean Sea in case of oligotrophic and coastal waters [29,35].

The QAA uses an empirical inversion of R_{rs} to retrieve absorption and then it solves total backscattering (b_b) analytically. b_{bp} is calculated by subtraction of pure seawater backscattering (b_{bw}) for an average temperature of 14 °C and an average salinity of 38 PSU [36]. b_{bp} is first estimated at a reference wavelength of $\lambda = 555$ nm and then the calculation is extended to other wavelengths by assuming a power law $b_{bp} = b_{bp}(\lambda_0)(\lambda/\lambda_0)^{-\eta}$ for the b_{bp} with a spectral slope.

$$\eta = p_1\left[1 - p_2 \exp\left(-p_3\frac{r_{rs}(443)}{r_{rs}(555)}\right)\right] \tag{1}$$

Equation (1) is widely used for QAA retrievals of b_{bp} at multiple wavelengths. Nevertheless, we use the in situ dataset presented here to evaluate the accuracy of analytical η. The functional form of Equation (1) is used and the default numerical coefficients $p_1 = 2.0$, $p_2 = 1.2$ and $p_3 = 0.9$ [22] are replaced by unknown variables established by non-linear regression. To this aim, we used the iterative bi-square method, which minimizes a weighted sum of squared errors, where the weight given to each data point decreases with the distance from the fitted curve [37]. This procedure makes the curve sensitive to the bulk of the data and the effect of outliers is reduced. The error function is minimized through the trust region algorithm [38]. In addition, the 95% confidence prediction bounds are also computed.

It is known that for oligotrophic waters the Raman scattering plays a significant role that is not accounted for in the semi-analytical R_{rs} modeling [39]. Therefore, a pertinent question is how much this phenomenon affects the semi-analytical b_{bp} retrievals. With this aim, Lee et al. [40] developed empirical correction formulas to compensate the Raman scattering on the R_{rs}. Here, we assess the effects of this compensation on the difference between the in situ b_{bp} and R_{rs}-derived b_{bp}. The statistical assessment was also replicated in other different cases: (i) validation of satellite CCI R_{rs} against in situ R_{rs}; and (ii) b_{bp} retrievals after the application of QAA to satellite R_{rs} (Raman corrected), were compared to in situ measurements at the different available wavelengths.

Estimated data y_i are compared to reference data x_i by using the following statistical indexes: relative bias (units of %), relative root-mean square error (RMS, units of %) and determination coefficient (r^2)

$$bias = 100\frac{1}{N}\sum_{i=1}^{N}\frac{y_i - x_i}{x_i} \tag{2a}$$

$$RMS = 100\sqrt{\frac{1}{N}\sum_{i=1}^{N}\left(\frac{y_i - x_i}{x_i}\right)^2} \tag{2b}$$

$$r^2 = \frac{\sum_{i=1}^{N}(x_i - \bar{x})(y_i - \bar{y})}{\sqrt{\sum_{i=1}^{N}(x_i - \bar{x})^2}\sqrt{\sum_{i=1}^{N}(y_i - \bar{y})^2}} \tag{2c}$$

2.2. In Situ Data

The in situ database is composed of three distinct datasets containing multi-spectral R_{rs} and b_{bp}: the recently updated global in situ database ([33] hereafter V19 dataset), an in situ dataset collected by the Italian National Research Council (CNR) during two field campaigns in the Mediterranean Sea ([29] hereafter CNR dataset) and the time-series of data acquired by the BOUSSOLE buoy in the northwestern Mediterranean Sea ([34,41]; hereafter BOU dataset [42]). The three in situ databases were quality-checked as described below. All the R_{rs} data were band-shifted to the CCI bands (those of the NASA SeaWiFS instrument, namely 412, 443, 490, 510, 555, and 670 nm). The band-shifting procedure [43] is a technique to compensate small band differences in multispectral R_{rs} data. It takes into account the spectral shape of the absorption and scattering that contribute to R_{rs} and constitutes a more accurate approach than a simple linear interpolation. Considering every wavelength an independent measurement, the final dataset accounts for a total of $N = 2881$ R_{rs} and b_{bp} co-located measurements around the global ocean (Figure 1). As shown in Figure 2, the total R_{rs} and b_{bp} spectra cover from oligotrophic open-ocean to more eutrophic coastal waters as the range of R_{rs} and b_{bp} values vary between 0–0.02 sr^{-1} and 10^{-4}–10^{-1} m^{-1} respectively.

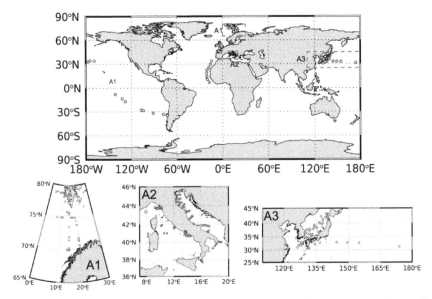

Figure 1. Geographical distribution of the in situ R_{rs} vs. b_{bp} matchups. Some areas (A1, A2, and A3) concentrate a high point density and are highlighted in zoomed maps. Pink, yellow, and green dots refer to V19, BOU, and CNR data, respectively.

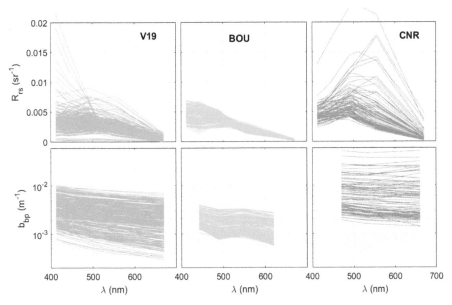

Figure 2. R_{rs} and b_{bp} spectra for the three-different datasets: V19, BOU, and CNR. Pink, yellow, and green lines refer to V19, BOU, and CNR data, respectively.

2.2.1. V19 Dataset

R_{rs} and IOPs, aggregated within ±6 nm, were downloaded. V19 is a global compilation of in situ data that was acquired from many sources (e.g., MOBY, AERONET-OC, SeaBASS, NOMAD, MERMAID, AMT, and many others), motivated by the validation of the ocean-color products from the

ESA OC-CCI products. Methodologies were implemented for homogenization, quality control and merging of all data. No changes were made to the original data, other than averaging of observations that were close in time and space, elimination of some points after quality control and conversion to a standard format [44].

In this study, data were selected only if valid and corresponding R_{rs} and b_{bp} measurements at all CCI bands were available. Such condition determines a total of $N = 319$ matchups. Remaining minor b_{bp} wavelength mismatches were removed by linear interpolation to the CCI bands. Although V19 is a merged dataset from multiple datasets, the condition we set for the matchup left data that were originally from the NOMAD dataset only.

2.2.2. BOU Dataset

The BOUSSOLE (*BOUee pour l'acquiSition d'une Série Optique a Long termE*) project started in 1999, and its activities are developed on a site located in the northwestern Mediterranean Sea (7°54′ E, 43°22′ N, Figure 1, panel A2). Essential information about the site characteristics, the measurement platform, and the instrumentation was previously documented [34,41,42]. The b_{bp} data were collected at 9 m nominal depth with a Hobilabs Hydroscat-4 (442, 488, 550, and 620 nm) and processed as in [45]. In addition, a quality control on b_{bp} was applied that required a spectral b_{bp} slope, calculated from every pair of two consecutive bands, within given bounds (more than −1 and less than 6). R_{rs} data were derived with a set of Satlantic 200-series multispectral radiometers ([46] and references therein). The R_{rs} is available at a varying number of the following bands, depending on the time period: 412.5, 442.5, 490, 510, 555, 560, 665, 670, and 681.25 nm. Since the application of the QAA requires R_{rs} at 443, 490, 555, and 670 nm, only R_{rs} records whose native bands matched those needed by the QAA algorithms were selected (within a ±6 nm range). Data at 412.5 nm and 442.5 nm were band-shifted to 412 nm and 443 nm [43], respectively. In the green region, if the R_{rs} at 555 nm was available, it was directly sampled and the R_{rs} at 560 nm was ignored. If the R_{rs} at 560 nm was available when the R_{rs} at 555 nm was missing, the R_{rs} at 560 nm was band-shifted to 555 nm. Similarly, in the red region, between the R_{rs} at 665 nm and 670 nm, preference to R_{rs} at 670 nm was given. R_{rs} data at 681.25 nm was not considered for the analysis. Data was generally available within two hours from the local noon. The time series at sub-daily resolution were reduced by calculating the daily medians.

2.2.3. CNR Dataset

Data belong to two field campaigns conducted in 2013 and 2015 in Italian seas, encompassing a high optical range between open and coastal waters. Measurements were performed between 8:30 h and 16:00 h UTC. IOPs and R_{rs} were collected sequentially at each station, with a maximum delay of ~1 h and ship drift of maximum few hundreds of meters.

Backscattering was measured with an ECO-VSF3, manufactured by WET Labs, Inc., at the wavelengths 470, 530, and 660 nm. This instrument measures the volume scattering function at three backward angles and calculates b_b by integration of a polynomial fit. Final data are the result of a binning across the first optical depth.

Radiometry was performed with OCR-507 radiometers, manufactured by Satlantic, Inc., measuring at the center bands 412, 443, 490, 510, 556, 665, and 865 nm. In-water upwelling radiance at nadir (L_u) sensor was mounted onto a free-falling T-shaped structure, with the multicast technique. Above-water downwelling irradiance (E_s) data were collected by a reference sensor, mounted at the top of the ship's deck. R_{rs} was computed using the SERDA software developed at CNR [47]. All the R_{rs} data were band-shifted to the CCI bands for consistency with the satellite R_{rs}. Further details about this dataset are provided in Pitarch et al. [29].

2.3. Satellite ESA OC-CCI R_{rs} Data

The ESA OC-CCI version 4.0 global daily R_{rs} data at 4 km resolution for the period 1997–2017 were downloaded [48]. CCI products are the result of the merging of SeaWiFS, MERIS, MODIS, and VIIRS data in which the inter-sensor biases are removed [49]. This version 4.0 includes the latest NASA reprocessing R2018.0 that mostly accounts for the aging of the MODIS sensor. ESA OC-CCI provides the daily R_{rs} data and associated uncertainty maps in terms of bias and RMS, which were generated with a procedure that included comparison to in situ data and optical water type analysis [48].

In this work, a conservative extraction procedure was followed, in which the center R_{rs} data within a 3×3 pixels box was extracted only if all the 9 pixels were not flagged, therefore minimizing possible land border, cloud or other environmental contaminations, and obtaining the highest quality of matchups. Finally, for each single R_{rs}, the bias was also extracted and then compensated pixel-by-pixel.

3. Results and Discussion

3.1. QAA Performance for b_{bp} Retrievals from In Situ Data

QAA-retrieved b_{bp} from in situ R_{rs} is here compared to in situ measured b_{bp}. Comparisons are made at the native bands of each in situ b_{bp} instrument for the cases of BOU and CNR and at the CCI bands for V19. Statistics are also presented for each band and dataset.

A first assessment consists of applying the QAA without performing the compensation for Raman scattering. Here, results show a general overestimation of around +43.4% for V19 (Table 1) that is not significant given the overall noise expressed by the RMS (152%). This high RMS is the likely consequence of the different protocols, instrumental and geophysical noises affecting all single contributors to the V19 dataset (Table 1). In the case of the BOU dataset, an overall overestimation of +49.2% is found for all the bands which is statistically significant given the related RMS (58.7%). On the other hand, the QAA applied to the CNR dataset showed the highest performances, with a bias of +3.3% and a RMS below 23%.

Table 1. Statistical descriptors of the difference between the QAA-derived b_{bp} and in situ b_{bp} for each dataset, without Raman scattering compensation. Figure A1 provides a graphical representation of this table.

	Band (nm)	Bias (%)	RMS (%)	r^2	N
V19	412	40.3	128.4	0.35	319
	443	42.7	129.4	0.37	319
	490	44.5	127.8	0.41	319
	510	45.0	127.1	0.42	319
	555	45.2	124.2	0.44	319
	670	43.1	114.2	0.47	319
	All	43.4	125.3	0.43	1914
BOU	442	44.5	50.7	0.73	172
	488	71.3	79.2	0.73	172
	550	29.0	36.5	0.78	172
	620	52.0	60.2	0.73	172
	All	49.2	58.7	0.75	688
CNR	470	11.8	25.1	0.88	93
	530	7.7	22.8	0.89	93
	660	−9.6	20.7	0.93	93
	All	3.3	22.9	0.88	279

To understand the importance of the Raman scattering correction in semi-analytical b_{bp} retrievals, the analysis is repeated with corrected R_{rs} [40]. The application of the Raman scattering correction reduces both bias and RMS nearly for all the b_{bp} at all bands (Table 2 and Figure 3). Indeed, for the V19 dataset, the bias decreases to 12% with respect to the retrievals obtained without correction of the Raman scattering (Table 2). The RMS reduction is around 34%. For the BOU data, the RMS and bias improve of about 11% and 12%, respectively. In the case of the CNR dataset, statistics show a modest increase in accuracy except for $\lambda = 660$ nm, which is likely influenced by chlorophyll-a fluorescence.

Although fluorescence peaks at around λ = 660 nm, the ECO-VSF 3 sensor, used to collect the CNR dataset, has a full width at half maximum (FWHM) of about 20 to 30 nm, so a fluorescence interference may not be excluded.

Overall, these results are somewhat expected as the Raman scattering correction produces a smaller effect in coastal waters [50], which represent a significant part of the CNR dataset with respect to the two other datasets (Figure 2). The overall statistics are in agreement with previous comparisons that showed negligible biases over noise at global scale [5] and at regional level [29]. Results in this section highlight the importance of applying the Raman scattering correction to the source R_{rs} prior to semi-analytical b_{bp} retrieval in order to increase the accuracy.

Table 2. Statistical descriptors of the difference between the b_{bp}-QAA derived and in situ b_{bp} for each dataset with Raman scattering compensation. Figure A2 provides a graphical representation of this table.

	Band (nm)	Bias (%)	RMS (%)	r^2	N
	412	28.5	94.6	0.45	319
	443	30.7	95.0	0.47	319
	490	32.2	93.4	0.50	319
V19	510	32.6	92.8	0.51	319
	555	32.7	90.4	0.52	319
	670	30.7	83.1	0.54	319
	All	31.2	91.6	0.52	1914
	442	33.0	40.1	0.73	172
	488	57.2	64.8	0.73	172
BOU	550	18.2	27.1	0.78	172
	620	39.0	47.8	0.73	172
	All	37.0	47.0	0.75	688
	470	6.5	22.6	0.88	93
CNR	530	2.5	21.3	0.89	93
	660	−14.2	23.0	0.93	93
	All	−1.73	22.3	0.89	279

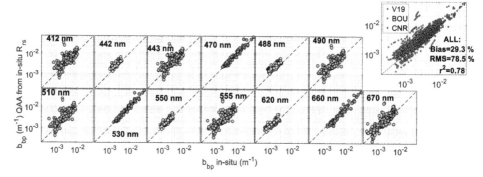

Figure 3. Scatter plots of QAA-derived b_{bp} vs. in situ b_{bp} data for each wavelength and dataset considered and for the merged dataset. Raman correction is applied to R_{rs}. The dashed line represents the 1:1 ratio. Pink, yellow and green dots refer to V19, BOU and CNR data, respectively.

3.2. Estimation of the b_{bp} Spectral Slope from R_{rs} Data

The in situ dataset described above is used (see Section 2.2) to assess the proposed relationship in the QAA and perform a model to data fit that is compared to the common QAA v6 equation [50]. Figure 4 shows a comparison of the independent variable (i.e., the blue-to-green band ratio $r_{rs}(443)/r_{rs}(555)$)

with respect to η derived from the in situ b_{bp}. Fitting a functional form of Equation (1) returns a curve ($p_1 = 2.2$, $p_2 = 0.9$ and $p_3 = 0.5$) and a 95% prediction interval, which is around ±1 wide, caused by the high scatter of the data cloud. The difference between η computed here and the one derived via QAA is much smaller than the width of the prediction interval, thus making them equivalent for prediction purposes. Therefore, by the principle of parsimony, the operational η functional form (dashed line in Figure 4) remains valid. However, one must keep in mind that the low predictive value of this relationship may result in b_{bp} extrapolations to bands outside the reference one (usually 555 nm) that accumulate significant errors. In particular, within a worst-case scenario, an error in η estimation, $\Delta\eta = 1$, will lead to a ~26% error when extrapolating b_{bp} from 555 nm to 412 nm.

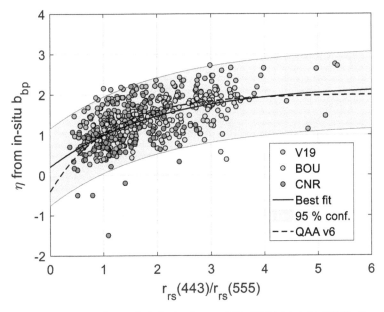

Figure 4. η calculation considering all the in situ data available: V19 (pink dots), BOU (yellow dots), and CNR (green dots). The solid curve is the best fit of Equation (1) to all the data ($p_1 = 2.2$, $p_2 = 0.9$ and $p_3 = 0.5$). The 95% confidence prediction bounds are represented by the grey shaded area. The dashed curve is the η estimation from R_{rs} as defined in Equation (1). Pink, yellow, and green dots refer to V19, BOU, and CNR data, respectively.

3.3. Validation of CCI R_{rs}

Prior to applying to satellite data an algorithm that has been developed with in situ data, assessing the quality of the satellite R_{rs} with respect to in situ measured R_{rs} is desirable in order to identify possible biases. Therefore, this section uses the in situ R_{rs} contained in the three datasets to evaluate the CCI R_{rs}. There is a total of 882 matchups for V19, 581 for BOU, and 252 for CNR. Good agreement between in situ values and the CCI R_{rs} products is found (Figure 5, Table 3) at all wavelengths, rather consistently with other previous results [5]. Overall, all datasets display similar performance, with negligible biases with respect to the overall noise expressed by the RMS. In the case of $\lambda = 670$, increased RMS is mostly due to the low values R_{rs}, except for CNR, that contains a higher data range. It is concluded that the CCI R_{rs} do not require adjustments at the studied wavelengths.

The magnitude of this RMS expresses a high bound for the overall uncertainty of the R_{rs} product as it is a measure of the errors in the comparison experiment, including those within the in situ data. The fraction of this error which is attributable to the satellite data only is likely to be much lower.

To have a measure of this fraction, a comparison to global in situ dataset with a traceable uncertainty budget would be desirable, though such option is presently not available.

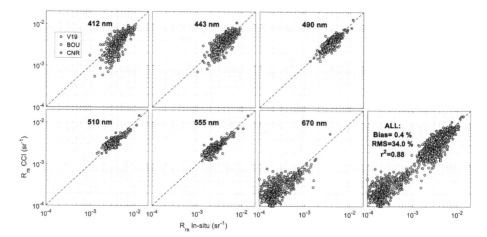

Figure 5. Scatter plots of CCI R_{rs} versus in situ R_{rs} for the six different wavelengths. The dashed line represents the 1:1 ratio. Pink, yellow, and green dots refer to V19, BOU, and CNR data, respectively.

Table 3. Statistical descriptors of the difference between satellite CCI R_{rs} and in situ R_{rs} for each dataset. Figure A3 provides a graphical representation of this table.

	Band (nm)	Bias (%)	RMS (%)	r^2	N
	412	−19.6	42.7	0.37	147
	443	−16.9	30.6	0.53	147
	490	−5.0	19.3	0.66	147
V19	510	−0.4	15.3	0.73	147
	555	−4.6	18.7	0.78	147
	670	28.4	117.9	0.47	147
	All	−3.0	54.2	0.73	882
	412	−4.0	22.5	0.50	96
	443	−3.7	23.9	0.63	97
	490	−1.9	11.1	0.66	97
BOU	510	−6.4	11.9	0.47	97
	555	9.5	16.0	0.64	97
	670	24.2	49.5	0.31	97
	All	3.0	26.0	0.89	581
	412	−10.7	24.8	0.42	42
	443	2.6	18.2	0.53	42
	490	−0.4	13.2	0.75	42
CNR	510	−3.7	14.9	0.81	42
	555	0.9	19.9	0.88	42
	670	−4.9	83.1	0.90	42
	All	−2.7	21.9	0.87	252

3.4. QAA Performance for b_{bp} Retrievals from CCI Data

After assessing the quality of the QAA retrievals with in situ b_{bp} and the quality of the CCI R_{rs} respect to in situ R_{rs}, the QAA is applied to the CCI R_{rs} to retrieve the b_{bp} that are then compared to the in situ data. In agreement with our findings in Section 3.1, CCI R_{rs} are corrected for Raman scattering. Results of this comparison are shown in Figure 6 and Table 4. For V19, biases are not significant (less than 30%) in comparison of RMS values (less than 60%). On the other hand, similarly to the statistics derived from Section 3.1, QAA-derived b_{bp}, as compared to the BOU data displays significant positive biases. Comparison with CNR data shows the highest performances, with bias of +2.7% and RMS of 48%. The conclusions from our analysis are consistent with previous comparisons to QAA, reporting negligible biases above noise level both at global and regional scales [25,30,51].

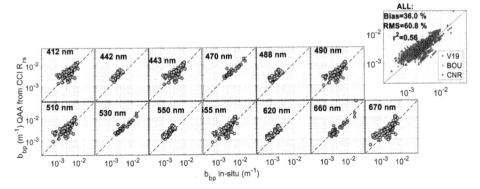

Figure 6. Scatter plots of QAA derived b_{bp} from CCI R_{rs} versus in situ b_{bp} data for each wavelength and dataset considered and for the merged data set. The dashed line represents the 1:1 ratio. Pink, yellow, and green dots refer to V19, BOU, and CNR data, respectively.

Table 4. Statistical descriptors of the difference between the QAA-derived b_{bp} from satellite CCI R_{rs} and in situ b_{bp} for each dataset with Raman scattering compensation. Figure A4 provides a graphical representation of this table.

	Band (nm)	Bias (%)	RMS (%)	r^2	N
	412	24.2	51.8	0.66	147
	443	26.8	53.9	0.67	147
	490	29.1	56.0	0.68	147
V19	510	29.9	56.8	0.67	147
	555	31.0	58.1	0.67	147
	670	31.6	60.7	0.62	147
	All	28.8	56.3	0.68	882
	442	56.6	62.7	0.67	97
	488	86.9	96.2	0.64	97
BOU	550	41.9	50.2	0.70	97
	620	66.8	75.3	0.69	97
	All	63.1	73.1	0.69	388
	470	10.1	52.9	0.48	42
CNR	530	7.8	54.9	0.46	42
	660	−9.6	33.5	0.63	42
	All	2.7	48.1	0.50	126

4. Conclusions

The main findings of this work and their relevance for ocean color studies are summarized here:

(1) Raman scattering compensation of R_{rs} prior to the application of the QAA significantly reduces errors in the retrieval of b_{bp} with respect to in situ b_{bp}. Inclusion of this processing step in operational schemes is recommended.

(2) The QAA-derived b_{bp} from in situ radiometry has negligible biases with respect to in situ b_{bp}.

(3) CCI R_{rs} shows low biases but higher RMS differences with respect to in situ data, that could be excessive for the monitoring of natural change over short periods. Here, the standardization of in

situ radiometry protocols is highly encouraged [52], in order to reduce the errors when in situ datasets formed by multiple contributors are merged and used for R_{rs} matchup analysis.

(4) In part as a consequence of the findings above, QAA-derived b_{bp} from CCI R_{rs} displays negligible biases respect to in situ b_{bp}, with moderately low RMS errors.

(5) The in situ radiometry-derived spectral backscattering slope (η) has low predictive value as compared to η derived from b_{bp} matchups. In this context, the impact of using the best fitted curve instead of the widely used expression [22] is negligible, thus validating the application of the latter without its retuning.

Notwithstanding these results, one future challenge should be to evaluate the impact of two other sources of inelastic scattering before the application of QAA on R_{rs}: (i) red fluorescence, caused by chlorophyll, that usually plays an important role around the peak close to 685 nm; and (ii) the blue fluorescence, caused by CDOM, that can be relevant close to the peak at 425 nm [53].

In addition, there is the need of increasing the amount of spatial and spectral coverage of high-quality in situ b_{bp} observations. As of today, available multispectral b_{bp} is limited to a small number of ship-borne data, or longer datasets but in fixed points (i.e., buoy). On the other hand, Biogeochemical-Argo floats cover large areas but their data are mainly given at a single band. Therefore, there is need to significantly increase the amount of b_{bp} data at multiple bands, seasons and geographical regions. New technological developments on autonomous platforms will aid to enhance data density across many water types, to extend the CCI uncertainty derivation approach to b_{bp} as well, thus allowing the mapping of uncertainties for every b_{bp} product.

Lastly, in situ b_{bp} measurements lag behind the standards on protocols and uncertainty characterization with respect to other quantities such as the radiometry [52]. Only when in situ uncertainty-characterized datasets, from instrument characterization to deployment [54], become available, more detailed algorithm validation could be performed and this will help to better evaluate the influence of optically active constituents (e.g., CDOM, chlorophyll).

Author Contributions: Conceptualization, J.P., M.B., and S.M.; Methodology, J.P., M.B., and S.M.; Data curation, J.P., E.O., G.V., S.C., V.V.; Writing-review and editing, all authors contributed equally. All authors have read and agreed to the published version of the manuscript.

Funding: J.P. has received partial funding from the 'Coastal Ocean Darkening' project funded by the Ministry for Science and Culture of Lower Saxony, Germany (VWZN3175). M.B. has a postdoctoral fellowship by the European Space Agency (ESA). This work was supported by the ESA Living Planet Fellowship Project PHYSIOGLOB: Assessing the inter-annual physiological response of phytoplankton to global warming using long-term satellite observations, 2018–2020.

Acknowledgments: This is a contribution to the ESA Ocean Colour Climate Change Initiative of the European Space Agency. ESA and CNES are thanked for funding the BOUSSOLE project. We are grateful to all the contributors to the V19 dataset. J.P. thanks CNR-ISMAR for the stay at the Rome CNR location in which this manuscript was elaborated. Finally, we wish to thank four anonymous reviewers for their criticisms and suggestions that helped the manuscript to be improved.

Conflicts of Interest: The authors declare no conflict of interest.

Appendix A

This appendix includes the graphical representation of the information in Tables 1–4, for a quicker interpretation of the results and the derived conclusions. The error bars are made by taking the mean bias in tables as central values and the standard deviation (σ) as bar width, calculated as $\sigma = \sqrt{RMS^2 - Bias^2}$. When error bars intersect the zero-difference line, the differences are assumed to be not significant.

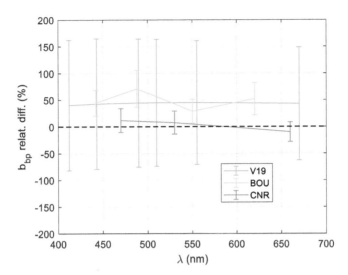

Figure A1. Relative differences between the QAA-derived b_{bp} and in situ b_{bp} for each dataset, without Raman scattering compensation. Data taken from Table 1. Pink, yellow, and green lines refer to V19, BOU, and CNR data, respectively.

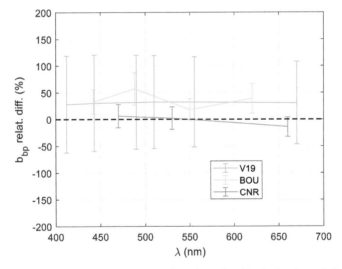

Figure A2. Relative differences between the b_{bp}-QAA derived and in situ b_{bp} for each dataset with Raman scattering compensation. Data taken from Table 2. Pink, yellow and green lines refer to V19, BOU, and CNR data, respectively.

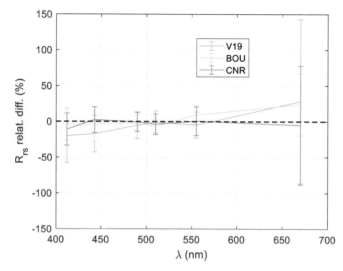

Figure A3. Relative differences between satellite CCI R_{rs} and in situ R_{rs} for each dataset. Data taken from Table 3. Pink, yellow, and green lines refer to V19, BOU, and CNR data, respectively.

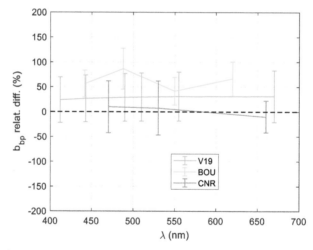

Figure A4. Relative difference between the QAA-derived b_{bp} from satellite CCI R_{rs} and in situ b_{bp} for each dataset with Raman scattering compensation. Data taken from Table 1. Pink, yellow, and green lines refer to V19, BOU, and CNR data, respectively.

References

1. Behrenfeld, M.J.; Boss, E.; Siegel, D.A.; Shea, D.M. Carbon-based ocean productivity and phytoplankton physiology from space. *Glob. Biogeochem. Cycles* **2005**, *19*. [CrossRef]
2. Bellacicco, M.; Volpe, G.; Colella, S.; Pitarch, J.; Santoleri, R. Influence of photoacclimation on the phytoplankton seasonal cycle in the Mediterranean Sea as seen by satellite. *Remote Sens. Environ.* **2016**, *184*, 595–604. [CrossRef]
3. Lee, Z.; Carder, K.L.; Arnone, R.A. Deriving inherent optical properties from water color: A multiband quasi-analytical algorithm for optically deep waters. *Appl. Opt.* **2002**, *41*, 5755–5772. [CrossRef] [PubMed]

4. Loisel, H.; Stramski, D.; Dessailly, D.; Jamet, C.; Li, L.; Reynolds, R.A. An Inverse Model for Estimating the Optical Absorption and Backscattering Coefficients of Seawater from Remote-Sensing Reflectance Over a Broad Range of Oceanic and Coastal Marine Environments. *J. Geophys. Res. Oceans* **2018**, *123*, 2141–2171. [CrossRef]

5. Brewin, R.J.W.; Sathyendranath, S.; Müller, D.; Brockmann, C.; Deschamps, P.-Y.; Devred, E.; Doerffer, R.; Fomferra, N.; Franz, B.; Grant, M.; et al. The Ocean Colour Climate Change Initiative: III. A round-robin comparison on in-water bio-optical algorithms. *Remote Sens. Environ.* **2015**, *162*, 271–294. [CrossRef]

6. Twardowski, M.S.; Boss, E.; Macdonald, J.B.; Pegau, W.S.; Barnard, A.H.; Zaneveld, J.R.V. A model for estimating bulk refractive index from the optical backscattering ratio and the implications for understanding particle composition in case I and case II waters. *J. Geophys. Res. Oceans* **2001**, *106*, 14129–14142. [CrossRef]

7. Slade, W.H.; Boss, E. Spectral attenuation and backscattering as indicators of average particle size. *Appl. Opt.* **2015**, *54*, 7264–7277. [CrossRef]

8. Organelli, E.; Dall'Olmo, G.; Brewin, R.J.W.; Tarran, G.A.; Boss, E.; Bricaud, A. The open-ocean missing backscattering is in the structural complexity of particles. *Nat. Commun.* **2018**, *9*, 5439. [CrossRef]

9. Stramski, D.; Kiefer, D.A. Light scattering by microorganisms in the open ocean. *Prog. Oceanogr.* **1991**, *28*, 343–383. [CrossRef]

10. Morel, A.; Ahn, Y.-H. Optics of heterotrophic nanoflagellates and ciliates: A tentative assessment of their scattering role in oceanic waters compared to those of bacterial and algal cells. *J. Mar. Res.* **1991**, *49*, 177–202. [CrossRef]

11. Stramski, D.; Boss, E.; Bogucki, D.; Voss, K.J. The role of seawater constituents in light backscattering in the ocean. *Prog. Oceanogr.* **2004**, *61*, 27–56. [CrossRef]

12. Thomalla, S.J.; Ogunkoya, A.G.; Vichi, M.; Swart, S. Using Optical Sensors on Gliders to Estimate Phytoplankton Carbon Concentrations and Chlorophyll-to-Carbon Ratios in the Southern Ocean. *Front. Mar. Sci.* **2017**, *4*. [CrossRef]

13. Loisel, H.; Bosc, E.; Stramski, D.; Oubelkheir, K.; Deschamps, P.-Y. Seasonal variability of the backscattering coefficient in the Mediterranean Sea based on satellite SeaWiFS imagery. *Geophys. Res. Lett.* **2001**, *28*, 4203–4206. [CrossRef]

14. Kostadinov, T.S.; Siegel, D.A.; Maritorena, S. Retrieval of the particle size distribution from satellite ocean color observations. *J. Geophys. Res. Oceans* **2009**, *114*. [CrossRef]

15. Kostadinov, T.S.; Milutinović, S.; Marinov, I.; Cabré, A. Carbon-based phytoplankton size classes retrieved via ocean color estimates of the particle size distribution. *Ocean Sci.* **2016**, *12*, 561–575. [CrossRef]

16. Behrenfeld, M.J.; O'Malley, R.T.; Boss, E.S.; Westberry, T.K.; Graff, J.R.; Halsey, K.H.; Milligan, A.J.; Siegel, D.A.; Brown, M.B. Revaluating ocean warming impacts on global phytoplankton. *Nat. Clim. Chang.* **2015**, *6*, 323–330. [CrossRef]

17. Bellacicco, M.; Volpe, G.; Briggs, N.; Brando, V.; Pitarch, J.; Landolfi, A.; Colella, S.; Marullo, S.; Santoleri, R. Global Distribution of Non-algal Particles From Ocean Color Data and Implications for Phytoplankton Biomass Detection. *Geophys. Res. Lett.* **2018**, *45*, 7672–7682. [CrossRef]

18. Martínez-Vicente, V.; Evers-King, H.; Roy, S.; Kostadinov, T.S.; Tarran, G.A.; Graff, J.R.; Brewin, R.J.W.; Dall'Olmo, G.; Jackson, T.; Hickman, A.E.; et al. Intercomparison of Ocean Color Algorithms for Picophytoplankton Carbon in the Ocean. *Front. Mar. Sci.* **2017**, *4*. [CrossRef]

19. Siegel, D.A.; Behrenfeld, M.J.; Maritorena, S.; McClain, C.R.; Antoine, D.; Bailey, S.W.; Bontempi, P.S.; Boss, E.S.; Dierssen, H.M.; Doney, S.C.; et al. Regional to global assessments of phytoplankton dynamics from the SeaWiFS mission. *Remote Sens. Environ.* **2013**, *135*, 77–91. [CrossRef]

20. Halsey, K.H.; Jones, B.M. Phytoplankton Strategies for Photosynthetic Energy Allocation. *Annu. Rev. Mar. Sci.* **2015**, *7*, 265–297. [CrossRef]

21. Mobley, C.D. *Light and Water: Radiative Transfer in Natural Waters*; Academic Press: Cambridge, MA, USA, 1994.

22. Lee, Z. Update of the Quasi-Analytical Algorithm (QAA_v6). Available online: http://www.ioccg.org/groups/Software_OCA/QAA_v6_2014209.pdf (accessed on 22 December 2019).

23. Huot, Y.; Morel, A.; Twardowski, M.S.; Stramski, D.; Reynolds, R.A. Particle optical backscattering along a chlorophyll gradient in the upper layer of the eastern South Pacific Ocean. *Biogeosciences* **2008**, *5*, 495–507. [CrossRef]

24. Brewin, R.J.W.; Dall'Olmo, G.; Sathyendranath, S.; Hardman-Mountford, N.J. Particle backscattering as a function of chlorophyll and phytoplankton size structure in the open-ocean. *Opt. Express* **2012**, *20*, 17632–17652. [CrossRef] [PubMed]

25. Mélin, F. Comparison of SeaWiFS and MODIS time series of inherent optical properties for the Adriatic Sea. *Ocean Sci.* **2011**, *7*, 351–361. [CrossRef]

26. Werdell, P.J.; Bailey, S.W. An improved in situ bio-optical data set for ocean color algorithm development and satellite data product validation. *Remote Sens. Environ.* **2005**, *98*, 122–140. [CrossRef]

27. Werdell, P.J.; Franz, B.A.; Lefler, J.T.; Robinson, W.D.; Boss, E. Retrieving marine inherent optical properties from satellites using temperature and salinity-dependent backscattering by seawater. *Opt. Express* **2013**, *21*, 32611–32622. [CrossRef] [PubMed]

28. Zheng, G.; Stramski, D.; Reynolds, R.A. Evaluation of the Quasi-Analytical Algorithm for estimating the inherent optical properties of seawater from ocean color: Comparison of Arctic and lower-latitude waters. *Remote Sens. Environ.* **2014**, *155*, 194–209. [CrossRef]

29. Pitarch, J.; Bellacicco, M.; Volpe, G.; Colella, S.; Santoleri, R. Use of the quasi-analytical algorithm to retrieve backscattering from in situ data in the Mediterranean Sea. *Remote Sens. Lett.* **2016**, *7*, 591–600. [CrossRef]

30. Bisson, K.M.; Boss, E.; Westberry, T.K.; Behrenfeld, M.J. Evaluating satellite estimates of particulate backscatter in the global open ocean using autonomous profiling floats. *Opt. Express* **2019**, *27*, 30191–30203. [CrossRef]

31. Sathyendranath, S.; Brewin, R.J.W.; Brockmann, C.; Brotas, V.; Calton, B.; Chuprin, A.; Cipollini, P.; Couto, A.B.; Dingle, J.; Doerffer, R.; et al. An Ocean-Colour Time Series for Use in Climate Studies: The Experience of the Ocean-Colour Climate Change Initiative (OC-CCI). *Sensors* **2019**, *19*, 4285. [CrossRef]

32. Bellacicco, M.; Cornec, M.; Organelli, E.; Brewin, R.J.W.; Neukermans, G.; Volpe, G.; Barbieux, M.; Poteau, A.; Schmechtig, C.; D'Ortenzio, F.; et al. Global Variability of Optical Backscattering by Non-algal particles from a Biogeochemical-Argo Data Set. *Geophys. Res. Lett.* **2019**, *46*, 9767–9776. [CrossRef]

33. Valente, A.; Sathyendranath, S.; Brotas, V.; Groom, S.; Grant, M.; Taberner, M.; Antoine, D.; Arnone, R.; Balch, W.M.; Barker, K.; et al. A compilation of global bio-optical in situ data for ocean-colour satellite applications—version two. *Earth Syst. Sci. Data* **2019**, *11*, 1037–1068. [CrossRef]

34. Antoine, D.; Chami, M.; Claustre, H.; d'Ortenzio, F.; Morel, A.; Bécu, G.; Gentili, B.; Louis, F.; Ras, J.; Roussier, E. *BOUSSOLE: A Joint CNRS-INSU, ESA, CNES, and NASA Ocean Color Calibration and Validation Activity*; NASA: Washington, DC, USA, 2006.

35. Bracaglia, M.; Volpe, G.; Colella, S.; Santoleri, R.; Braga, F.; Brando, V.E. Using overlapping VIIRS scenes to observe short term variations in particulate matter in the coastal environment. *Remote Sens. Environ.* **2019**, *233*, 111367. [CrossRef]

36. Zhang, X.; Hu, L.; He, M.-X. Scattering by pure seawater: Effect of salinity. *Opt. Express* **2009**, *17*, 5698–5710. [CrossRef] [PubMed]

37. Huber, P.J.; Ronchetti, E.M. *Robust Statistics*; John Wiley & Sons: New York, NY, USA, 1981.

38. Moré, J.J.; Sorensen, D.C. Computing a Trust Region Step. *SIAM J. Sci. Stat. Comput.* **1983**, *4*, 553–572. [CrossRef]

39. Westberry, T.K.; Boss, E.; Lee, Z. Influence of Raman scattering on ocean color inversion models. *Appl. Opt.* **2013**, *52*, 5552–5561. [CrossRef]

40. Lee, Z.; Hu, C.; Shang, S.; Du, K.; Lewis, M.; Arnone, R.; Brewin, R. Penetration of UV-visible solar radiation in the global oceans: Insights from ocean color remote sensing. *J. Geophys. Res. Oceans* **2013**, *118*, 4241–4255. [CrossRef]

41. Antoine, D.; Guevel, P.; Desté, J.-F.; Bécu, G.; Louis, F.; Scott, A.J.; Bardey, P. The "BOUSSOLE" Buoy—A New Transparent-to-Swell Taut Mooring Dedicated to Marine Optics: Design, Tests, and Performance at Sea. *J. Atmos. Ocean. Technol.* **2008**, *25*, 968–989. [CrossRef]

42. Antoine, D.; d'Ortenzio, F.; Hooker, S.B.; Bécu, G.; Gentili, B.; Tailliez, D.; Scott, A.J. Assessment of uncertainty in the ocean reflectance determined by three satellite ocean color sensors (MERIS, SeaWiFS and MODIS-A) at an offshore site in the Mediterranean Sea (BOUSSOLE project). *J. Geophys. Res. Oceans* **2008**, *113*. [CrossRef]

43. Mélin, F.; Sclep, G. Band shifting for ocean color multi-spectral reflectance data. *Opt. Express* **2015**, *23*, 2262–2279. [CrossRef]

44. Valente, A.; Sathyendranath, S.; Brotas, V.; Groom, S.; Grant, M.; Taberner, M.; Antoine, D.; Arnone, R.; Balch, W.M.; Barker, K.; et al. Inherent optical properties and diffuse attenuation coefficient aggregated within +/−6 nm of SeaWiFS, MODIS-AQUA, VIIRS, OLCI and MERIS bands, corrected Version 2019-06-12. In *A Compilation of Global Bio-Optical in Situ Data for Ocean-Colour Satellite Applications—Version Two*; Valente, A., Ed.; Pangaea: Bremen, Germany, 2019. [CrossRef]

45. Antoine, D.; Siegel, D.A.; Kostadinov, T.; Maritorena, S.; Nelson, N.B.; Gentili, B.; Vellucci, V.; Guillocheau, N. Variability in optical particle backscattering in contrasting bio-optical oceanic regimes. *Limnol. Oceanogr.* **2011**, *56*, 955–973. [CrossRef]

46. Organelli, E.; Bricaud, A.; Gentili, B.; Antoine, D.; Vellucci, V. Retrieval of Colored Detrital Matter (CDM) light absorption coefficients in the Mediterranean Sea using field and satellite ocean color radiometry: Evaluation of bio-optical inversion models. *Remote Sens. Environ.* **2016**, *186*, 297–310. [CrossRef]

47. Volpe, G.; Colella, S.; Brando, V.E.; Forneris, V.; La Padula, F.; Di Cicco, A.; Sammartino, M.; Bracaglia, M.; Artuso, F.; Santoleri, R. Mediterranean ocean colour Level 3 operational multi-sensor processing. *Ocean Sci.* **2019**, *15*, 127–146. [CrossRef]

48. Jackson, T.; Chuprin, A.; Sathyendranath, S.; Grant, M.; Zühlke, M.; Dingle, J.; Storm, T.; Boettcher, M.; Fomferra, N. Ocean Colour Climate Change Initiative (OC_CCI)—Interim Phase. Product User Guide, D3.4 PUG. 2019. Available online: https://esa-oceancolour-cci.org/sites/esa-oceancolour-cci.org/alfresco.php?file=a68aa514-3668-4935-9235-fca10f7e8bee&name=OC-CCI-PUG-v4.1-v1.pdf (accessed on 22 December 2019).

49. Mélin, F.; Chuprin, A.; Grant, M.; Jackson, T.; Sathyendranath, S. *Ocean Colour Climate Change Initiative (OC_CCI)—Phase Two*; Ocean Colour Data Bias Correction and Merging D2.6; Plymouth Marine Laboratory: Plymouth, UK, 2016; Volume 35.

50. Lee, Z.; Huot, Y. On the non-closure of particle backscattering coefficient in oligotrophic oceans. *Opt. Express* **2014**, *22*, 29223–29233. [CrossRef]

51. Mélin, F.; Zibordi, G.; Berthon, J.-F. Assessment of satellite ocean color products at a coastal site. *Remote Sens. Environ.* **2007**, *110*, 192–215. [CrossRef]

52. Zibordi, G.; Voss, K.J.; Johnson, B.C.; Mueller, J.L. (Eds.) Protocols for Satellite Ocean Colour Data Validation: In Situ Optical Radiometry (Volume 3.0). In *IOCCG Ocean Optics and Biogeochemistry Protocols for Satellite Ocean Colour Sensor Validation*; IOCCG Protocol Series; IOCCG: Dartmouth, NS, Canada, 2019; Volume 3. [CrossRef]

53. Haltrin, V.I.; Kattawar, G.W.; Weidemann, A.D. Modeling of elastic and inelastic scattering effects in oceanic optics. In Proceedings of the Ocean Optics XIII (1997), Halifax, NS, Canada, 6 February 1997.

54. Dall'Olmo, G.; Westberry, T.K.; Behrenfeld, M.J.; Boss, E.; Slade, W.H. Significant contribution of large particles to optical backscattering in the open ocean. *Biogeosciences* **2009**, *6*, 947–967. [CrossRef]

Letter

The Difference of Sea Level Variability by Steric Height and Altimetry in the North Pacific

Qianran Zhang [1], Fangjie Yu [1,2,*] and Ge Chen [1,2]

1   College of Information Science and Engineering, Ocean University of China, Qingdao 266100, China;
     zqr@stu.ouc.edu.cn (Q.Z.); gechen@ouc.edu.cn (G.C.)
2   Laboratory for Regional Oceanography and Numerical Modeling, Qingdao National Laboratory for Marine
     Science and Technology, Qingdao 266200, China
*   Correspondence: yufangjie@ouc.edu.cn; Tel.: +86-0532-66782155

Received: 4 December 2019; Accepted: 22 January 2020; Published: 24 January 2020

Abstract: Sea level variability, which is less than ~100 km in scale, is important in upper-ocean circulation dynamics and is difficult to observe by existing altimetry observations; thus, interferometric altimetry, which effectively provides high-resolution observations over two swaths, was developed. However, validating the sea level variability in two dimensions is a difficult task. In theory, using the steric method to validate height variability in different pixels is feasible and has already been proven by modelled and altimetry gridded data. In this paper, we use Argo data around a typical mesoscale eddy and altimetry along-track data in the North Pacific to analyze the relationship between steric data and along-track data (SD-AD) at two points, which indicates the feasibility of the steric method. We also analyzed the result of SD-AD by the relationship of the distance of the Argo and the satellite in Point 1 (P_1) and Point 2 (P_2), the relationship of two Argo positions, the relationship of the distance between Argo positions and the eddy center and the relationship of the wind. The results showed that the relationship of the SD-AD can reach a correlation coefficient of ~0.98, the root mean square deviation (RMSD) was ~1.8 cm, the bias was ~0.6 cm. This proved that it is feasible to validate interferometric altimetry data using the steric method under these conditions.

Keywords: steric height; sea level variability; interferometric altimeter validation

1. Introduction

Since the first radar altimetric satellites were launched in the 1970s, altimetric data have been widely used for understanding the ocean, including eddies and sea level change [1]. Since 1992, more than two radar altimetry datasets were able to be used at the same time. However, even when combining the data, it is difficult to observe ocean dynamic variability less than an ~150 km scale because of the traditional altimetry only has the along-track data [2]. Compared with traditional radar altimetry, new-generation interferometric altimetry, which is onboard satellites such as the Surface Water and Ocean Topography (SWOT) or "Guanlan" mission [3,4], can yield high spatial resolutions over two swaths to obtain sea level variability in two dimensions with unprecedented spatial resolution in the swath and give finer observations of the ocean phenomenon [5]. Meanwhile, to improve data accuracy, the calibration and validation (Cal/Val) of altimetry data are important. Fixed platforms, such as Harvest, Corsica, Gavdos and the Bass Strait, are widely used to validate radar altimetry along-track data [6–9]; this approach extrapolated the onshore sea surface height (SSH) out to the offshore nadir point with an accuracy of (1.88 ± 0.20) cm and a standard deviation of 3.3 cm, which suggested that the approach presented was feasible in absolute altimeter Cal/Val [10]. However, the validation of interferometric altimetry is not only for along-track data; it is a validation of spatial resolution such as validate sea level variability between the different pixels at the same time in a swath. Now, the research on the Cal/Val methods of interferometric altimetry mainly focuses on the offshore method [11] and the steric method [12,13]. There is no perfect solution. The offshore method was developed to extend the single-point approach to a wider regional scale. In this paper, the study is based on the steric method, which uses the observed data to calculate the steric height and validate the sea level variability. Wang et al. [13] proved that using the steric method to validate the interferometric altimetry data by 20 glider and mooring data in the swath area was feasible based on the model data. However, the model data and the theoretical analysis are not sufficient to prove that it is feasible to validate sea level variability between different pixels in a swath by the steric method.

The variability in sea surface height observed by an altimeter has two causes: the steric signal and the non-steric height (NSH) signal [14]. The sea surface height (SSH) variability is largely dominated by the steric signal within a small area [14–16]. Meanwhile, the previous studies have greatly contributed to the relationship between the steric height (SH) and the sea level. Roemmich et al. [17] used the 10-yr SH increase and the 22-yr SSH increase data for analyzing the interannual trends in the South Pacific and found an increase difference of approximately 3 cm. Meyssignac et al. [18] studied the relationship between sea level and SH in different areas and proved that trends were largely dominated by SH except in high-latitude areas (> 60°N and < 55°S) and some shallow shelf seas. For seasonal and inter-annual period variability, Dhomps et al. [19] showed that sea level variability is clearly dominated by baroclinic motion. Regarding wind and rainfall, Perigaud et al. [20] used the model to analyze the influence of rainfall anomalies on sea level variability due to the SH. Song et al. [21] analyzed the wind force influence on the sea level variability and proved that the annual SSH variability is mostly due to steric changes. Therefore, the SH variability is an important part of sea level variability [22].

However, studying long-term scale and large areas of the sea level variability were not enough to explain the feasibility of using the steric method to validate the sea level variability in two dimensions. In this paper, we used field observations to obtain the relationship between steric data and along-track data (SD-AD) at two points. There were 17 Argo stations data for captured the mesoscale eddy and the altimetric along-track data to analyze the feasibility of using the SH to validate the interferometric altimetry data. We used the experimental results to analyze the SD-AD under different factors. The altimetry data, Argo data and the methods are introduced in Section 2. The results are presented in Section 3. The discussion is presented in Section 4 and is followed by conclusions.

2. Materials and Methods

2.1. Argo Profile

Argo, as the broad-scale global array of profiling floats, has already grown to be a major component of the ocean observing system. The Argo data can be used for measuring high-vertical-resolution temperature-salinity-pressure (T-S-P) profiles that, in turn, can be used to compute ocean density and SH for ocean monitoring and study [23]. The Argo was deployed by the rate of about 800 per year since 2000. Now, the global array has about 3000 floats and distributed approximately every 3 degrees (300 km). Floats can be observed at the depth up to 2000 m every 10 days, with 4–5-year lifetimes for individual instruments. In this paper, the data were downloaded from the French Research Institute for Exploitation of the Sea (IFREMER) Global Data Assembly Centres (GDAC) Server (ftp.ifremer.fr/ifremer/argo/), which has passed through real-time quality control procedures, and most of the data have passed through delayed-mode procedures.

In order to find the Argo data with a larger number in a small area to match the satellite data, we used data from 17 Argo stations (numbers 2901550 to 2901566), which capture the eddy south of the Kuroshio extension located east of Japan, as shown in Figure 1a. The eddy was detected by the Okubo–Weiss (OW) method using the AVISO gridded data [24], and the eddy information is also obtained from the Okubo-Weiss method like the eddy radius. The eddy passing area was getting from the eddy radius and the eddy center position. In this paper, the Argo was in operation for more than one year, starting in March 2014 and the longest running time reaching June 2015. The profile sampling interval was one day, an anticyclonic eddy (AE) moved through the Argo track comprising more than

3000 hydrographic profiles following the AE [25]. The T-S-P profiles measured pressure at depths up to 1000 m, and the profiles were divided into more than 350 levels.

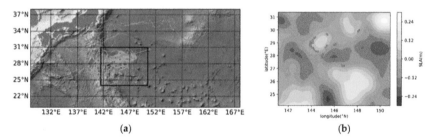

(a)   (b)

Figure 1. Argo and eddy data. (**a**) The coral lines are the Argo data with numbers from 2901550 to 2901566, the red line is the track of the eddy center, and the green area is the eddy passing area. The base map shows the depth of the ocean, and the sea level anomaly (SLA) of the black box is shown in Figure 1 (**b**). (**b**) The Argo position on May 15, 2014 and the base map show the performance of SLA.

2.2. Altimetry Data

In this paper, the altimetry along-track data were provided by Copernicus Marine Environment Monitoring Service (CMEMS) project, which includes the SLA and absolute dynamic topography (ADT) [26]. As shown in Figure 2, the different databases have the different reference level. The SSH is the sea level relative to the reference ellipsoid, which was obtained by the satellite altitude minus the altimetric range observed by the altimeter; the SLA was the sea level relative to the mean sea surface height (MSSH), and the ADT was the sea surface height above the Geoid [27]. Moreover, the reference ellipsoid is a regular ellipsoid that fits the earth with the equatorial radius and flattening coefficient. The geoid is a gravity equipotential surface that would correspond with the ocean surface if ocean was at rest, and the MSSH is the long-term sea surface height average. The altimetry data that we used were near real-time reanalysis along-track data with delivery times up to 6 months. Figure 3 shows the along-track data of which the spans of the two data were 14 km. With an objective to match the time of the Argo data, we selected the Jason-2 data from 2014 and 2015, which was derived from a satellite mission from 2008 to 2017.

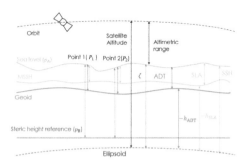

Figure 2. The relationships of sea level variability under different reference levels. The sea surface height (SSH) measured by the altimeter was the sea level reference for the reference ellipsoid, the sea level anomaly (SLA) was the sea level reference for the mean sea surface height (MSSH) and the absolute dynamic topography (ADT) was the sea surface height above the Geoid. The steric height (SH) was caused by the change of the density between the sea level and the reference level.

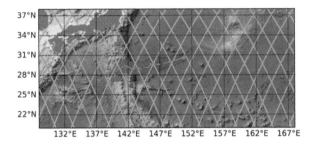

Figure 3. Satellite along-track data and eddy track data. The coral points represent the satellite position. The red points are the eddy center, and the red star is the start position of the eddy. The green areas are the eddy passing area.

2.3. Wind Data

In this paper, the wind data were provided by the European Centre for Medium Weather Forecasting (ECMWF) ERA-Interim data. The data assimilation system used to produce ERA-Interim is based on a 2006 release of the IFS (Cy31r2). The system includes a four-dimensional variational analysis (4D-Var) with a 12-hour analysis window. The spatial resolution of the data set is approximately 80 km (T255 spectral) on 60 vertical levels from the surface up to 0.1 hPa. In order to match the Argo data, we select the ERA-Interim data from 2014 and 2015, and the data were gridded data which size of 0.125 degrees, with one datum every 6 hours.

2.4. Method

In this paper, our aim was to determine the feasibility of using the SH to validate sea level variability between different pixels in the swath of the interferometric altimeter, and find the relationship between the height variability of the altimetry data and the SH variability at two points. To ensure the influencing factors of the relationship between the SD-AD, we conducted the theoretical analysis first.

As far as physical oceans are concerned, the hydrostatic equation is as follows:

$$\frac{\partial p}{\partial z} = -\rho g; \ \rho = \rho_0 + \rho\prime \tag{1}$$

where p is the pressure, ρ is the potential density, ρ_0 is the reference potential density (1027.5 kgm^{-3}), $\rho\prime$ is the potential density anomaly and g is the gravity of Earth. The equation also can be written as follows:

$$p_A - p_B = -\rho_0 g(\zeta + h) - \int_{-h}^{\zeta} \rho\prime g dz \tag{2}$$

where p_B is the ocean bottom pressure, p_A is the atmospheric surface pressure, ζ is the SSH referenced to $z = 0$ and $-h$ is the depth of the ocean. After rearrangement, one obtains:

$$\zeta = \frac{p'_B}{\rho_0 g} - \frac{p_A}{\rho_0 g} - \int_{-h}^{0} \frac{\rho\prime}{\rho_0} dz \tag{3}$$

where $p'_B = p_B - \rho_0 g(\zeta + h)$ represents the bottom pressure anomaly and the term $\int_0^{\zeta} \frac{\rho\prime}{\rho_0} dz$ is neglected because $\zeta \ll h$. Therefore, the variable of ζ at two points is:

$$D_\zeta = \zeta_1 - \zeta_2 = \frac{p'_{B1} - p'_{B2}}{\rho_0 g} - \frac{p_{A1} - p_{A2}}{\rho_0 g} - \int_{-h}^{0} \frac{\rho\prime_1 - \rho\prime_2}{\rho_0} dz \tag{4}$$

In this paper, we used the SLA and the ADT to represent the ζ as the satellite data and the p_2 was the MSSH and geoid, respectively.

SH is not exactly the height (in meters) of an isobaric surface above the geopotential surface, which is Φ. The Φ is:

$$\Phi = \int_0^z g dz \tag{5}$$

To calculate geostrophic currents, oceanographers use a modified form of the hydrostatic equation. The vertical pressure gradient (1) is written as follows:

$$\frac{\delta p}{\rho} = \alpha \delta p = -g \delta z; \alpha \delta p = \delta \Phi \tag{6}$$

where Φ is the geopotential surface, $\Phi = \int_0^z g dz$, α is the specific volume and $SH = \Phi/g$ (in SI units) has almost the same numerical value as height in metres.

At Point 1 (P_1), the SH is:

$$SH = \frac{1}{g} \int_{p_B}^{p_A} \alpha(35, 0, p) dp + \frac{1}{g} \int_{p_B}^{p_A} \alpha' dp \tag{7}$$

where α (35, 0, p) is the specific volume of sea water with a salinity of 35, temperature of 0 °C and pressure p. In this paper, in order to ensure the consistency of the depth in different profiles, 900 meters is selected as the p_2 value in the SH equation. The second term α' is the specific volume anomaly. Therefore, the variable of *SH* at two points is:

$$D_{SH} = SH_1 - SH_2 = \frac{1}{g} \int_{p_B}^{p_A} \alpha'_1 - \alpha'_2 dp = -\int_{-h}^{\zeta} \frac{\rho\prime_1 - \rho\prime_2}{\rho_0} dz \tag{8}$$

Thus, the difference between the SH and the satellite data is caused by the bottom pressure variable and the atmospheric surface pressure variable.

In order to improve the reliability of the results, we chose the SH and steric height anomaly (SHA) as the Argo data source to compare the satellite data (ADT and SLA) to study the feasibility of using the steric method to validate the interferometric altimetry. The mean steric height (MSH) was calculated by the World Ocean Atlas 2013 version 2 monthly fields (WOA 13). Thus, the difference between the two points is as follows:

$$D_{SLA} = SLA_1 - SLA_2 = SSH_1 - MSSH_1 - (SSH_2 - MSSH_2) = SSH_1 - SSH_2 - MSSH_1 + MSSH_2 \tag{9}$$

$$D_{ADT} = ADT_1 - ADT_2 = SSH_1 - Geoid_1 - (SSH_2 - Geoid_2) = SSH_1 - SSH_2 - Geoid_1 + Geoid_2 \tag{10}$$

$$D_{SH} = SH_1 - SH_2 = SSH_1 - NSH_1 - (SSH_2 - NSH_2) = SSH_1 - SSH_2 - NSH_1 + NSH_2 \tag{11}$$

$$D_{SHA} = SHA_1 - SHA_2 = SH_1 - MSH_1 - (SH_2 - MSH_2) = SSH_1 - SSH_2 - NSH_1 + NSH_2 - MSH_1 + MSH_2 \tag{12}$$

The difference between the two databases is mainly caused by the barotropic and the difference of the different reference planes. In this paper, the two databases were the satellite data and the Argo data. The satellite data include the SLA and the ADT, which got from the satellite along-track data, and the Argo data include the SH and the SHA. In order to eliminate the impact, we conducted the following study. We initially utilized the satellite along-track data position to match the Argo position based on distance and time; then, we obtained two points as a set of data for which the space-time interval between the Argo position and the satellite position are within a certain range. The set of data seemed to represent P_1 in equations 9–12, and the time matching was done again for each set of data to find point 2 (P_2). Then, we obtained a group of data that included the two satellite positions and two Argo position. According to the time information of this group of data, matching the eddy which was detected by the OW method like the eddy radius. As shown in Figure 4, each group of data had five points, including two Argo data points (blue cross-shaped symbol) as P_1 and P_2, respectively, two

AVISO data points (cyan dots) as P_1 and P_2, respectively, and an eddy centre point (red dot). Thus, we can got four height variables between P_1 and P_2: D_{SH} and D_{SHA}, which were from the Argo points, and D_{SLA} and D_{ADT}, which were from the altimetry data.

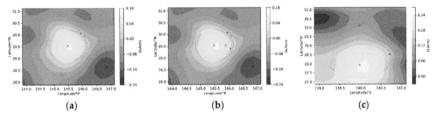

| (a) | (b) | (c) |

Figure 4. The grouped data. Each map has a group of data in which the blue cross-shaped points are the Argo positions, the cyan dot points are along-track positions, the red star is the eddy center and the green circle areas are the eddy areas. The maps (**a**) indicate data groups No. 6, the maps (**b**) indicate data groups No. 13 and the maps (**c**) indicate data groups No. 24. They will be discussed in the next section.

In the next section, we discuss the distance between the Argo position and the along-track position and the influence of barotropy to analyzing the impact of different factors on the SD-AD data.

3. Results

3.1. Distances between the Argo Positions and the Along-Track Positions

With the distance between an Argo position and the satellite position being small, the Argo position and the satellite position can be approximately considered as lying on the same point, and the representative error is reduced. The result is shown in Figure 5 when we used the SH as the Argo database. With the selected data based on the distance between the Argo position and the satellite position less than 0.2 degrees in the latitude and longitude direction and the time being less than four hours, the trends are show in Figure 5a. The maximum difference of the SD-AD reach near 20 cm, and when the satellite database selects different, the SLA performs better than the ADT. To output a better result using the SH to validate the altimetry data, we used more restrictive conditions. As shown in Figure 5b, 26 groups of data were used for the analysis with the distance between the Argo position and the satellite position less than 0.15 degrees in the latitude and longitude direction and the time being less than four hours.

We analyzed the result of the SD-AD by using three different evaluation data; the results are shown in Table 1. When we see the result of the SD-AD under the control condition is 0.2 degrees and four hours, the results indicate that the results of the SD-AD by SLA are better than the result by ADT. When the control condition up to 0.15 degrees, the results of the SD-AD have greatly improved, especially in the bias. The data analysis under the control condition is 0.15 degrees, indicating that for the root mean square deviation (RMSD) and the correlation coefficient, the SLA was more suitable than the ADT, but were different when the bias was used as an evaluation condition. Furthermore, the difference between the SD-AD by ADT and SLA are more similar from the control condition of 0.2 degrees to 0.15 degrees. For the interferometric altimeter, the satellite data can be divided into 5 km * 5 km grids; thus, the observation equipment used for validation can be integrated into the data less than 5 km away from the satellite observation. In addition, the results of using the steric method to validate the sea level variability between different pixels in the interferometric altimetry swath are better than this test because in reality, the limited conditions are stricter than those in this experiment. However, to avoid the samples being too sparse, the distance between the Argo position and the satellite position less than 0.15 degrees in the latitude and longitude direction was chosen as the screening condition in this experiment.

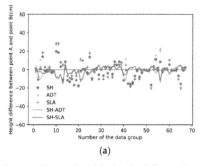 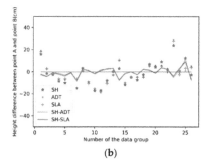

(a)   (b)

Figure 5. The results of the height difference between the P_1 and P_2 by using the SH as the Argo data. The blue star is the height difference between SH at P_1 and P_2, the cyan dots are the height difference between ADT at P_1 and P_2, and the red crosses are the height difference between SLA at P_1 and P_2. The cyan line was the height difference between SD-AD (SH-ADT) at two points, and the red line was the height difference between SD-AD (SH-SLA) at two points. (**a**) The height difference between P_1 and P_2 where the distance between the Argo position and the satellite position less than 0.2 degrees in the latitude and longitude direction with a time difference less than four hours; (**b**) the height difference between the P_1 and P_2 where the distance between the Argo position and the satellite position less than 0.15 degrees in the latitude and longitude direction with a time difference less than four hours.

Table 1. The result of the SD-AD by the Argo data selecting the SH.

Control Condition	Root Mean Square Deviation (RMSD)		Bias		Correlation Coefficient	
0.2 degrees	ADT	SLA	ADT	SLA	ADT	SLA
4 hours	5.25 cm	4.42 cm	1.54 cm	1.18 cm	0.8858	0.9034
0.15 degrees	ADT	SLA	ADT	SLA	ADT	SLA
4 hours	4.07 cm	3.87 cm	0.54 cm	0.57 cm	0.9241	0.9292

When we used the SHA as the Argo database, we conducted the same experiment; the results are shown in Figure 6 and the evaluation data are show in Table 2. Compared with the control condition, it was 0.2 degrees in the latitude and longitude direction; the RMSD, bias and correlation coefficients are greatly improved when the control condition is 0.15 degrees in the latitude and longitude direction. Compared with the SD-AD of the Argo data selecting SH, the result of the SD-AD by the SHA decreased significantly. Moreover, the trend of SD-AD was similar between SD-AD results with two different Argo databases (SH and SHA), although the increase by the SHA is much larger than using the SH as the Argo database. The main reason is that the results of SD-AD by SHA are worse under the control condition of 0.2 degrees. Meanwhile, with the limit condition of the distance between the altimetry and Argo positions decreasing, regardless of the difference between the SHA and SH or the difference between the ADT and SLA, the SD-AD had a significant reduction.

Table 2. The result of the SD-AD by the Argo data select the SHA.

Control Condition	RMSD		BIAS		Correlation Coefficient	
0.2 degrees	ADT	SLA	ADT	SLA	ADT	SLA
4 hours	7.79 cm	6.56 cm	2.46 cm	2.09 cm	0.7393	0.7994
0.15 degrees	ADT	SLA	ADT	SLA	ADT	SLA
4 hours	5.57 cm	5.42 cm	0.54 cm	0.57 cm	0.8862	0.8941

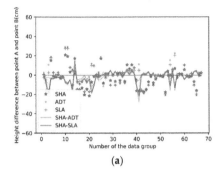
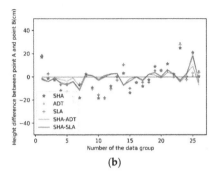

(a)   (b)

Figure 6. The results of the height difference between the P_1 and P_2 by using the SHA as the Argo data. The blue stars are the height difference between SH at P_1 and P_2, the cyan dots are the height difference between ADT at P_1 and P_2, and the red crosses are the height difference between SLA at P_1 and P_2. The cyan line was the height difference between SD-AD (SHA-ADT) at two point, and the red line was the height difference between SD-AD (SHA-SLA) at two point. The faded lines repeating lines from Figure 5 (for SH). (**a**) The height difference between P_1 and P_2 where the distance between the Argo position and the satellite position less than 0.2 degrees in the latitude and longitude direction with a time difference less than four hours; (**b**) the height difference between the P_1 and P_2 where the distance between the Argo position and the satellite position less than 0.15 degrees in the latitude and longitude direction with a time difference less than four hours.

To obtain better SD-AD results, we used more factors to analyze the results. At P_1, there is a distance between the Argo and the satellite. At P_2 there is still a distance between the Argo and the satellite. Figure 7 shows the relationships between the SD-AD as a function of the distance difference between these two distances. Regardless of the SH or SHA, the relationship of the SD-AD had a similar trend; moreover, no matter for the SH and SHA, the ADT data performed better than the SLA data when the distance difference was smaller. However, the results after linear regression were different in terms of the slope, the SHA slope larger than the SH one, thus, the SHA datasets had a clearer trend. In Figure 7b, with the distance different between the distance of the Argo and the satellite in P_1 and P_2 being larger than 13 km, the results became larger than 3 cm for the SHA case. However, there are some special set of data that the SD-AD performed poorly even when the distance difference between P_1 and P_2 was not very large, such as for No. 13. As shown in Figure 4b, the distance difference between P_1 and P_2 were similar; at this time, the directions of the Argo positions relative to the altimetry positions became another contributing factor. At P_1 and P_2, the angle difference between Argo position and satellite positions is close to 180 degrees. However, the sea level variability in the small range always occurred in the same direction. At P_1 and P_2, compared with the position of the Argo point, in which the satellite position in the different direction, the sea level variability was different. Therefore, making the distance difference between P_1 and P_2 similar and the direction between the Argo points and altimetry data points similar are an important condition when using the SD to validate the altimetry data.

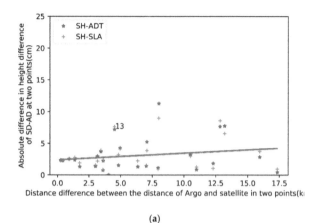

(a)

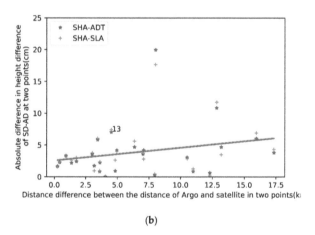

(b)

Figure 7. The relationship of the SD-AD as a function of the distance difference between the distance of the Argo and the satellite in P_1 and P_2, where (**a**) shows the relationship between the SH and the altimetry data; the blue stars show the absolute difference in the height difference between SD-AD (SH-ADT) at two points, and the red crosses show the absolute difference in the height difference between SD-AD (SH-SLA) at two points. (**b**) Shows the relationship between the SHA and the altimetry data; the blue stars show the absolute difference in the height difference between SD-AD (SHA-ADT) at two points, and the red crosses show the absolute difference in the height difference between SD-AD (SHA-SLA) at two points.

3.2. Barotropic Influence

Variability in the SSH can be decomposed into two contributions: barotropic and baroclinic [28]. Steric data can respond well to baroclinic changes, so in order to better reveal the effects of SD-AD, it is necessary to analyze the barotropic influence. In this paper, we analyzed the SD-AD by using three different approaches.

3.2.1. Distance between Two Points

The results of the SD-AD as a function of the distance between two Argo points are shown in Figure 8a,c. We can see that the distribution of the height difference in two points between the positive and negative values is not consistent. And as the distance between the two Argo points became shorter, the results of the SD-AD under the satellite data selecting the ADT data or the SLA data are more similar. This proved that the geoid and MSSH were similar in a small range. With the increase in distance between the two Argo points, it is clear that the SD-AD increased and the result of the SD-AD has a significant linear relationship either selecting the ADT data or the SLA data as the satellite data (Figure 8b,d). When a linear regression analysis was performed on the distance between two Argo points, regarding the result of the SD-AD for the SH or SHA, the ADT data performed well when the distance between two Argo closer, despite the result after linear regression are basically consistent.

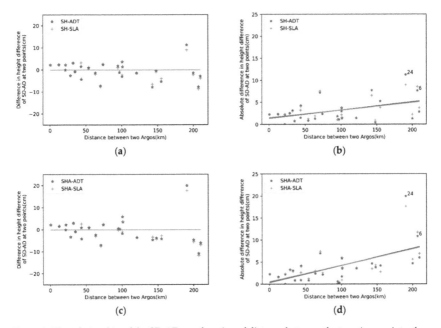

Figure 8. The relationship of the SD-AD as a function of distance between the two Argo points; the y axes of (**a**) and (**c**) are the difference at two points (A and B) of Argo data minus satellite data; the y axes of (**b**) and (**d**) represent the SD-AD. (**a**) and (**b**) are the result of the SD-AD by using the Argo data to select the SH; the blue stars show the difference in the height difference between SD-AD (SH-ADT) at two points, and the red crosses show the difference in the height difference between SD-AD (SH-SLA) at two points. (**c**) and (**d**) are the result of the SD-AD by using the Argo data to select the SHA; the blue stars show the difference in the height difference between SD-AD (SHA-ADT) at two points, and the red crosses show the difference in the height difference between SD-AD (SHA-SLA) at two point.

As shown in Figure 8, when the distance was large enough, the SD-AD data indicated that the values were not good enough to use the SH data to validate the altimetry data, as shown in group data No. 24 and group data No. 6. As shown in Figure 4, the distances between the two Argo points and the corresponding along-track points were 190 km and 207 km, which were too long to assume that the barotropic motion was similar. The trend of the SHA was clearer, which we can see in Figure 8d; with a distance larger than 120 km, it is hard to find a good result of the SD-AD by using the Argo data to select the SHA, which means that the non-steric influence significantly impacts the relationship between the steric height and the sea level.

3.2.2. Distances between the Argo Points and the Eddy Centre

There is a strong correlation between the mesoscale eddy and baroclinic instability [29]. Therefore, we used the sum of the distance of the Argo point and the eddy center in P_1 and P_2 to analyze the influence of the SD-AD. As shown in Figure 9, with the increase of the sum of the distance between an Argo point and the eddy center at two points, the SD-AD also shows an upward trend, and the selected ADT database as the altimetry data was better than the SLA database at smaller distances. The slope of the linear regression analysis result is similar between the SH database and SHA database but different in terms of the ADT and SLA.

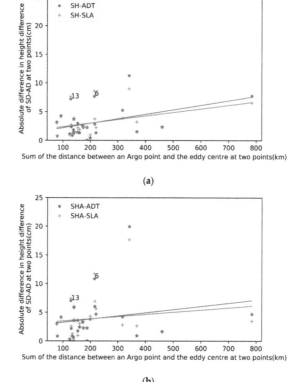

Figure 9. Shows the relationships of the SD-AD as a function of the sum of the distance between an Argo point and the eddy center at two points. (**a**) Shows the relationship between the SH and the altimetry data; the blue stars show the absolute difference in the height difference between SD-AD (SH-ADT) at two points, and the red crosses show the absolute difference in the height difference between SD-AD (SH-SLA) at two points. (**b**) The relationship between the SHA and the altimetry data; the blue stars show the absolute difference in the height difference between SD-AD (SHA-ADT) at two points, and the red crosses show the absolute difference in the height difference between SD-AD (SHA-SLA) at two points.

In Figure 9, with the sum of the distance between an Argo point and the eddy center at two points being larger, the difference between the ADT and the SLA was larger, thus, the MSSH and the geoid had greater differences under this condition. When we analyze the SD-AD data with distances smaller than ~200 km, the results perform well, except for group data No. 13 and group data No. 6; the reason why was introduced in the previous section. Moreover, it was also proven that the SD-AD is the result of the multiple factors. Therefore, when the distance between the Argo points and the eddy center is smaller than 220 km, the results of the SD-AD are relatively concentrated, simply because the mesoscale eddy is a typical baroclinic phenomenon, the steric data can better represent sea level variations.

3.2.3. Wind Speeds

Barotropic ocean response to wind stress forcing is important for understanding sea level variability [30]. In this paper, considering that wind-forced motion is a type of barotropic motion and the wind force are related to the wind speed [31,32], we tried to find the relationship of the SD-AD by the wind speed. The range of the wind speed was from 1 m/s to 9.6 m/s for all points, and the maximum value of the wind speed difference was 1.3 m/s. As we can see from Figure 10, using the SHA provides clearer results than the SD-AD using SH. After the linear regression analysis accounting for the wind factor was performed, the SLA data initially performed better than the ADT data regardless of SH or SHA, although the slope of the SHA was greater than that of the SH. When the wind speed difference between two points was smaller than 1 m/s, the SD-AD results performed well in most cases.

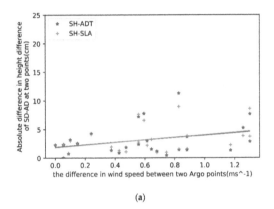

(a)

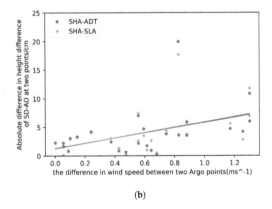

(b)

Figure 10. Shows the relationships of the SD-AD as a function of the difference in wind speed between two Argo points. (**a**) The relationship between the SH and the altimetry data; the blue stars show the absolute difference in the height difference between SD-AD (SH-ADT) at two points, and the red crosses show the absolute difference in the height difference between SD-AD (SH-SLA) at two point. (**b**) The relationship between the SHA and the altimetry data, the blue stars show the absolute difference in the height difference between SD-AD (SHA-ADT) at two points, and the red crosses show the absolute difference in the height difference between SD-AD (SHA-SLA) at two points.

3.3. Results under the Conditions

Considering the previous analysis in this paper, we used four conditions to limit the results to prove the influence of these factors, which included the distance difference between the distance of the Argo and the satellite in P_1 and P_2 being less than ~13 km and an angle difference of less than 120 degrees, the distance between two Argo points was less than ~120 km, the sum of the distance of the Argo point and the eddy center in P_1 and P_2 were less than ~220 km, and the difference in wind speed between two Argo points was less than ~1 m/s. There were seven groups of data after data control was applied which shown in Figure 11. The results are showed in Table 3. These values represent a further improvement with respect to previous analyses, and the results showed that sea level variability dominates by the baroclinicity in the small range. The results of these analyses proved that it is feasible to use the steric method to validate altimetry data.

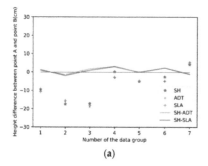 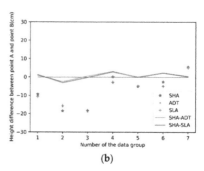

(a) (b)

Figure 11. The SD-AD results after screening with the influence factor; (**a**) is the SH data selected as the steric data, where the blue stars are the height difference between SH at P_1 and P_2, the cyan dots are the height difference between ADT at P_1 and P_2 and the red crosses are the height difference between SLA at P_1 and P_2. The cyan line was the difference in the height difference between SD-AD (SH-ADT) at two points, and the red line was the difference in the height difference between SD-AD (SH-SLA) at two points. (**b**) is the SHA data selected as the steric data, where the blue stars are the height difference between SHA at P_1 and P_2, the cyan dots are the height difference between ADT at P_1 and P_2 and the red crosses are the height difference between SLA at P_1 and P_2. The cyan line shows the difference in the height difference between SD-AD (SHA-ADT) at two points, and the red line shows the difference in the height difference between SD-AD (SHA-SLA) at two points.

Table 3. The result of the SD-AD by four control conditions.

Argo Database	RMSD		BIAS		Correlation Coefficient	
	ADT	SLA	ADT	SLA	ADT	SLA
SH	1.79 cm	1.76 cm	0.81 cm	0.69 cm	0.9785	0.9780
	ADT	SLA	ADT	SLA	ADT	SLA
SHA	1.76 cm	1.89 cm	0.66 cm	0.55 cm	0.9841	0.9828

4. Discussion

We analyzed the relationships of the SD-AD from multiple perspectives. The first is the distance between the Argo position and the altimetry position. As the distance between the Argo position and altimetry data position decreases, the Argo data can better validate the result of the satellite data. However, when we validate the sea level variability between the different pixels in the swath of the interferometric altimeter, the influence of the distance between the Argo point and the altimetry position can be neglected, because the pixel from the interferometric altimeter is sufficiently smaller. Meanwhile, considering that the ocean variability has the same trend in a small area, the direction between the Argo and altimeter measurements sites becomes another important factor as shown in Figure 4b. Thus, the effect of direction cannot be ignored when validating the interferometric altimetry data.

Barotropic influence is another important factor. Most of the ocean barotropic motions, such as barotropic tides and pressure- and wind-forced barotropic motions, drove mass variability over all frequency band studied here and yielded particularly strong variables at periods of 3–4 days, primarily at mid-high latitudes [33]. However, there are still some barotropic motions that are temporary and can cause the SH data to vary substantially. We used three different factors to analyze the SD-AD, deepening our understanding of the relationship of the SD-AD, which proved that the SD-AD is influenced by multiple factors, which include the distance difference between the distance of the Argo and the satellite in P_1 and P_2 and angle difference, the distance between two Argo points, the sum of the distance of the Argo point and the eddy center in P_1 and P_2 and the difference in wind speed between two Argo points.

Regardless of the changing factors, the slope of the satellite data selecting the ADT or SLA is similar under each factor except for the eddy center factor. Moreover, when the limit conditions were 0.2 degrees, the result of the SLA database was obviously superior to that of the ADT database. With the limit conditions becoming 0.15 degrees, the result of the ADT database performance improved. As shown in Figures 7–10, except for the influence of wind, the ADT data were better than the SLA data when the limit was more restrictive.

When comparing the SHA and the SH, when the limit conditions were 0.2 degrees, the result of the SH database was obviously superior to that of the SHA database. As shown in Figures 7–10, except for the influence of the eddy center, the slope of the SHA was larger than that of the SH which showed that the variability was more obvious and the SHA was influenced more by those factors. When the experimental conditions from the distance between the Argo position and the satellite position less than 0.2 degrees in the latitude and longitude direction and the time being less than four hours to the last 7 group data, the result of the RMSD, bias and the correlation coefficient are obviously better when the SHA is selected as the Argo data.

5. Conclusions

In this study, we evaluated the feasibility of using a steric method to validate the sea level variability between different pixels in a swath from multiple perspectives. We analyzed the influence of the SD-AD data under four different influence factors, including the distance difference between the distance of the Argo and the satellite in P_1 and P_2 and angle difference, the distance between two Argo points, the sum of the distance of the Argo point and the eddy center in P_1 and P_2 and the difference in wind speed between two Argo points. The highlights of this study are listed below:

1. The feasibility of validating the altimetry swath data by using the steric method. In this paper, we used in-situ observation data to analyze the feasibility of using a steric method to validate the interference altimetry sea level variability in different pixels. The result showed that when considering the distance difference between the distance of the Argo and the satellite in P_1 and P_2 and angle difference, the distance between two Argo points, the sum of the distance of the Argo point and the eddy center in P_1 and P_2 and the difference in wind speed between two Argo points, the relationship of the SD-AD has a highly corrected coefficient of 0.98, the RMSD was ~1.8 cm and the bias was ~0.6 cm. This proved that it is feasible to validate interferometric altimetry data using the steric method under these conditions.

2. As we can see from Figures 4–10, when the distance difference between the distance of the Argo and the satellite in P_1 and P_2 were less than ~13 km, and in the same direction, the distance between two Argo points was less than ~120 km, the sum of the distance of the Argo point and the eddy center in P_1 and P_2 less than 220 km, difference in wind speed between two Argo points were less than ~1 m/s and the non-steric influence had a significant reduction. The relationship between the steric data and sea level data had a highly corrected coefficient of 0.98. This proved

that using the steric method to validate the sea level variability in different pixels is feasible, and the relationship needs to be studied in more detail in the future.

Author Contributions: Q.Z., F.Y. and G.C. designed the study, Q.Z. and F.Y. conducted the analysis, Q.Z. wrote the original draft, all authors contributed to final review and editing. All authors have read and agreed to the published version of the manuscript.

Funding: This research was funded by the following programs: (1) the Qingdao National Laboratory for Marine Science and Technology, grant number QNLM2016ORP0105; (2) the National key research and development program of China, grant number 2016YFC1402608, 2016YFC1401008; (3) the Marine S&T Fund of Shandong Province for Pilot National Laboratory for Marine Science and Technology (Qingdao), grant number 2018SDKJ0102-7.

Acknowledgments: We are thankful to the AVISO and IFREMER for providing the data.

Conflicts of Interest: The authors declare no conflict of interest.

Abbreviations

The list of the abbreviation:

ADT	Absolute Dynamic Topography
AE	Anticyclonic Eddy
AVISO	Archiving, Validation and Interpretation of Satellite Oceanographic
Cal/Val	Calibration and Validation
CMEMS	Copernicus Marine Environment Monitoring Service
ECMWF	European Centre for Medium Weather Forecasting
GDAC	Global Data Assembly Centers
IFREMER	French Research Institute for Exploitation of the Sea
MSSH	Mean Sea Surface Height
MSH	Mean Steric Height
NSH	Non-Steric Height
OW	Okubo–Weiss method
P_1	Point 1
P_2	Point 2
RMSD	Root Mean Square Deviation
SD-AD	Steric Data and Along-track Data
SH	Steric Height
SHA	Steric Height Anomaly
SLA	Sea Level Anomaly
SSH	Sea Surface Height
SWOT	Surface Water and Ocean Topography
T-S-P	Temperature-Salinity-Pressure
WOA 13	World Ocean Atlas 2013
4D-Var	Four-Dimensional Variational Analysis

References

1. Johnson, D.R.; Thompson, J.D.; Hawkins, J.D. Circulation in the Gulf of Mexico from Geosat altimetry during 1985–1986. *J. Geophys. Res. Ocean.* **1992**, *97*, 2201. [CrossRef]
2. Fu, L.-L.; Ubelmann, C. On the Transition from Profile Altimeter to Swath Altimeter for Observing Global Ocean Surface Topography. *J. Atmos. Ocean. Technol.* **2014**, *31*, 560–568. [CrossRef]
3. Chen, G.; Tang, J.; Zhao, C.; Wu, S.; Yu, F.; Ma, C.; Xu, Y.; Chen, W.; Zhang, Y.; Liu, J.; et al. Concept Design of the "Guanlan" Science Mission: China's Novel Contribution to Space Oceanography. *Front. Mar. Sci.* **2019**, *6*, 194. [CrossRef]
4. Morrow, R.; Fu, L.-L.; Ardhuin, F.; Benkiran, M.; Chapron, B.; Cosme, E.; d'Ovidio, F.; Farrar, J.T.; Gille, S.T.; Lapeyre, G.; et al. Global Observations of Fine-Scale Ocean Surface Topography With the Surface Water and Ocean Topography (SWOT) Mission. *Front. Mar. Sci.* **2019**, *6*, 232. [CrossRef]
5. Fu, L.L.; Ferrari, R. Observing Oceanic Submesoscale Processes From Space. *Eos Trans. Am. Geophys. Union* **2013**, *89*, 488. [CrossRef]

6. Bonnefond, P.; Exertier, P.; Laurain, O.; Ménard, Y.; Orsoni, A.; Jan, G.; Jeansou, E. Absolute Calibration of Jason-1 and TOPEX/Poseidon Altimeters in Corsica Special Issue: Jason-1 Calibration/Validation. *Mar. Geod.* **2010**, *26*, 261–284. [CrossRef]

7. Christensen, E.J.; Haines, B.J.; Keihm, S.J.; Morris, C.S.; Norman, R.A.; Purcell, G.H.; Williams, B.G.; Wilson, B.D.; Born, G.H.; Parke, M.E.; et al. Calibration of TOPEX/POSEIDON at Platform Harvest. *J. Geophys. Res.* **1994**, *99*, 24465. [CrossRef]

8. Mertikas, S.P.; Daskalakis, A.; Tziavos, I.N.; Vergos, G.S.; Frantzis, X.; Tripolitsiotis, A.; Partsinevelos, P.; Andrikopoulos, D.; Zervakis, V. Ascending and Descending Passes for the Determination of the Altimeter Bias of Jason Satellites using the Gavdos Facility. *Mar. Geod.* **2011**, *34*, 261–276. [CrossRef]

9. Watson, C.; Coleman, R.; White, N.; Church, J.; Govind, R. Absolute Calibration of TOPEX/Poseidon and Jason-1 Using GPS Buoys in Bass Strait, Australia Special Issue: Jason-1 Calibration/Validation. *Mar. Geod.* **2010**, *26*, 285–304. [CrossRef]

10. Liu, Y.; Tang, J.; Zhu, J.; Lin, M.; Zhai, W.; Chen, C. An improved method of absolute calibration to satellite altimeter: A case study in the Yellow Sea, China. 2014, 33, 103–112. *Acta Oceanol. Sinica* **2014**, *33*, 103–112. [CrossRef]

11. Jan, G.; Ménard, Y.; Faillot, M.; Lyard, F.; Jeansou, E.; Bonnefond, P. Offshore Absolute Calibration of Space-Borne Radar Altimeters. *Mar. Geod.* **2004**, *27*, 615–629. [CrossRef]

12. Li, Z.; Wang, J.; Fu, L.L. An Observing System Simulation Experiment for Ocean State Estimation to Assess the Performance of the SWOT Mission: Part 1—A Twin Experiment. *J. Geophys. Res. Ocean.* **2019**, *124*, 4838–4855. [CrossRef]

13. Wang, J.; Fu, L.-L.; Qiu, B.; Menemenlis, D.; Farrar, J.T.; Chao, Y.; Thompson, A.F.; Flexas, M.M. An Observing System Simulation Experiment for the Calibration and Validation of the Surface Water Ocean Topography Sea Surface Height Measurement Using In Situ Platforms. *J. Atmos. Ocean. Technol.* **2018**, *35*, 281–297. [CrossRef]

14. Wahr, J.; Smeed, D.A.; Leuliette, E.; Swenson, S. Seasonal variability of the Red Sea, from satellite gravity, radar altimetry, and in situ observations. *J. Geophys. Res. Ocean.* **2015**, *119*, 5091–5104. [CrossRef]

15. Guinehut, S.; Le Traon, P.-Y.; Larnicol, G. What can we learn from Global Altimetry/Hydrography comparisons? *Geophys. Res. Lett.* **2006**, *33*, 10. [CrossRef]

16. Tapley, B.D.; Srinivas, B.; Ries, J.C.; Thompson, P.F.; Watkins, M.M. GRACE measurements of mass variability in the Earth system. *Science* **2004**, *305*, 503–505. [CrossRef]

17. Roemmich, D.; Gilson, J.; Sutton, P.; Zilberman, N. Multidecadal Change of the South Pacific Gyre Circulation. *J. Phys. Oceanogr.* **2016**, *46*, 1871–1883. [CrossRef]

18. Meyssignac, B.; Piecuch, C.G.; Merchant, C.J.; Racault, M.F.; Palanisamy, H.; Macintosh, C.; Sathyendranath, S.; Brewin, R. Causes of the Regional Variability in Observed Sea Level, Sea Surface Temperature and Ocean Colour Over the Period 1993–2011. In *Integrative Study of the Mean Sea Level and Its Components*; Springer: Cham, Germany, 2017; pp. 187–215.

19. Dhomps, A.L.; Guinehut, S.; Traon, P.Y.L.; Larnicol, G. A global comparison of Argo and satellite altimetry observations. *Ocean. Sci.* **2011**, *7*, 175–183. [CrossRef]

20. Perigaud, C. Influence of interannual rainfall anomalies on sea level variations in the tropical Indian Ocean. *J. Geophys. Res.* **2003**, *108*, C10. [CrossRef]

21. Song, Y.T.; Qu, T. Multiple Satellite Missions Confirming the Theory of Seasonal Oceanic Variability in the Northern North Pacific. *Mar. Geod.* **2011**, *34*, 477–490. [CrossRef]

22. Shao, Q.; Zhao, J. Comparing the steric height in the Nordic Seas with satellite altimeter sea surface height. *Acta Oceanol. Sin.* **2015**, *34*, 32–37. [CrossRef]

23. Dandapat, S.; Chakraborty, A. Mesoscale Eddies in the Western Bay of Bengal as Observed From Satellite Altimetry in 1993–2014: Statistical Characteristics, Variability and Three-Dimensional Properties. *IEEE J. Sel. Top. Appl. Earth Obs. Remote. Sens.* **2016**, *9*, 5044–5054. [CrossRef]

24. Henson, S.A.; Thomas, A.C. A census of oceanic anticyclonic eddies in the Gulf of Alaska. *Deep. Sea Res. Part I* **2008**, *55*, 163–176. [CrossRef]

25. Xu, L.; Li, P.; Xie, S.P.; Liu, Q.; Liu, C.; Gao, W. Observing mesoscale eddy effects on mode-water subduction and transport in the North Pacific. *Nat. Commun.* **2016**, *7*, 10505. [CrossRef] [PubMed]

26. Lázaro, C.; Juliano, M.F.; Fernandes, M.J. Semi-automatic determination of the Azores Current axis using satellite altimetry: Application to the study of the current variability during 1995–2006. *Adv. Space Res.* **2013**, *51*, 2155–2170.

27. Rio, M.H. A mean dynamic topography computed over the world ocean from altimetry, in situ measurements, and a geoid model. *J. Geophys. Res.* **2004**, *109*, C12. [CrossRef]

28. Baker-Yeboah, S.; Watts, D.R.; Byrne, D.A. Measurements of Sea Surface Height Variability in the Eastern South Atlantic from Pressure Sensor–Equipped Inverted Echo Sounders: Baroclinic and Barotropic Components. *J. Atmos. Ocean. Technol.* **2009**, *26*, 2593–2609. [CrossRef]

29. Badin, G.; Williams, R.G.; Holt, J.T.; Fernand, L.J. Are mesoscale eddies in shelf seas formed by baroclinic instability of tidal fronts? *J. Geophys. Res. Ocean.* **2009**, *114*, C10. [CrossRef]

30. Piecuch, C.G.; Calafat, F.M.; Dangendorf, S.; Jordà, G. The Ability of Barotropic Models to Simulate Historical Mean Sea Level Changes from Coastal Tide Gauge Data. *Surv. Geophys.* **2019**, *40*, 1399–1435. [CrossRef]

31. Fourniotis, N.T.; Horsch, G.M. Three-dimensional numerical simulation of wind-induced barotropic circulation in the Gulf of Patras. *Ocean. Eng.* **2010**, *37*, 355–364. [CrossRef]

32. Güting, P.M.; Hutter, K. Modeling wind-induced circulation in the homogeneous Lake Constance using. *Aquat. Sci.* **1998**, *60*, 266–277.

33. Savage, A.C.; Arbic, B.K.; Richman, J.G.; Shriver, J.F.; Alford, M.H.; Buijsman, M.C.; Farrar, J.T.; Sharma, H.; Voet, G.; Wallcraft, A.J. Frequency content of sea surface height variability from internal gravity waves to mesoscale eddies. *J. Geophys. Res. Ocean.* **2017**, *122*, 2519–2538. [CrossRef]

Article

Multi-Sensor Observations of Submesoscale Eddies in Coastal Regions

Gang Li [1], Yijun He [1,2], Guoqiang Liu [1,3,*], Yingjun Zhang [4], Chuanmin Hu [4] and William Perrie [3,5,6]

1   School of Marine Sciences, Nanjing University of Information Science and Technology, Nanjing 210044, China; cafferylg@gmail.com (G.L.); yjhe@nuist.edu.cn (Y.H.)
2   Laboratory for Regional Oceanography and Numerical Modeling, Qingdao National Laboratory for Marine Science and Technology, Qingdao 266237, China
3   Fisheries and Oceans Canada, Bedford Institute of Oceanography, Dartmouth, NS B2Y 4A2, Canada; william.perrie@dfo-mpo.gc.ca
4   College of Marine Science, University of South Florida, St. Petersburg, FL 33701, USA; yingjunzhang@mail.usf.edu (Y.Z.); huc@usf.edu (C.H.)
5   Department of Oceanography, Dalhousie University, Halifax, NS B3H 4R2, Canada
6   Department of Engineering Mathematics and Internetworking, Dalhousie University, Halifax, NS B3H 4R2, Canada
*   Correspondence: Guoqiang.Liu@dfo-mpo.gc.ca

Received: 13 January 2020; Accepted: 19 February 2020; Published: 21 February 2020

Abstract: The temporal and spatial variation in submesoscale eddies in the coastal region of Lianyungang (China) is studied over a period of nearly two years with high-resolution (0.03°, about 3 km) observations of surface currents derived from high-frequency coastal radars (HFRs). The centers and boundaries of submesoscale eddies are identified based on a vector geometry (VG) method. A color index (CI) representing MODIS ocean color patterns with a resolution of 500 m is used to compute CI gradient parameters, from which submesoscale features are extracted using a modified eddy-extraction approach. The results show that surface currents derived from HFRs and the CI-derived gradient parameters have the ability to capture submesoscale processes (SPs). The typical radius of an eddy in this region is 2–4 km. Although no significant difference in eddy properties is observed between the HFR-derived current fields and CI-derived gradient parameters, the CI-derived gradient parameters show more detailed eddy structures due to a higher resolution. In general, the HFR-derived current fields capture the eddy form, evolution and dissipation. Meanwhile, the CI-derived gradient parameters show more SPs and fill a gap left by the HFR-derived currents. This study shows that the HFR and CI products have the ability to detect SPs in the ocean and contribute to SP analyses.

Keywords: high-frequency radar; MODIS ocean color patterns; submesoscale eddies

1. Introduction

Submesoscale processes (SPs) are frequently observed as eddies, filaments and fronts, where the approximate scale ranges are 0.1–10 km in the horizontal, 0.01–1 km in the vertical, and between hours and days in time over open oceans [1]. The most fundamental characteristic difference between submesoscale and mesoscale is that the Rossby (Ro) and Richardson numbers (Ri) are both close to $O(1)$ for SPs. SPs have large vertical velocities which are typically several times larger than those associated with mesoscale processes; thus, they play a critical role in vertical transport [2,3], enhance bio-physical interactions, and provide a possible link between mixing and dissipation. Submesoscale buoyancy flux in SPs enhances seasonal restratification and the mixing of the mixed layer [4]. Submesoscale restratification can weaken submesoscale turbulence and enhance surface inertia-gravity waves [5]. The generation mechanisms of SPs may be summarized as: strain-induced frontogenesis, baroclinic instabilities in the mixed layer, turbulent thermal winds and topographic wakes [1,6,7]. Idealized numerical models under relevant theoretical frameworks may efficiently be employed in studies of SPs, which require high-resolution data of less than one hour in time and $O(1)$ km in space [8,9].

Compared with numerical model simulations, the in situ observations at these scales are limited greatly by their ability to resolve the detailed horizontal and vertical structures of SPs [10–16]. However, in recent years, high-precision observations have mainly focused on some special submesoscale events and phenomena, due to increased sampling rates in time and space [6,11]. Remote-sensing observations can also be used to investigate SPs, such as surface currents derived by high-frequency radar [2,17–19], geostationary ocean color imagery (GOCI)-derived chlorophyll surface concentration maps, and colored dissolved organic matter [20], which are available at hourly and kilometer-scale resolutions. HFRs can provide information on ocean surface currents over a large, horizontal distance with high spatial-temporal resolution, which can reach up to 200 m in space (e.g., SeaSonde [21]) and 20 min in time. Meanwhile, satellite images can rarely observe SPs continuously because of their limitation in sampling frequency, contamination from high-frequency motions and incoherent internal tides. L. Pomale-Velázquez et al. [22] analyzed the characterization of mesoscale eddies and submesoscale eddies using satellite images (i.e., satellite-measured sea surface height anomalies, satellite-derived geostrophic velocity components, satellite chlorophyll-a (Chl-a) and available floating algae index images). HFRs were used to measure surface velocities off the coast of southwestern Puerto Rico, and it was found that they are capable of identifying submesoscale eddies lasting for 3 days. However, satellite products cannot capture these eddies for long time periods. Moreover, because certain satellite images, for example high-resolution radar images, show variable anomalies instantaneously, they can be used to obtain some characteristics of submesoscale eddies instantaneously. Although HFRs have previously been applied to study coastal submesoscale currents in other coastal regions, this is the first application of HFRs in characterizing submesoscale eddies in the Lianyungang area (Figure 1a); thus, we are able to improve our understanding of issues related to oceanography, relevant to port planning and fisheries management in Haizhou Bay.

In this area, there is one flow existing in the coastal regions: the Yellow Sea Coastal Current coming from the northwest [23]. The tidal current is characterized by regular semidiurnal currents (the M2 tide is the main constituent). The tidal current shows a rectilinear motion, where the direction of the flood (ebb) tide tends to be southwestward (northeastward) and winter tide currents are generally stronger than summer tide currents. In winter (summer), the major axis of the M2 tidal ellipse is 30.85 cm/s (28.53 cm/s) which shows that the local current is affected by the tidal currents [24]. The dominant bathymetric feature is the shallow coastal area, with a maximum water depth of 33 m. There is a shallow bank in the center of the region and a channel located between the bank and the 10 m isobath. The interaction between the tidal current and the bathymetry is the main driver for the generation of submesoscale eddies and filaments. Vortex stretching on the ebb and flood tides is considered to be closely linked to eddy generation and the local wind-driven currents also affect the distribution and movement of submesoscale eddies [2,25].

In order to detect submesoscale eddies, methods based on their geometric properties are applied. Traditionally, methods for automatic eddy detection have been mostly based on physical or geometric properties of the flow field [26,27]. Methods based on physical properties mostly identify eddies by comparing specified physical parameters, including pressure, the magnitude of sea level anomalies, vorticity, and various quantities derived from the velocity gradient tensor [26,28–30]. Methods based on the geometric properties include the winding-angle method [27,31,32], the vector geometry (VG) method [33], vector pattern matching [34] and the Clifford convolution [35]. Compared with methods based on physical properties, methods based on the geometric properties can better detect eddies and tend to consider a non-vortical structure as an eddy [36].

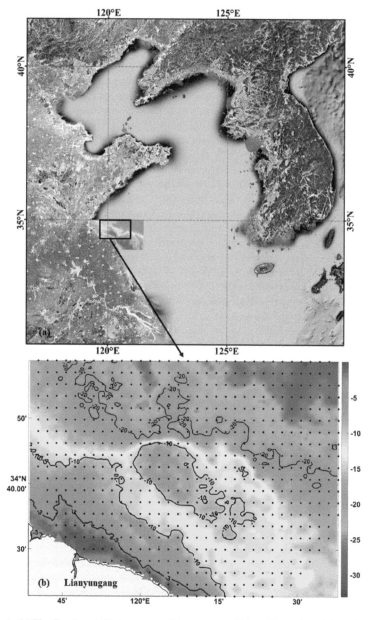

Figure 1. (**a**) The Google satellite image near Lianyungang, China. The region of the color image is the coverage region of Moderate Resolution Imaging Spectroradiometer (MODIS) color patterns. The region of the black solid frame is the coverage region of High-Frequency coastal Radars (HFRs) (90.8 × 66.7 km); (**b**) The bathymetry of the study area (units: m). The background color shows the topography. The black contours are the isobaths, which show the 3, 10, and 20 m bathymetric contours. The black dots are vector points derived from the HFRs. The two green triangles are the locations of the two HFRs. The bathymetry data is from Global Ocean and Land Terrain Models' (GEBCO's) gridded bathymetric dataset, which is a global terrain model for ocean and land at 15 arc-second intervals.

Dong et al. [37] used satellite-measured SST images to obtain the characteristics of ocean mesoscale eddies, which extracted most mesoscale signals limited by the coarse resolution of Remote Sensing System SST production. Here, Moderate Resolution Imaging Spectroradiometer (MODIS) ocean color patterns of resolution up to 500 m are applied to extract submesoscale signals, based on a similar approach proposed by Dong et al. [37]. A color index (CI, units: non-dimensional) represents the MODIS ocean color patterns, which are derived from an empirical approach proposed by Hu [38]. The empirical approach has been proven to validly remove sun glint and clouds and increase the data coverage. The CI has a significant correlation with the MODIS band-ratio Chl-a (<1 mg m^{-3}) and is a good index to represent coastal color changes. Liu et al. [39] pointed out that mesoscale anticyclonic eddies in the Gulf of Mexico Loop Current system were accompanied by low CI values which indicated clear waters. Local CI increases (decreases) when submesoscale cyclonic (anticyclonic) eddies occur, while local SST decreases (increases) when submesoscale cyclonic (anticyclonic) eddies occur. This could help us understand the process by which eddies drive the upwelling of colder, nutrient rich waters that boost plankton production, the aggregation and retention of particles, etc. Because SST and CI have opposite responses to cyclonic and anticyclonic eddies, CI-derived gradient parameters are calculated by a modified method based on an automated approach to extract submesoscale signals. Zhang et al. [40] used two thresholds to determine the candidate boundaries for eddy detection. These two thresholds were determined after trial-and-error ananlyses of 625 sets of threshold values, which may not be applicable for other regions. Our modified method focuses on the gradient information only, without the need to use thermal wind approximation, such as in Dong et al. [37]. This is because the thermal wind approximation is not applicable for submesoscale eddies. In this study, the modified method of calculating gradient parameters goes beyond originally intended limits of the applicability of Dong et al. [37]. With the availability of HFR ground-truth data, this method is evaluated when scaled down to submesoscale. The gradient parameters are used to indicate the relative changes in CI values across eddy features; they contain meridional and zonal values to show eddy types (anticyclonic or cyclonic).

In this study, the VG method is modified to extract submesoscale signals from MODIS CI fields and to identify submesoscale eddies from HFR-derived currents fields and CI-derived gradient parameters. The primary goal of this paper is to analyze the coastal surface currents measured by HFRs with high-resolution data and MODIS ocean color patterns. The paper is organized as follows: in Section 2 we describe the dataset, including the HFR data and the high-resolution MODIS ocean color patterns, the eddy detection method, and the dynamics quantities. Section 3 presents results concerning our ability to extract submesoscale signals by CI-derived gradient parameters, and the eddy properties from HFR-derived current fields and MODIS CI-derived gradient parameters. Discussion is provided in Section 4. Final summary and conclusions are presented in Section 5.

2. Materials and Methods

2.1. HFR-Derived Currents

Surface currents were measured from June 2017 to March 2019 for the coastal area of Huang Sea, close to the city of Lianyungang, China, using two HFRs. The temporal sampling rate is about 20 min. The final radial current maps recorded simultaneously from two HFRs were synthesized to construct total vector maps on a Cartesian grid. The composite radar system mapped the Lianyungang coastal surface currents on 34 × 21 grids covering the surface area from 119.64° to 120.63° E and from 34.38° to 34.98° N, with a spacing of 0.03° (about 3 km) in both directions. Figure 1b shows the observational range covered by the HFRs. The depth of the study area varied from about 1 to 33 m.

The numbers of observed data are shown in Figure 2. Part of the data are missing because of terrible weather and equipment failures. Although the temporal coverage rate is not always continuous, only a small number of submesoscale eddies are observed to survive for more than one day [13,22]. Therefore, the data are sufficient to show the entire submesoscale eddy processes.

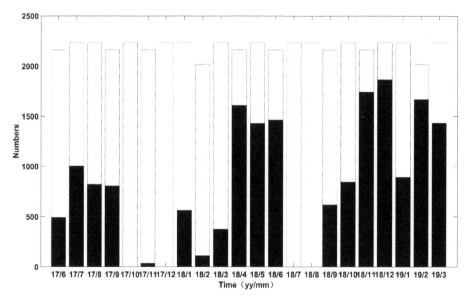

Figure 2. Number of HFR-derived surface current field records for each month. The data are recorded once every 20 min. Shaded areas indicate the number of data for each month.

2.2. MODIS Ocean Color Patterns

The MODIS ocean color patterns obtained from MODIS/Aqua satellite measurements over our study area (119.64°–120.63°E, 34.38°–34.98°N) have a spatial resolution of 0.004° × 0.004° and a daily temporal sampling rate (under cloud-free conditions) for the period from June 2017 to March 2019. Its original spatial resolution is about 500 m, but a 3 × 3 median filter was applied to remove speckle noise. Details about MODIS color index (CI) can be seen in Hu [38]. Briefly, CI is defined as the reflectance magnitude at 555 nm relative to two neighboring wavelengths (469 and 645 nm). For most ocean waters (chlorophyll concentration < 1 mg m^{-3}) CI is highly correlated with MODIS chlorophyll data product (Chl-a) derived from the NASA default algorithms. For more productive waters, CI can be used as an index to show relative color patterns. The CI algorithm takes a similar band-subtraction approach to the OCI algorithm, but the former uses 469, 555, 645 nm band combinations while the latter uses 443, 547, 667 nm band combinations. The former bands do not saturate in severe sun glint, thick aerosols, and thin clouds, but the latter bands do saturate. The advantage of CI over MODIS Chl-a is that, because of the three land bands used to derive CI, CI does not saturate over sun glint, thin clouds, or thick aerosols, thereby providing more spatial and temporal coverage than MODIS chlorophyll to reveal the relative ocean color patterns. As shown in Figure 3, CI patterns from three subsequent images show the evolution of eddy patterns.

The CI dataset has advantages in research activities related to SPs because of its higher resolution and coverage compared to other MODIS SST or Chl-a datasets. The area surrounded by dotted lines in Figure 3 shows decreases in CI, corresponding to anticyclonic eddies shown in Figure 4.

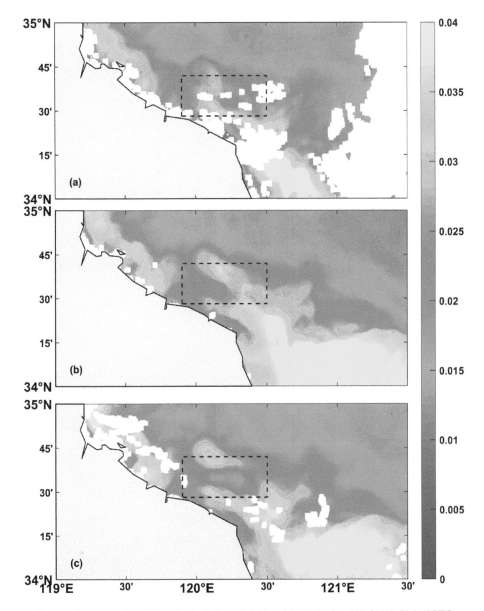

Figure 3. Three examples of MODIS color index (CI) fields at (**a**) 0535 UTC on 10 Jul 2017, (**b**) 0440 UTC on 11 Jul 2017 and (**c**) 0525 UTC on 12 Jul 2017. The submesoscale anticyclonic eddies are indicated by the black dashed lines. The values on the legend represent the relative magnitudes of CI, with higher values indicating more turbid waters. Note that, for the three days shown here, MODIS Chl-a images do not show valid data in most areas.

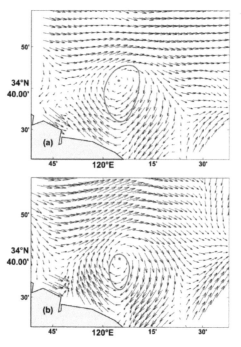

Figure 4. Examples of the eddy detection algorithm. Surface current velocity fields are obtained with HFRs at (**a**) 11:40 UTC on 10 July 2017 and (**b**) 12:40 UTC on 10 July 2017. The blue solid lines are eddy boundaries and the blue dots are centers of anticyclonic eddies, both estimated from the vector geometry (VG) method.

MODIS ocean color patterns do not only focus on the local CI changes in time, but may also be used to estimate divergence and convergence through the CI-derived gradient fields. In our study, we define a modified method to extract CI gradient parameters from the MODIS CI data based on the method proposed by Dong et al. [37].

A CI gradient parameter $V_{CI} = \left(U_x, U_y\right)$ is defined as

$$U_x = G_y; U_y = -G_x \text{ in the northern hemisphere,} \tag{1}$$

$$U_x = G_y; U_y = G_x \text{ in the southern hemisphere.} \tag{2}$$

The meridional and zonal derivatives for each datapoint, G_x and G_y, are calculated from a Sobel gradient operator [41] in order to smooth the images as, follows

$$G_y = \begin{bmatrix} +1 & +2 & +1 \\ 0 & 0 & 0 \\ -1 & -2 & -1 \end{bmatrix} * A, \tag{3}$$

$$G_x = \begin{bmatrix} +1 & 0 & -1 \\ +2 & 0 & -2 \\ +1 & 0 & -1 \end{bmatrix} * A \tag{4}$$

where * denotes the 2-D convolution operation and A is a 2-D matrix of the CI data.

Since the rotation of cyclonic (anticyclonic) eddies is counterclockwise (clockwise) in the northern hemisphere, the CI values remain higher (lower) for cyclonic (anticyclonic) eddies. The parameter

$V_{CI}(U_x, U_y)$ contains the gradient information of the CI data. Figure 5 shows examples of the gradient parameters derived from the CI data. The highest (lowest) CI values correspond to the centers of the cyclonic (anticyclonic) eddies.

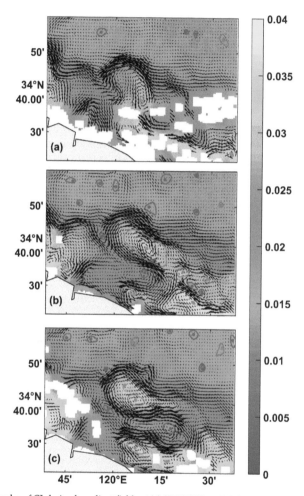

Figure 5. Examples of CI-derived gradient fields at (**a**) 0535 UTC on 10 Jul 2017, (**b**) 0440 UTC on 11 July 2017 and (**c**) 0525 UTC on 12 July 2017. Vectors are derived from the CI patterns. The blue (red) dots and solid lines are anticyclonic (cyclonic) eddy centers and boundaries identified by the VG method. Values on the legend represent relative magnitudes of CI. Higher values indicate more turbid waters.

2.3. Eddy Detection Algorithm

In this study, we have used the eddy detection algorithm based on the VG method proposed by Nencioli et al. [33]. This method is suitable for measuring the velocity field. It has been successfully applied to different study regions, such as the leeward side of Hawaii [42], the subtropical zonal band of the North Pacific Ocean [43], the Taiwan Strait [18], and the Kuroshio Extension Region [44]. We invite the reader to refer to Nencioli et al. [33] for more details.

There are two key steps in using the VG-based method to define an eddy structure:

(a): In order to find the center of an eddy with four constraints, first, the u(v) component of the HF current velocity along the north–south (east–west) section must reverse sign across the eddy center, and its magnitude must increase away from the center, where the u and v components are the meridional and zonal components of the HF current velocity vector, respectively. Second, the center is where the velocity is at a local minimum. Finally, the velocity's change in direction must exhibit a constant sense of rotation and the directions of the two neighboring velocities must lie within the same quadrant of the two adjacent quadrants to the eddy center. The detection algorithm of eddy centers is applied to the HFR-derived, and CI-derived, velocity fields;

(b): In order to obtain the eddy boundary, the outermost enclosed streamline is chosen as the eddy boundary around the center. The streamline at one position (i, j) is computed as

$$\psi(i, j) = \int_{(0,0)}^{(i,j)} -vdx + udy, \tag{5}$$

where u and v are meridional and zonal components of the velocity. This approach is consistent with the criterion requiring that the values of the velocity components should remain increasing along the eddy boundary. In Figure 4, the blue dots and lines indicate, respectively, the detected centers and the boundaries of anticyclonic eddies. The radius is defined as the mean distance between the center and the perimeter of a given shape.

2.4. Diagnostic

The VG method for the eddy detection algorithm was applied to both the HFR-derived current fields and CI-derived gradient parameters to detect eddy structures. Several eddy properties were calculated after one eddy structure was identified. To understand the surface flow kinematics in the study region, various diagnostic values were calculated. The eddy kinetic energy (EKE) at each point was computed using the classical relation

$$\text{EKE} = \frac{1}{2}(u^2 + v^2), \tag{6}$$

where u and v are the velocities derived with HFRs. In addition, the eddy vorticity(ζ), the shearing deformation rate (S_s), the stretching rate (S_n), the total deformation rate (S) and the divergence (ψ) were calculated using the following relations:

$$\zeta = \frac{\partial v}{\partial x} - \frac{\partial u}{\partial y} \tag{7}$$

$$S_s = \frac{\partial v}{\partial x} + \frac{\partial u}{\partial y} \tag{8}$$

$$S_n = \frac{\partial u}{\partial x} - \frac{\partial v}{\partial y} \tag{9}$$

$$S = \sqrt{S_s^2 + S_n^2} \tag{10}$$

$$\psi = \frac{\partial u}{\partial x} + \frac{\partial v}{\partial y} \tag{11}$$

3. Results

3.1. Features of Submesoscale Eddies (Eddy Properties) from HFR Observations

Submesoscale eddies were obtained from the HFR velocity fields, with a total of 1102 cyclonic eddies and 886 anticyclonic eddies identified, based on the VG eddy detection method. Submesoscale eddies cover most of the study region. Their number density is shown in Figure 6, which indicates that the occurrence of eddies mostly happens in the center of the study region and also in the southwestern coastal part. Moreover, the region where eddies frequently appear is off the cape at (39°40′N, 119°50′E) and the seamount in the center. In addition, a small number of eddies are detected in the study region as a result of being close to the outer edge of the HFR observation region. However, the identified eddies in the region do not show significant monthly variability due to differences in the validated data obtained each month (Figure 7).

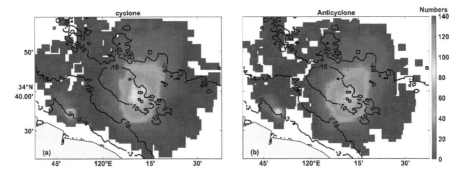

Figure 6. The total number of cyclonic (**a**) and anticyclonic (**b**) eddies derived by HFRs current fields based on the VG method during the observation period. The total number of eddies in each grid is increased if the eddy exists in the grid.

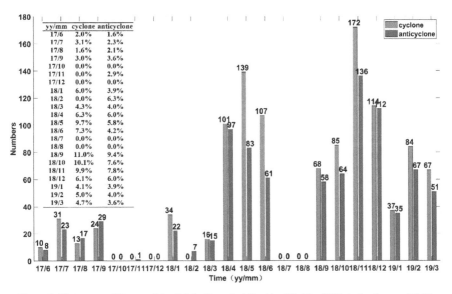

Figure 7. Histogram of the monthly distribution of eddies identified by HFR-derived current fields. The table shows the ratio of the cyclonic (anticyclonic) eddy number to the total observation records each month.

The submesoscale characteristics have a radius of 1.3 to 8 km and an asymmetric normalized vorticity with a maximum between 0.3 and 1.5 (Figure 8). There is little difference between the size distribution of cyclonic and anticyclonic eddies. Figure 8 shows that eddies with a radius between 2 and 4 km are frequent, whereas eddies with a radius larger than 10 km are very infrequent. The mean radius values are 2.70 and 2.59 km for the anticyclonic and cyclonic eddies, respectively. Table 1 lists the mean values of the eddies' kinematic parameters. It illustrates that the average vorticity of anticyclonic eddies is larger than that of cyclonic eddies; moreover, the instability of anticyclonic eddies is also larger, according to the values of total deformation. Moreover, since surface current fields in this region are influenced by tides and topography, the vorticity of the surface current field is not large compared with western boundary region currents [45]. In addition, no significant difference is found for the radii of anticyclonic and cyclonic eddies.

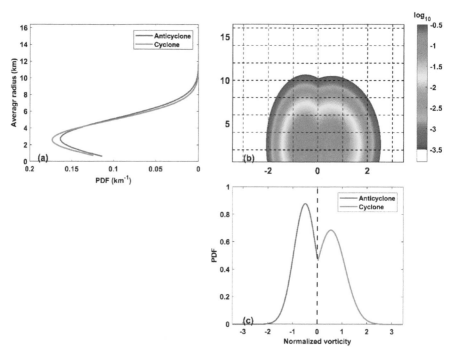

Figure 8. Probability density functions (PDFs) of the various characteristics of the identified submesoscale eddies from HFR-derived velocity fields. (a,c): 1-D PDFs for radius and vorticity; (b): joint PDF for radius vorticity.

Table 1. Mean values of eddy kinematics.

	Mean	Std dev	Min	Max
Cyclonic eddies				
Vorticity	2.84	1.97	0.31	9.92
Shearing deformation	0.30	1.46	−6.80	7.44
Stretching deformation	0.10	2.39	−11.94	10.42
Total deformation	2.13	2.14	0.16	13.34
Divergence	0.20	3.15	−15.20	15.40
Anticyclonic eddies				
Vorticity	−3.30	2.06	−9.92	−0.25
Shearing deformation	−0.44	1.67	−7.73	5.31
Stretching deformation	0.32	2.52	−6.70	11.77
Total deformation	2.50	2.22	0.15	12.86
Divergence	0.25	3.36	−7.72	19.73

units: $10^{-5} s^{-1}$.

3.2. Ability of Submesoscale Signals Extracted by CI-Derived Gradient Parameters

The CI-derived gradient parameters were calculated by the method described in Section 2. Figure 5 shows some examples of the gradient parameters at three different times. According to the location surrounded by the black dashed lines in Figure 3, the local CI is changed from large to small values. When anticyclonic eddies lasted for a short period, as shown in Figure 4, the anticyclonic eddies are found near the location of the black dashed lines in Figure 3; the size of anticyclonic eddies in the CI images is smaller than that from HFRs. In addition, a big difference is that there is one eddy that does exist in the center of HFR map and some other small eddies in this region. This illustrates that CI images could find the submesoscale signals due to physical processes, but they also contain other submesoscale information. The gradient parameters capture the submesoscale signals due to submesoscale anticyclonic eddies, and also other signals from other physical processes or biological processes. These processes mostly result from the sinking process accompanying anticyclonic eddies shown in Figure 3. The size and location of the anticyclonic eddies in CI gradient parameter fields have little difference from those in HFR-derived current fields because: (1) the Chl-a field aroused by the physical eddy field tends to be weaker than the current field and appears to be shifted by the force of the advection term; (2) the two datasets have different spatial resolutions and different properties (CI and velocity vectors). In addition, our method relies on a single image and gradient information, which is different from other studies about tracer-derived velocity products [46–48]. Nevertheless, some limitations do exist in both our method and the CI imagery when compared with HFR imagery. In fact, it is known in fluid mechanics that (a) submesoscale patterns could have taken a long time to emerge due to purely biological processes; (b) submesoscale patterns could be an example of "fossil turbulence" [49]. In both cases, the eddy patterns may only show up in the CI imagery as opposed to the HFR imagery. On the other hand, submesoscale eddies can exist without any color signature. Even with these known limitations, as long as eddy patterns are captured in the CI imagery, the CI-derived gradient parameters can successfully capture submesoscale signals.

Therefore, the CI-derived gradient parameters show similar eddy structures and provide even more submesoscale eddies. Although these gradient parameters do not provide realistic velocity fields, they are evidence of SPs and can also provide more detailed information about the eddy structures. This enhances the observation and analysis of submesoscale phenomena, implying that a CI (MODIS color pattern) with a resolution up to 500 m is a good tool to extract submesoscale signals from relatively larger observation areas, compared with HFR datasets.

3.3. Characteristics of Submesoscale Eddies (Eddy Properties) from CI-derived Gradient Fields

The results indicate that more submesoscale eddies can be detected with MODIS CI patterns with a resolution of 500 m than with the HFR data with a resolution of 3 km, although the former have a much lower temporal frequency (i.e., at most once per day versus every 20 min from HFR). A total of 4990 anticyclonic eddies and 4968 cyclonic eddies were identified with the abovementioned eddy detection algorithm from CI-derived gradient fields, which is more than four times those identified from the HFR velocity fields. The monthly distribution of the number of anticyclonic and cyclonic eddies is shown in Figure 9. We don't have sufficient CI data to report results during summer due to frequent clouds.

The CI-derived parameters (units: s^{-1}) aren't the actual velocities, and so the vorticity of the CI parameters only focuses on its relative value. In Figure 10a, the mean radius values for submesoscale eddies are 1.85 and 1.56 km for the anticyclonic and cyclonic eddies, respectively. The mean radius values are smaller than those computed by the HFR-derived eddies. This indicates that eddy structures with a higher resolution are more suitable to detect SPs, and therefore better able to observe submesoscale signals. The strength of anticyclonic and cyclonic eddies is almost equal, as shown in Figure 10c.

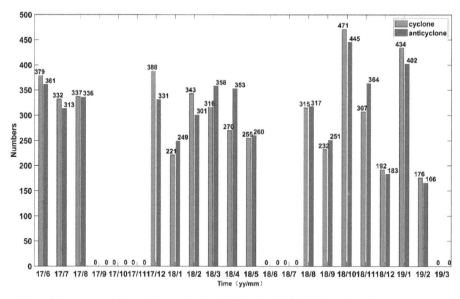

Figure 9. Histogram of the monthly distribution of eddies identified by CI-derived gradient parameters.

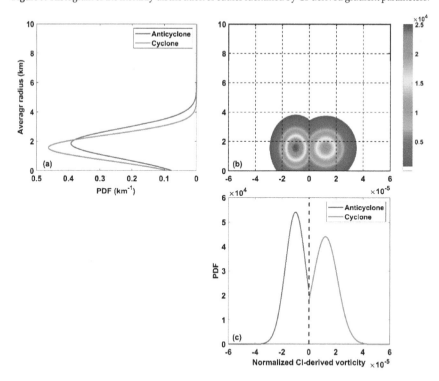

Figure 10. Probability density functions (PDFs) of the various characteristics of the identified submesoscale eddies from the CI-derived parameters. (**a,c**), 1-D PDFs for radius and vorticity; (**b**), joint PDF of radius vorticity.

4. Discussion

The small sizes of the eddies detected from both HFR-derived currents and CI-derived gradient parameters are different from other coastal submesoscale eddies using HFRs [13,33]. Here, although eddies have high relative vorticities, they are weaker than the submesoscale island wake at Green Island in the Kuroshio, as simulated by Liu and Chang [45], who found that values for the eddy vorticity can reach up to $10f$ (f is the Coriolis parameter), and the wake exhibits frequently detached recirculation, containing upwelling and von Kármán vortex streets. In our study region, cyclonic (anticyclonic) eddies were often accompanied by an increase (decrease) in CI (representing Chl-a and other in-water constituents). Jang et al. [16] reported similar studies about eddy properties from HFRs but didn't consider coastal responses of Chl-a because part of their study region did not have valid Geostationary Ocean Color Imager (GOCI) images. In this study, it is demonstrated that the efficient way to understand coastal SPs is via the application of HFR-derived currents and CI-derived gradient parameters.

The potential mechanism for the generation of submesoscale eddies in coastal areas can be inferred from geophysical factors, tides and wind stress. Firstly, a primary source for eddies' generation consists of geophysical factors, including coastline changes and bottom topography. Topographic features, such as islands, headlands and seamounts, generally exhibit a turbulent–drag interaction between currents and the topography. The formation and separation of bottom boundary layers affects the generation and injection of vertical vorticity. In this study area, multiple regions of eddies are mainly located along the coastline and near the seamount. The seamount changes the local vorticity and thus increases the probability of submesoscale eddies. The current fields near the coastline become unstable due to lateral boundaries, which also contribute to the generation of submesoscale eddies. Secondly, eddy generation is one possible cause of asymmetric advection by the rotating tidal flow [5]. The major axis of the M2 semidiurnal current is ~29 cm/s near Lianyungang, which greatly influences the local currents. Thirdly, local eddies respond to the wind direction and the wind stress curl, thereby providing some indication of when and where eddies may be formed, evolve and ultimately dissipate [2]. Detailed discussions of these mechanisms will be presented in future papers, using numerical simulations.

In summary, although HFRs have some limitations due to their limited coastal coverage, they do meet the requirements for the spatio-temporal resolutions of SPs. Moreover, the gradient parameters derived from daily CI images have the ability to fill the gap left by HFRs, with the advantages of high space coverage. The latter method therefore shows potential for further research on SPs of global oceans.

5. Conclusions

In our study, CI-derived gradient parameters are calculated by a modified method from MODIS CI data. Two kinds of dataset, HFR-derived current fields and CI-derived gradient parameters, were applied to identify submesoscale eddies using the VG eddy detection algorithm, and thus to analyze the eddy properties. Both the HFR-derived velocity fields and the CI-derived gradient parameters are able to efficiently observe submesoscale eddies with data resolutions of 3 km and 500 m, respectively. The CI gradient parameters are able to capture more detailed submesoscale eddy features due to the higher spatial resolution of the CI images (500 m). However, the eddies' distribution and sizes are found to have some differences because of different sampling rates, observational variables, response delay, fossil turbulence, and possibly colorless eddies. In conclusion, the study reveals that HFR surface currents and MODIS CI are good options to investigate SPs.

In the coastal Lianyungang region, a total of 1102 cyclonic eddies and 886 anticyclonic eddies were recorded from the HFR-derived velocity fields, and a total of 4990 anticyclonic eddies and 4968 cyclonic eddies were recorded from the MODIS CI-derived gradient parameters, based on the VG eddy detection algorithm. The typical radius of eddies derived from the HFR velocity fields was 2.70 and 2.59 km for the anticyclonic and cyclonic eddies, respectively. The typical radius of eddies derived from the CI images is slightly smaller than those found from the HFR velocity fields. Thus, an advantage of high-resolution data is that they can resolve the eddy structures in more detail.

This study is the first to extract submesoscale signals from MODIS CI data, with a resolution of 500 m, to investigate submesoscale coastal eddies. Comparing HFR surface currents with MODIS CI data, they can be mutually verified and they are complementary. Submesoscale eddies in CI images are detected near the location, where there are indeed some anticyclonic or cyclonic eddies from recent HFR records, revealing their consistency. So HFR data significantly prove the ability of submesoscale signals extracted from CI-derived gradient parameters, while CI data enhance the spatial resolution and the scope of the observations. Based on the VG eddy detection algorithm, submesoscale eddies near the coastal region of the Huang Sea, close to the city of Lianyungang, China have been well analyzed. This methodology is potentially useful for investigations of ecosystem distributions in coastal biological communities in other areas of the global ocean.

Author Contributions: G.L. (Guoqiang Liu) and Y.H. contributed to the idea of this study; Y.Z. and C.H. preprocessed CI data; G.L. (Gang Li) collected HFR-derived currents and analyzed data; G.L. (Gang Li) wrote the paper; Y.H., C.H. and W.P. helped with manuscript preparation and revision. All authors have read and agree to the published version of the manuscript.

Funding: This work was supported in part by the National Natural Science Foundation under Grant 41620104003, the National Key Research and Development Program under Grant 2016YFC1401002, the National Key Research and Development Program of China (2016YFC1401407) and by the Canada Government Research Initiatives Program (GRIP) in Climate Change Impacts and Ecosystem Resilience.

Acknowledgments: The authors would like to thank Jiangsu Research Center for Ocean Survey Technology for HFR products, thank Changming Dong for the VG eddy detection algorithm and thank Zhenyu Xu for image processing in Figure 1.

Conflicts of Interest: The authors declare no conflict of interest

References

1. McWilliams, J.C. Submesoscale currents in the ocean. *Proc. R. Soc. A* **2016**, *472*, 32. [CrossRef] [PubMed]
2. Kirincich, A. The Occurrence, Drivers, and Implications of Submesoscale Eddies on the Martha's Vineyard Inner Shelf. *J. Phys. Oceanogr.* **2016**, *46*, 2645–2662. [CrossRef]
3. Kim, S.Y.; Cornuelle, B.D.; Terrill, E.J. Decomposing observations of high-frequency radar-derived surface currents by their forcing mechanisms: Decomposition techniques and spatial structures of decomposed surface currents. *J. Geophys. Res.* **2010**, *115*. [CrossRef]
4. Du Plessis, M.; Swart, S.; Ansorge, I.J.; Mahadevan, A. Submesoscale processes promote seasonal restratification in the Subantarctic Ocean. *J. Geophys. Res.* **2017**, *122*, 2960–2975. [CrossRef]
5. Rocha, C.B.; Gille, S.T.; Chereskin, T.K.; Menemenlis, D. Seasonality of submesoscale dynamics in the Kuroshio Extension. *Geophys. Res. Lett.* **2016**, *43*, 304–311. [CrossRef]
6. Callies, J.; Ferrari, R.; Klymak, J.M.; Gula, J. Seasonality in submesoscale turbulence. *Nat. Commun.* **2015**, *6*, 6862. [CrossRef]
7. Thomas, L.N.; Tandon, A.; Mahadevan, A. Submesoscale processes and dynamics. *Ocean Model. Eddy. Regime* **2008**, 17–38. [CrossRef]
8. Mahadevan, A.; Tandon, A. An analysis of mechanisms for submesoscale vertical motion at ocean fronts. *Ocean Model.* **2006**, *14*, 241–256. [CrossRef]
9. Tulloch, R.; Smith, K.S. A theory for the atmospheric energy spectrum: Depth-limited temperature anomalies at the tropopause. *Proc. Natl. Acad. Sci. USA* **2006**, *103*, 14690. [CrossRef]
10. Buckingham, C.E.; Naveira Garabato, A.C.; Thompson, A.F.; Brannigan, L.; Lazar, A.; Marshall, D.P.; George Nurser, A.J.; Damerell, G.; Heywood, K.J.; Belcher, S.E. Seasonality of submesoscale flows in the ocean surface boundary layer. *Geophys. Res. Lett.* **2016**, *43*, 2118–2126. [CrossRef]
11. D'Asaro, E.; Lee, C.; Rainville, L.; Harcourt, R.; Thomas, L. Enhanced Turbulence and Energy Dissipation at Ocean Fronts. *Science* **2011**, *332*, 318. [CrossRef] [PubMed]
12. Haza, A.C.; Özgökmen, T.M.; Griffa, A.; Molcard, A.; Poulain, P.-M.; Peggion, G. Transport properties in small-scale coastal flows: Relative dispersion from VHF radar measurements in the Gulf of La Spezia. *Ocean Dyn.* **2010**, *60*, 861–882. [CrossRef]

13.   Kim, S.Y. Observations of submesoscale eddies using high-frequency radar-derived kinematic and dynamic quantities. *Cont. Shelf Res.* **2010**, *30*, 1639–1655. [CrossRef]

14.   Lekien, F.; Coulliette, C. Chaotic stirring in quasi-turbulent flows. *Philos. Trans. R. Soc. A* **2007**, *365*, 3061–3084. [CrossRef]

15.   Shcherbina, A.Y.; D'Asaro, E.A.; Lee, C.M.; Klymak, J.M.; Molemaker, M.J.; McWilliams, J.C. Statistics of vertical vorticity, divergence, and strain in a developed submesoscale turbulence field. *Geophys. Res. Lett.* **2013**, *40*, 4706–4711. [CrossRef]

16.   Yoo Jang, G.; Kim Sung, Y.; Kim Hyeon, S. Spectral Descriptions of Submesoscale Surface Circulation in a Coastal Region. *J. Geophys. Res.* **2018**. [CrossRef]

17.   Kim, S.Y.; Terrill, E.J.; Cornuelle, B.D.; Jones, B.; Washburn, L.; Moline, M.A.; Paduan, J.D.; Garfield, N.; Largier, J.L.; Crawford, G.; et al. Mapping the U.S. West Coast surface circulation: A multiyear analysis of high-frequency radar observations. *J. Geophys. Res.* **2011**, *116*. [CrossRef]

18.   Lai, Y.; Zhou, H.; Yang, J.; Zeng, Y.; Wen, B. Submesoscale Eddies in the Taiwan Strait Observed by High-Frequency Radars: Detection Algorithms and Eddy Properties. *J. Atmos. Ocean. Technol.* **2017**, *34*, 939–953. [CrossRef]

19.   Soh, H.S.; Kim, S.Y. Diagnostic Characteristics of Submesoscale Coastal Surface Currents. *J. Geophys. Res.* **2018**, *123*, 1838–1859. [CrossRef]

20.   Choi, J.-K.; Park, Y.J.; Ahn, J.H.; Lim, H.-S.; Eom, J.; Ryu, J.-H. GOCI, the world's first geostationary ocean color observation satellite, for the monitoring of temporal variability in coastal water turbidity. *J. Geophys. Res.* **2012**, *117*. [CrossRef]

21.   Garfield, N.; Hubbard, M.; Pettigrew, J. Providing SeaSonde high-resolution surface currents for the America's Cup. In Proceedings of the 2011 IEEE/OES 10th Current, Waves and Turbulence Measurements (CWTM), Monterey, CA, USA, 20–23 March 2011; pp. 47–49.

22.   Pomales-Velázquez, L.; Morell, J.; Rodriguez-Abudo, S.; Canals, M.; Capella, J.; Garcia, C. Characterization of mesoscale eddies and detection of submesoscale eddies derived from satellite imagery and HF radar off the coast of southwestern Puerto Rico. In Proceedings of the OCEANS 2015-MTS/IEEE Washington, Monterey, CA, USA, 19–22 October 2015; pp. 1–6.

23.   Hu, J.; Wang, X.H. Progress on upwelling studies in the China seas. *Rev. Geophys.* **2016**, *54*, 653–673. [CrossRef]

24.   Huang, W.; Wang, W.; Zhang, W.; Zhang, J. Analysis of Characteristics of the Tides and the Tidal Currents in Adjacent Waters of Lianyungang Port. *Zhejiang Hydrotech.* **2012**, *3*, 1–5. [CrossRef]

25.   Chavanne, C.; Flament, P.; Gurgel, K.-W. Interactions between a Submesoscale Anticyclonic Vortex and a Front. *J. Phys. Oceanogr.* **2010**, *40*, 1802–1818. [CrossRef]

26.   Jeong, J.; Hussain, F. On the identification of a vortex. *J. Fluid Mech.* **1995**, *285*, 69–94. [CrossRef]

27.   Sadarjoen, I.A.; Post, F.H. Geometric Methods for Vortex Extraction. In Proceedings of the Data Visualization '99, Vienna, Austria, 26–28 May 1999; pp. 53–62.

28.   Chaigneau, A.; Pizarro, O. Eddy characteristics in the eastern South Pacific. *J. Geophys. Res.* **2005**, *110*. [CrossRef]

29.   McWilliams, J.C. The vortices of two-dimensional turbulence. *J. Fluid Mech.* **1990**, *219*, 361–385. [CrossRef]

30.   Morrow, R.; Donguy, J.-R.; Chaigneau, A.; Rintoul, S.R. Cold-core anomalies at the subantarctic front, south of Tasmania. *Deep Sea Res. Part I* **2004**, *51*, 1417–1440. [CrossRef]

31.   Ari Sadarjoen, I.; Post, F.H. Detection, quantification, and tracking of vortices using streamline geometry. *Comput. Graph.* **2000**, *24*, 333–341. [CrossRef]

32.   Sadarjoen, I.A.; Post, F.H.; Bing, M.; Banks, D.C.; Pagendarm, H. Selective visualization of vortices in hydrodynamic flows. In Proceedings of the Visualization '98 (Cat. No.98CB36276), Triangle Park, NC, USA, 18–23 October 1998; pp. 419–422.

33.   Nencioli, F.; Dong, C.; Dickey, T.; Washburn, L.; McWilliams, J.C. A Vector Geometry–Based Eddy Detection Algorithm and Its Application to a High-Resolution Numerical Model Product and High-Frequency Radar Surface Velocities in the Southern California Bight. *J. Atmos. Ocean. Technol.* **2010**, *27*, 564–579. [CrossRef]

34.   Heiberg, E.; Ebbers, T.; Wigstrom, L.; Karlsson, M. Three-dimensional flow characterization using vector pattern matching. *IEEE Trans. Vis. Comput. Graph.* **2003**, *9*, 313–319. [CrossRef]

35.   Ebling, J.; Scheuermann, G. Clifford convolution and pattern matching on vector fields. In Proceedings of the IEEE Visualization, VIS 2003, Seattle, WA, USA, 19–24 October 2003; pp. 193–200.

36.   Chaigneau, A.; Gizolme, A.; Grados, C. Mesoscale eddies off Peru in altimeter records: Identification algorithms and eddy spatio-temporal patterns. *Prog. Oceanogr.* **2008**, *79*, 106–119. [CrossRef]

37. Dong, C.; Nencioli, F.; Liu, Y.; McWilliams, J.C. An Automated Approach to Detect Oceanic Eddies From Satellite Remotely Sensed Sea Surface Temperature Data. *IEEE Geosci. Remote Sens. Lett.* **2011**, *8*, 1055–1059. [CrossRef]

38. Hu, C. An empirical approach to derive MODIS ocean color patterns under severe sun glint. *Geophys. Res. Lett.* **2011**, *38*. [CrossRef]

39. Liu, Y.Y.; Weisberg, R.H.R.H.; Hu, C.C.; Kovach, C.C.; RiethmüLler, R.R. Evolution of the Loop Current System During the Deepwater Horizon Oil Spill Event as Observed with Drifters and Satellites. *Monit. Modeling Deep. Horiz. Oil Spill* **2011**, *195*, 91–101.

40. Zhang, Y.; Hu, C.; Liu, Y.; Weisberg, R.H.; Kourafalou, V.H. Submesoscale and Mesoscale Eddies in the Florida Straits: Observations from Satellite Ocean Color Measurements. *Geophys. Res. Lett.* **2019**, *46*, 13262–13270. [CrossRef]

41. Sobel, I.; Feldman, G. A 3 × 3 Isotropic Gradient Operator for Image Processing. In *Pattern Classification and Scene Analysis*; Duda, R., Hart, P., Eds.; John Wiley & Sons: Stanford, CA, USA, 1968; pp. 271–272.

42. Dong, C.; Mavor, T.; Nencioli, F.; Jiang, S.; Uchiyama, Y.; McWilliams, J.C.; Dickey, T.; Ondrusek, M.; Zhang, H.; Clark, D.K. An oceanic cyclonic eddy on the lee side of Lanai Island, Hawai'i. *J. Geophys. Res.* **2009**, *114*. [CrossRef]

43. Liu, Y.; Dong, C.; Guan, Y.; Chen, D.; McWilliams, J.; Nencioli, F. Eddy analysis in the subtropical zonal band of the North Pacific Ocean. *Deep Sea Res. Part I* **2012**, *68*, 54–67. [CrossRef]

44. Ji, J.; Dong, C.; Zhang, B.; Liu, Y.; Zou, B.; King, G.P.; Xu, G.; Chen, D. Oceanic Eddy Characteristics and Generation Mechanisms in the Kuroshio Extension Region. *J. Geophys. Res.* **2018**, *123*, 8548–8567. [CrossRef]

45. Liu, C.-L.; Chang, M.-H. Numerical Studies of Submesoscale Island Wakes in the Kuroshio. *J. Geophys. Res.* **2018**. [CrossRef]

46. Yang, H.; Choi, J.-K.; Park, Y.-J.; Han, H.-J.; Ryu, J.-H. Application of the Geostationary Ocean Color Imager (GOCI) to estimates of ocean surface currents. *J. Geophys. Res.* **2014**, *119*, 3988–4000. [CrossRef]

47. Crocker, R.I.; Matthews, D.K.; Emery, W.J.; Baldwin, D.G. Computing Coastal Ocean Surface Currents From Infrared and Ocean Color Satellite Imagery. *IEEE Trans. Geosci. Remote Sens.* **2007**, *45*, 435–447. [CrossRef]

48. Kozlov, I.E.; Artamonova, A.V.; Manucharyan, G.E.; Kubryakov, A.A. Eddies in the Western Arctic Ocean From Spaceborne SAR Observations Over Open Ocean and Marginal Ice Zones. *J. Geophys. Res.* **2019**, *124*, 6601–6616. [CrossRef]

49. Leung, P.-T.; Gibson, C.H. Turbulence and fossil turbulence in oceans and lakes. *Chin. J. Oceanol. Limnol.* **2004**, *22*, 1–23. [CrossRef]

Article

Estimation of Hourly Sea Surface Salinity in the East China Sea Using Geostationary Ocean Color Imager Measurements

Dae-Won Kim [1], Young-Je Park [2], Jin-Yong Jeong [2] and Young-Heon Jo [1,*]

[1] Department of Oceanography, Pusan National University, Geumjeong-Gu, Busan 46241, Korea; daewon@pusan.ac.kr

[2] Korea Institute of Ocean Science and Technology, Busan 49111, Korea; youngjepark@kiost.ac.kr (Y.-J.P.); jyjeong@kiost.ac.kr (J.-Y.J.)

* Correspondence: joyoung@pusan.ac.kr; Tel.: +82-51-510-3372

Received: 23 January 2020; Accepted: 22 February 2020; Published: 25 February 2020

Abstract: Sea surface salinity (SSS) is an important tracer for monitoring the Changjiang Diluted Water (CDW) extension into Korean coastal regions; however, observing the SSS distribution in near real time is a difficult task. In this study, SSS detection algorithm was developed based on the ocean color measurements by Geostationary Ocean Color Imager (GOCI) in high spatial and temporal resolution using multilayer perceptron neural network (MPNN). Among the various combinations of input parameters, combinations with three to six bands of GOCI remote sensing reflectance (Rrs), sea surface temperature (SST), longitude, and latitude were most appropriate for estimating the SSS. According to model validations with the Soil Moisture Active Passive (SMAP) and Ieodo Ocean Research Station (I-ORS) SSS measurements, the coefficient of determination (R^2) were 0.81 and 0.92 and the root mean square errors (RMSEs) were 1.30 psu and 0.30 psu, respectively. In addition, a sensitivity analysis revealed the importance of SST and the red-wavelength spectral signal for estimating the SSS. Finally, hourly estimated SSS images were used to illustrate the hourly CDW distribution. With the model developed in this study, the near real-time SSS distribution in the East China Sea (ECS) can be monitored using GOCI and SST data.

Keywords: sea surface salinity estimation; Changjiang diluted water; neural network; GOCI application; ocean color

1. Introduction

The Changjiang Diluted Water (CDW) discharge influences the maritime environment, not only in the East China Sea (ECS), but also in Korean coastal regions including the Jeju Island coastal regions (Figure 1) [1–9]. Especially, the CDW is strengthened by high precipitation in the East China region in the summer season [10–12]. During summer season, high solar radiation heats the surface water and enhances the upper ocean stratification via high sea surface temperature (SST) and low sea surface salinity (SSS) [9]; this can damage aquaculture via anomalous marine environmental conditions. For example, in the summer of 2016, an anomalously large CDW caused approximately $9 million in economic damage to Korean aquaculture by enhancing the stratification via low SSS and high SST [13] (pp. 1–2). In the ECS, regular SSS observation has been consistently conducted at stationary points, the Ieodo Ocean Research Station (I-ORS) (Figure 1), and the shipboard observation by Korean National Institute of Fisheries Science serial oceanographic observation. However, such spatio-temporally limited surveys make it difficult to determine the exact spatial and temporal CDW distribution and its paths.

Accordingly, satellite sensor measurements are useful to understand the features of the CDW and its extension into the surrounding regions. The space borne SSS datasets are currently available through the Soil Moisture and Ocean Salinity (SMOS) and Soil Moisture Active Passive (SMAP) missions, which have global coverage. However, these two sensors have limitations for observing near-real time SSS variations in coastal regions because those microwave measurements have coarse spatial resolutions (30–100 km), low revisit frequencies (three days or longer), and experience land contamination [14–19]. Moreover, SMAP product has the issues of radio-frequency interference (RFI) in the ECS [20]. Therefore, an accurate investigation of the SSS near coastal regions is needed to understand the variations and distribution of the CDW [21].

To continuously monitor the SSS distribution along coastal regions, ocean color measurements have been used in various coastal regions [22–28]. Some studies have primarily applied the relationship between the SSS and the colored dissolved organic matter (CDOM) in different regions, such as the Clyde Sea [23], the ECS [6,25,26], and Osaka Bay in Japan [21]. A large portion of the CDOM is caused by river plumes originating from terrestrial sources and is associated with fresh water. The freshwater from river plumes has a low reflectance for blue wavelengths due to the absorption of blue wavelengths by the CDOM; this feature allows for the detection of CDOM using optical sensors. The relationship between the CDOM and the salinity indicates that low salinity conditions are expected when the CDOM concentration is high. However, the relationships between the CDOM optical properties and the SSS reported in previous studies have varied in space and time due to local river characteristics and their seasonalities [6,29]. In addition, the relationship between the CDOM optical properties and the salinity can be different even in the same region because the CDOM can be changed, not only by salinity, but also by other factors, such as the chlorophyll concentration and suspended sediment type differences [18,22,24].

To overcome such limitations, Geiger et al. [18] used a neural network (NN) approach to estimate the SSS in the coastal Mid-Atlantic region using marine environmental variables and ocean color measurements. The strengths of a NN approach are that the assumptions made by calculating the CDOM from optical signals are not required, and that the inclusion of other parameters, such as SST, is possible. Indeed, they demonstrated that the SST is an essential parameter to estimate the SSS in four different coastal regions (the coastal Mid-Atlantic region, the Chesapeake and Delaware bays, and the Hudson Estuary). In addition, Chen and Hu [19] estimated the SSS from ocean color and SST measurements using a multilayer perceptron neural network (MPNN) in the northern Gulf of Mexico after extensive evaluations testing many other empirical approaches (i.e., principle component analysis, multi-nonlinear regression, decision tree, random forest, and supporting vector machines). They demonstrated the robustness of their MPNN model for estimating the SSS using the input errors. Because the CDW is highly correlated with the SST in the ECS in the summer season [9], an MPNN approach was used in this study instead of a CDOM approach in order to use the SST measurement. The importance of SST is discussed via an evaluation of the model performance in this study.

The highest spatial and temporal resolution ocean color sensor measurements available in the ECS are those of the Geostationary Ocean Color Imager (GOCI) with spatial and temporal resolutions of 500 m × 500 m and eight times per day, respectively [30,31]. Compared to polar orbiting satellite ocean color products, GOCI has an advantage for monitoring the CDW distribution because it observes the same area eight times per day with high spatial resolution. Therefore, GOCI data were used to develop the SSS estimation model and produce SSS images in the ECS. Liu et al. [27], Nakada et al. [21], and Sun et al. [28] attempted to analyze the SSS using GOCI remote sensing reflectance (Rrs) measurements in the Bohai Sea, the Osaka Bay, and the Southern Yellow Sea (SYS), respectively. Liu et al. [27] developed a simple multi-linear statistical regression model to estimate the SSS using GOCI data. Their study demonstrated the ability of GOCI to observe the hourly SSS variability validated by the buoy measured SSS. Nakada et al. [21] estimated the SSS using a CDOM optical absorption coefficient from GOCI. Using a time series of in-situ SSS data from an automated observation system, they improved their satellite-derived SSS accuracy. Their study also compared a plume area before and after a typhoon period. Sun et al. [28] estimated the SSS for the month of August from 2011 to 2018 in the SYS and achieved a root mean square error (RMSE) of 0.29 psu.

Nevertheless, the algorithms discussed in these studies were specific to each study region and require further validations to apply to other regions. Moreover, the three studies described above analyzed the same ocean color satellite data (GOCI) to develop their algorithms with different wavelength GOCI bands (i.e., Liu et al. [27] used Rrs490, Rrs555, and Rrs660; Nakada et al. [21] used Rrs412 and Rrs555; and Sun et al. [28] used Rrs490 and Rrs555).

Regardless of the method and sensor, previous studies used different wavelengths either in different regions or in the same region. Therefore, this study aims to test various combinations of wavelengths to develop a new model for estimating SSS using the GOCI data in the ECS. Accordingly, eight combinations of input parameters using GOCI Rrs bands 1–6, three band ratios (band 1/band 4, band 2/band 4, and band 2/band 4), and the SST were analyzed. In addition, the SMAP-derived SSS products were used as an output parameter due to the lack of available in-situ data. The accuracy of the SMAP SSS products in the ECS was evaluated using the I-ORS SSS measurements (the result is described in Section 2.1.2). Moreover, a sensitivity analysis to evaluate the importance of the input parameters was conducted to determine the relationship between each input parameter and the predicted SSS, as shown in Section 3.3. Finally, hourly SSS images generated by the proposed model are discussed in Section 4.

2. Materials and Methods

2.1. Materials

2.1.1. GOCI Rrs Data

Ocean color data from GOCI can be downloaded from the Korea Ocean Satellite Center website (http://kosc.kiost.ac.kr/). GOCI has six visible bands (412 nm, 443 nm, 490 nm, 555 nm, 660 nm, and 680 nm) and two near-infrared bands (745 nm and 865 nm). Because the two near-infrared bands are primarily used for atmospheric correction to GOCI, the six visible Rrs bands were used in this study. According to previous studies, not only the visible bands but also the GOCI band ratios (412 nm/555 nm (R1), 443 nm/555 nm (R2), and 490 nm/555 nm (R3)) can be used to estimate the SSS using ocean color measurements [18,26,32]. Three band ratios are used representing for the spectral shape in the blue–green portion of the spectrum, which varies with CDOM concentration [18,26,33]. GOCI L1B image covers Korea, Japan, the eastern coast of China, and parts of the northern coast of Taiwan (2500 km × 2500 km) and is composed of 16 subimages called slots [34]. A part of GOCI images, slot number 9 and 10 covering ECS, Chanjiang River mouth, Jeju Island, and the south coast of Korea are used in this study.

2.1.2. I-ORS Data

The I-ORS is located at the boundary between the Yellow Sea and the ECS (125.18°E 32.12°N), and the CDW frequently passes through this location (Figure 1). Due to the favorable geographical location of the I-ORS, various oceanic, metrological, and coastal studies have been conducted using measurements at I-ORS [35,36]. The I-ORS salinity data used in this study were measured at various depths (3 m, 5 m, 10 m, 15 m, 20 m, 30 m, and 40 m) with sampling intervals of 1 min. To match the SMAP data, 3-m depth salinity data were used after daily averaging (or sampling). During the study period (the year of 2016), 141 days of SSS data were available (April–October). These data were used to evaluate the accuracy of the SSS data from SMAP and the performance of the developed algorithm.

2.1.3. Group for High Resolution Sea Surface Temperature (GHRSST) SST Data

In addition to the GOCI Rrs data, SST data were also used as input variables for the algorithm to estimate the SSS. SST data have been used in previous studies to estimate the SSS using the NN approach [18,19]; however, the influence of the SST on the SSS is not clearly understood. Therefore, algorithms trained with or without SST were compared to evaluate the impact of the SST on the SSS.

In this study, the GHRSST level 4 data were used to train the NN model. This measurement produces a 0.01° × 0.01° spatial resolution and a daily temporal resolution. The GHRSST data are available on the GHRSST website (https://www.ghrsst.org/). The GHRSST level 4 gridded products are generated by combining complementary satellite and in-situ observations within optimal interpolation systems. The differences in the validation results when training with or without SST are discussed in this paper.

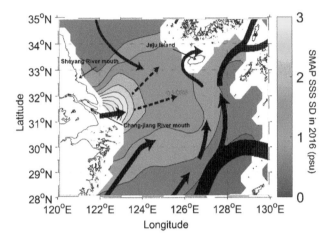

Figure 1. Soil Moisture Active Passive (SMAP) standard deviation (SD) of sea surface salinity (SSS) in 2016. Black arrows indicate typical ocean currents during summer season from 2002 to 2005 in this region (provided by Korea Hydrographic and Oceanographic Agency). Location of Ieodo Ocean Research Station (I-ORS) is marked with a solid red circle. Regions of more than 1 psu of SD were chosen to train multilayer perceptron neural network (MPNN) algorithm.

2.1.4. SMAP Data

Even though the primary mission of SMAP is to measure the soil moisture, the L-band radiometer of the SMAP platform can be used to measure the SSS in the global ocean. In this study, the version 4.0 of SMAP level 3 was used (http://www.remss.com/missions/smap/salinity/). The RFI was filtered in the version 4.0 of SMAP product. The SMAP level 3 product is provided with a 25-km spatial resolution and a daily temporal resolution (8-day running average). The near-polar orbits of SMAP cover the global oceans in 3 days with a repeat cycle of 8 days. To verify the SMAP performance, a validation with 8-day running averaged I-ORS data was conducted (Figure 2). The scatter diagram in Figure 2 shows the relationship between the SSS from I-ORS and that from SMAP with a coefficient of determination (R^2) of 0.71 and an RMSE of 1.02 psu. The SSS from SMAP agreed well with the I-ORS measurement compared to previous results (RMSE of approximately 3 psu near the Changjiang River mouth) obtained by Wu et al. [37]. Using this relationship, the SMAP SSS in the ECS could be correlated to the output parameter of the MPNN model despite it not being an in-situ measurement.

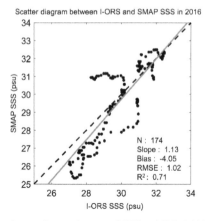

Figure 2. Scatter diagram between I-ORS and SMAP SSS in 2016.

2.2. Methods

2.2.1. GOCI Data Preprocessing

To reduce the noise contamination of the GOCI Rrs data, two analyses were conducted. First, Lee et al.'s [38] suspended particulate matter (SPM) speckle removing method was applied. This methodology effectively eliminated unexpected noise and some extremely high or low SPM values. Second, a novel quality assurance (QA) system was included [39]. The QA system was applied by measuring the Rrs data quality as a score between 0 and 1, with 1 indicating a perfect Rrs spectrum and 0 indicating an unusable Rrs spectrum. In this study, a QA score under 0.5 was treated as an error and removed. Following these two steps, the GOCI error was significantly decreased for the GOCI SSS images compared to the error obtained before preprocessing the SSS images.

2.2.2. MPNN

When developing the MPNN, a back-propagation learning technique with adaptive moment estimation (Adam) optimization and Rectified Linear Unit (ReLU) activation function were implemented in TensorFlow [40]. Figure 3 shows an example of an MPNN schematic diagram, which is composed of the inputs and the hidden and output layers. The training process was conducted 5000 times to improve the accuracy by adjusting the weights and biases of connections by changing the numbers of neurons and hidden layers. Table 1 shows detailed information concerning the training of the MPNN. All datasets were randomly divided into training (80%) and validation (20%) datasets in the training period. R^2 and RMSE were calculated using the validation dataset, which was an independent from the training dataset. All input parameters were normalized before train.

Since detection of low salinity water mass is important, the network was trained and validated with as many relatively low salinity conditions as possible. In the summer of 2016, a large amount of freshwater from the Changjiang River plume was observed [9,13,41]. Therefore, the year 2016 was selected as a study period when relatively low SSS covered wide regions in the ECS. The first attempt to train the MPNN was made using all pixels of slots 9 and 10 of the GOCI products. However, the training result showed a bias toward relatively high SSS ranges (33–35 psu), which are normal salinity conditions. Therefore, the training datasets were re-chosen, where the standard deviation (SD) of the SMAP SSS was more than 1 psu to prevent the MPNN from biasing relatively high SSS. Figure 1 shows the SD of the SMAP SSS in 2016 and reveals the frequently CDW-affected regions.

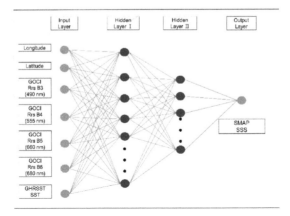

Figure 3. An example of a schematic diagram of the MPNN using D3 (longitude, latitude, Rrs490, Rrs555, Rrs660, Rrs680, and SST), consisting of input, hidden, and output layers. The number of neurons were different with dataset (Table 1).

Table 1. Detail training information of optimized multilayer perceptron neural network (MPNN).

Input dataset	Train R^2	Train RMSE (psu)	Validation R^2	Validation RMSE (psu)
Bands 3 to 6 (D1)	0.75	1.73	0.68	1.73
Bands 3 to 6 and R3 (D2)	0.87	1.10	0.67	1.73
Bands 3 to 6 and SST (D3)	0.85	1.15	0.81	1.30
Bands 3 to 6, R3, and SST (D4)	0.87	1.10	0.76	1.51
Bands 1 to 6 (D5)	0.81	1.44	0.67	1.74
Bands 1 to 6, R1, R2, and R3 (D6)	0.79	1.50	0.63	1.92
Bands 1 to 6 and SST (D7)	0.85	1.28	0.79	1.39
Bands 1 to 6, R1, R2, R3, and SST (D8)	0.85	1.15	0.81	1.33

In this study, eight different combinations of input parameters (longitude, latitude, Rrs bands 1 (412 nm)–6 (680 nm), R1, R2, R3, and SST) were used to develop the MPNN model, and schematic diagram of one example network is illustrated in Figure 3. The eight input datasets D1–D8 are defined as follows: D1 consists of Rrs bands 3–6; D2 consists of Rrs bands 3–6 and R3; D3 consists of Rrs bands 3–6 and SST; D4 consists of Rrs bands 3–6, R3, and SST; D5 consists of Rrs bands 1–6; D6 consists of Rrs bands 1–6, R1, R2, and R3; D7 consists of Rrs bands 1–6 and SST; and D8 consists of Rrs bands 1–6, R1, R2, and R3, and SST, and all datasets include the longitude and latitude (Table 1). Out of daily SMAP SSS images from January 1, 2016, to December 31, 2016, every four-day images (92 images) are used for training and the remaining 274 images were used to validate the output of the MPNN.

3. Results

3.1. Model Performance Compared to SMAP SSS

Figure 4 shows scatter diagrams between the MPNN results and the SMAP SSS for validation data with the eight different input datasets (D1–D8). Noticeable differences are observed with and without SST in the input datasets. In the case of input datasets without SST (Figure 4a,b,e and f), R^2 is lower and RMSE is larger than for input datasets with SST (Figure 4c,d,g and h) by approximately 0.13 and 0.40 psu, respectively. Conversely, the inclusion of band ratios (Figure 4b,d,f and h) did not improve the model performance compared to the inclusion of SST. It appears that the SST is significantly related to the SSS in this region. A comparison between the four-band group (Figure 4a–d) and the six-band group (Figure 4e–h) performances reveal no significant differences. This result indicates that the blue-wavelength (bands 1 and 2 of GOCI) spectral signals are not important parameters for estimating the SSS. Of the eight different scatter diagrams in Figure 4, the D3 dataset result shows the best performance (R^2 of 0.81 and RMSE of 1.30 psu).

In addition, Figures 5 and 6 show maps of R^2 and RMSE between the predicted SSS and the SMAP SSS. Notably, in these two figures, the overall performance increased when including SST; this is shown clearly in Figure 5. Without SST (D1, D2, D5, and D6), the R^2 values in the western and eastern areas were under 0.8 and 0.4, respectively. However, including SST, the R^2 values of the overall regi- on were approximately 0.8, except near the Sheyang River mouth region (in the northwestern part of the study area, see Figure 1). The eastern part of the study area includes the Kuroshio Current, which has high SST and SSS. Moreover, the Kuroshio Current system is known to have a high SST and SSS water type. Therefore, including the SST could improve the model for the SSS estimation in the overall region, especially in the eastern part of the study area.

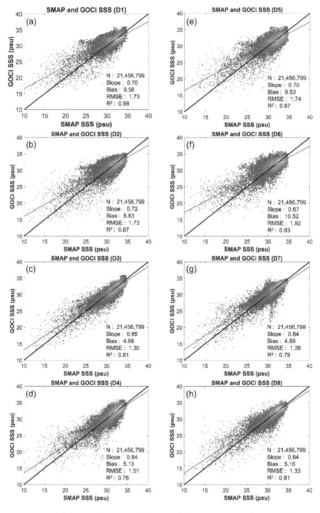

Figure 4. Density scatter diagram between SMAP and Geostationary Ocean Color Imager (GOCI) SSS estimated by (**a**) D1, (**b**) D2, (**c**) D3, (**d**) D4, (**e**) D5, (**f**) D6, (**g**) D7, and (**h**) D8 input dataset. Color indicates scatter density.

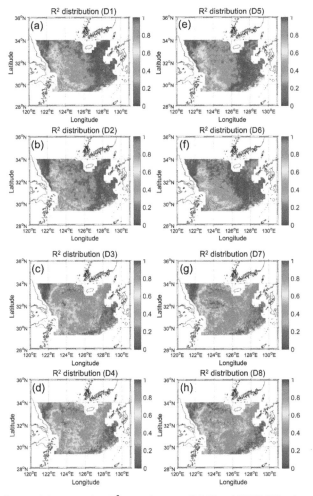

Figure 5. Coefficient of determination (R^2) maps between SMAP and GOCI SSS estimated by (**a**) D1, (**b**) D2, (**c**) D3, (**d**) D4, (**e**) D5, (**f**) D6, (**g**) D7, and (**h**) D8 input dataset.

More detailed differences are found in Figure 6. Every panel in Figure 6 has relatively high RMSE values between the SMAP and predicted SSS near the Changjiang River mouth region. However, the areas near the Sheyang River mouth region, the southwestern area of Jeju Island, and the Kuroshio Current region (the southeastern part of the study area) show different RMSE features in the eight RMSE maps. In the cases that do not include the SST in the input datasets (Figure 6a,b,e and f), relatively high RMSE values are seen in the southwestern area of Jeju Island and the Kuroshio Current region. Conversely, the RMSE is decreased when including the SST as an input parameter in this area (Figure 6c,d,g, and h). Of the four RMSE maps including SST (Figure 6c,d,g, and h), the D3 dataset result shows the relatively good RMSE distribution (Figure 6c). Near the Sheyang River mouth region, the four-band group (Figure 6a,b,c and d) and six-band group (Figure 6e,f,g and h) show opposite features when including the band ratios and SST. The four-band group shows small changes in the RMSE due to the inclusion of just the band ratio (R3) or the SST (Figure 6b,c, respectively); however, when including both the band ratio and the SST, the RMSE is increased notably in this region (Figure 6d). Conversely, the inclusion of the three band ratios and the SST decreased the RMSE of the six-band group (Figure 6h), while the individual parameters (the band ratios or SST) increased the RMSE in this region (Figure 6f,g, respectively). Similar to the spatial pattern changes of the R^2 image caused by different input datasets (Figure 5), the RMSE decreased overall in the study area when using the SST as an input parameter. The decrease in the RMSE was remarkable west of the Jeju Island area and the Kuroshio Current region. The RMSE results when including the SST and band ratios show an opposite effect for the four-band group (Figure 6c,d) and the six-band group (Figure 6g,h). In Figure 6d, RMSE increased compared to Figure 6c near the Sheyang River mouth and in the eastern parts of the study regions. Conversely, the SST and the three band ratios improve the MPNN performance in the six-band group (Figure 6g,h) near the Sheyang River mouth and around the 124°E 32.5°N area. By including SST in the input dataset, the MPNN performances (R^2 and RMSE) improved significantly in the overall study region (Figures 4–6). Therefore, SST is an important parameter for estimating the SSS in this study.

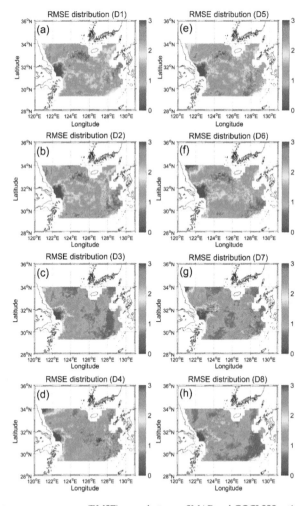

Figure 6. Root mean square error (RMSE) maps between SMAP and GOCI SSS estimated by (**a**) D1, (**b**) D2, (**c**) D3, (**d**) D4, (**e**) D5, (**f**) D6, (**g**) D7, and (**h**) D8 input dataset.

3.2. Model Performance Compared to I-ORS SSS

Figure 7 shows scatter plots of the validation with the I-ORS SSS measurements in 2016. Of the 141 days from April 30 to October 1, 8 days were validated due to the GOCI data availability over the I-ORS during this period. Even though only 8 days were validated, the performance of the MPNN varied with the eight different input datasets. The ranges of R^2 and RMSE were 0.58–0.94 and 0.25–0.88 psu, respectively. Of the eight datasets, D3, D4, and D7 had R^2 values of more than 0.90 and RMSE values of less than 0.35 psu (Figure 7c,d and g). Even though the D7 dataset results show the best performance for I-ORS (Figure 7g), the R^2 and RMSE distributions of the overall study area indicate that the MPNN model with D3 performs better than that with D7. The developed MPNN model has an impressive performance compared to previous studies that estimated the SSS using GOCI. R^2 value of the SSS estimation model from Liu et al. [27] was 0.795. Nakada et al. [21] estimated the SSS using GOCI in Osaka Bay and achieved R^2 of 0.57 and RMSE of 1.5 psu. Even though the SSS estimation model developed by Sun et al. [28] had a relatively low RMSE (0.29 psu), its R^2 value was 0.62. In comparison to these three studies, the performance of the MPNN model with D3 has a relatively high R^2 (0.92) and low RMSE (0.30 psu). Therefore, the MPNN model using the GOCI and SST data successfully estimated the SSS near the I-ORS regions.

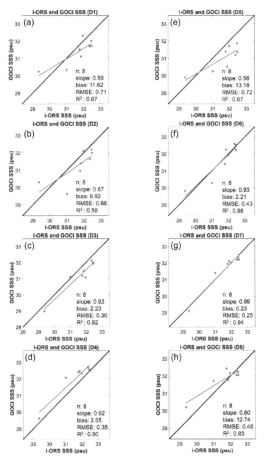

Figure 7. Scatter diagrams between I-ORS and GOCI SSS estimated by (**a**) D1, (**b**) D2, (**c**) D3, (**d**) D4, (**e**) D5, (**f**) D6, (**g**) D7, and (**h**) D8 input dataset.

3.3. Importance of Input Parameters

The MPNN validation results were described using four different approaches (validation with SMAP, R^2 and RMSE maps, and validation with I-ORS). Consequently, the MPNN training with the D3 dataset (longitude, latitude, Rrs490, Rrs555, Rrs660, Rrs680, and SST) was used to determine the SSS estimation model. Even though the D7 result was better than the D3 result for the validation with I-ORS (Figure 7), the D7 result had a relatively high RMSE near the Sheyang River mouth and the ECS (around 124°E 32.3°N) compared to the D3 result in Figure 6. To understand the importance of the input parameters for the SSS estimation, a sensitivity analysis was conducted. Instead of analyzing the sensitivity analysis for the entire region, the location of I-ORS was chosen because the CDW passes through this location frequently. Figure 8 shows the sensitivity analysis results for the five input parameters at the I-ORS location. The analysis was conducted by changing one parameter from the minimum to the maximum of its real value while fixing the other four parameters at their mean values. For example, the blue line indicates the sensitivity of Rrs490 for the SSS estimation when the other parameters were constant (at their mean values at the I-ORS, i.e., Rrs490: 0.0110, Rrs555: 0.0107, Rrs660: 0.0046, Rrs680: 0.0043, and SST: 19.97°C). According to Figure 8, a relatively high SSS (>34 psu) was associated with high Rrs660 values (orange line) and a relatively low SSS (<28 psu) was associated with high Rrs680 (red line); the middle range of the SSS (28–34 psu) was adjusted using all of the input parameters. The SSS variability shows positive relationships with Rrs490 and Rrs660 and negative relationships with Rrs555, Rrs680, and SST. The sensitivity analysis results show the significance of the red-wavelength spectral signal at I-ORS. This result is different from those of other previous studies that did not include red-wavelength signals in their algorithms [6,21,23,25–28]. In addition to the significance of SST, the importance of the red-wavelength spectrum signal for the SSS estimation is implied from this result.

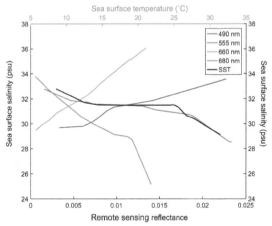

Figure 8. Sensitivity analysis results at the I-ORS location. Each line reveal the influence of input parameter on SSS variability with fixing the other parameters as mean value of real data(Rrs490: 0.0110, Rrs555: 0.0107, Rrs660: 0.0046, Rrs680: 0.0043, and SST: 19.97 °C).

3.4. Validation of the GOCI SSS

The model developed in this study enables the SSS distribution to be monitored with high spatial and temporal resolution using GOCI and SST measurements in the ECS. Figure 9 shows the monthly distribution of the SSS in 2016 as estimated by SMAP. The monthly SMAP SSS map shows that the SSS distribution in the ECS and the adjacent seas around the Korean Peninsula are primarily affected by the CDW in the summer season. However, SMAP SSS has coarse spatial scale (25 km) SSS distribution despite CDW extension revealed more detail shape and pattern in the past studies [26,42]. Conversely, the monthly GOCI SSS maps represent the SSS variability at small scales (500 m) as well (Figure 10). In this figure, intricate relatively low SSS features are observed even in the monthly averaged SSS images. In addition, Figure 11 shows a time series of three different SSS measurements at the I-ORS location. The SMAP and I-ORS SSS represent relatively low SSS conditions during the summer season. Even though the GOCI SSS did not observe this location during July and August, the overall pattern reveals a good agreement with the SMAP and I-ORS SSS. In addition, the model shows a reasonably good performance at the I-ORS location and is capable of detecting and monitoring the SSS in the ECS.

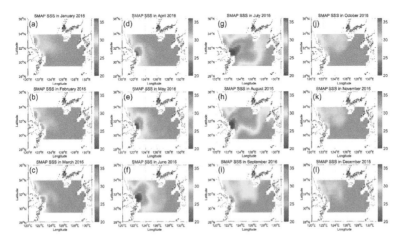

Figure 9. Monthly SSS map of SMAP in (**a**) January, (**b**) February, (**c**) March, (**d**) April, (**e**) May, (**f**) June, (**g**) July, (**h**) August, (**i**) September, (**j**) October, (**k**) November, and (**l**) December 2016. The Changjiang Diluted Water (CDW) was extended to Jeju Island in the summer season.

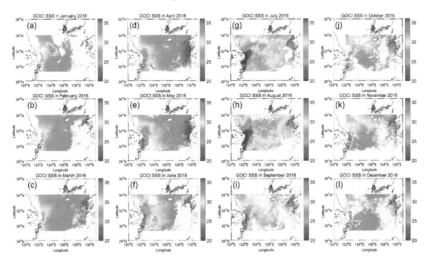

Figure 10. Monthly SSS map of GOCI in (**a**) January, (**b**) February, (**c**) March, (**d**) April, (**e**) May, (**f**) June, (**g**) July, (**h**) August, (**i**) September, (**j**) October, (**k**) November, and (**l**) December. General SSS patterns are similar to those of monthly SMAP images, but more specific features are revealed due to high spatial resolution.

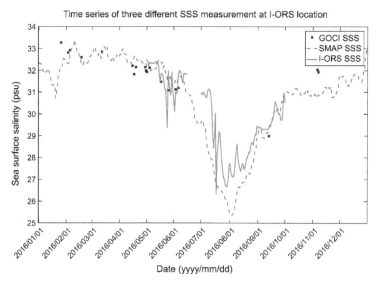

Figure 11. Time series of three different SSS measurement at I-ORS location. Overall patterns of three measurement are very similar.

4. Discussion

Eight different MPNN models were trained and validated in this study. The model developed using the D3 dataset showed the optimum performance. The overall model performances were an R^2 of 0.81 and an RMSE of 1.30 psu for SSS ranges between 1 psu and 36 psu (N = 21,269,124). In comparison to other studies, the MPNN model developed in this study shows a similar performance to the results from Chen and Hu [19] (R^2 of 0.86 and RMSE of 1.2 psu in the northern Gulf of Mexico) and a better performance than that of Geiger et al. [18] (RMSE of 1.40–2.29 psu). In Figure 5c, the R^2 values demonstrates the good agreement of the model with the SMAP SSS, except in the northwestern part of the study region (the Sheyang River mouth region). The relatively low R^2 in this region was caused by the lack of available GOCI data (less than 20 days in one year). Moreover, relatively fewer observations (approximately five days) were used for training in this region. In addition, Figure 5c–h revealed anomalously high or low values of R^2 in the southeastern part of the study region (the northwest Pacific region). The northwest Pacific region is known to have heavily clouded areas; unsurprisingly, these cloudy conditions led to a lack of observations in this region. The RMSE distribution also shows good agreement of the MPNN model in the ECS (Figure 6c). The reason for the relatively high RMSE near the Changjiang River mouth is that the SSS fluctuated widely over this region (from approximately 1 psu to 35 psu).

Finally, the validation with the I-ORS SSS measurement yielded an R^2 of 0.92 and an RMSE of 0.30 psu. Because the I-ORS is located between the Changjiang River mouth and Jeju Island, the SSS estimation model accuracy in this area is important to monitor the CDW extension to the Korean coastal region, including Jeju Island. The predicted SSS shows good agreement with the I-ORS data and supports the model accuracy near the I-ORS region. In addition, a sensitivity analysis was conducted to understand the importance of the input parameters for estimating the SSS at the I-ORS site (Figure 8). Each input parameter played a role in the MPNN model. Of these parameters, Rrs660 was associated with relatively high SSS (>34 psu) and Rrs680 was associated with relatively low SSS (<28 psu). The red wavelengths of Rrs are important in this model. This is an important result because many previous studies have estimated the SSS in the coastal region without using the red spectral signal but rather using the blue and green spectral signals [6,21,23,25–28]. The relationship between the optical features and the SSS differs near the Changjiang River mouth region and Jeju Island because the CDOM or suspended particles can settle with distance from the river mouth. Even if the same CDOM signal is observed at different locations, the SSS is not necessarily the same. Therefore, SSS estimations using simple equations, primarily derived from a CDOM approach, are likely inappropriate for estimating the SSS over wide areas. Red-wavelength signals have been found to be highly related to turbid water in various studies [43–49]. In addition, the chlorophyll concentration (Chl) has been found to be highly linked to the CDW summer distributions during the period of 1998–2007 [4]. Moreover, the red wavelength of the spectrum (675 nm) is the second peak of the Chl absorption spectra [50]. Therefore, the relationship between Chl in turbid and fresh water might facilitate the SSS estimation in the MPNN model. However, it is difficult to determine the influence of Chl using only one wavelength signal (680 nm). Therefore, the importance of Chl for estimating SSS needs to be verified in a future study. In addition, the relationship between the SSS and the red-wavelength signal is not clear. The influence of Chl in the CDW, and its effect on the SSS requires further study. Figure 12 shows the hourly SSS distribution near the Changjiang River mouth region on April 30, 2016.

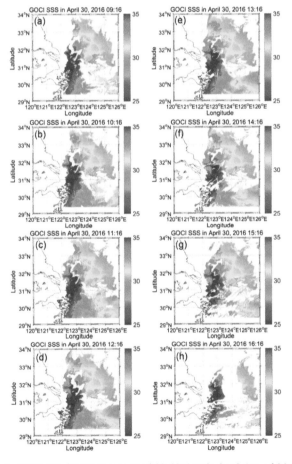

Figure 12. Hourly GOCI SSS distribution in April 30, 2016 at the local time of (**a**) 09:16, (**b**) 10:16, (**c**) 11:16, (**d**) 12:16, (**e**) 13:16, (**f**) 14:16, (**g**) 15:16, and (**h**) 16:16. Such hourly GOCI SSS measurements permit the SSS monitoring with high spatial and temporal resolutions.

In order to estimate hourly GOCI SSS using neural network, SST has to be observed at the same time as the important input as demonstrated in this study. However, such SST measurement from satellite sensor is not available during this study period. For the future hourly SSS estimation using GOCI and SST, hourly SST in the ECS can be used, which has been observed by the Advanced Meteorological Imager (AMI) sensor on Geostationary Earth Orbit (GEO)-Korea Multi-Purpose Satellite (KOMPSAT)-2A. Therefore, hourly SSS can be achieved based on the similar method suggested in this study using the hourly SST data from GEO-KOMPSAT-2A (GK-2A) and ocean color measurements from the GEO-KOMPSAT-2B (GK-2B). The former was launched on December 5, 2018, and the latter was launched February 19, 2020.

5. Conclusions

This study is the first attempt to estimate the SSS using GOCI with NN in the ECS. Since CDW volume is large during the summer season, the ocean upper layer in ECS is very stably stratified by low SSS and high SST. The previous studies for estimating SSS in the ECS using CDOM approach, did not consider the relationship between SSS and SST [6,25,26]. In this study, SST was used as one of input parameter, and the results indicated that the SST was crucial factor to obtain better SSS estimations. Therefore, the utilization of SST is highly recommended for estimating the coastal salinity using satellite ocean color measurements. Finally, hourly SSS distributions can be estimated using the MPNN model. Figure 12 shows the hourly SSS distribution near the Changjiang River mouth region on April 30, 2016. The eight panels show the continuous SSS distribution over eight hours. Hourly GOCI SSS data can be used for various fields. For example, hourly SSS values will be very useful information are needed to avoid the influence of low SSS at aquaculture sites. A typical aquaculture site is approximately 2 km long and coastal water typically moves at approximately 10 cm s^{-1}. At this rate, low SSS water would expand an aquaculture site within approximately 5.6 h. With that, aqua farmers may know its spatial trajectories and time to reach their site. In addition, SSS is an important parameter for not only marine physical conditions (e.g., density), but also the marine biogeochemical environment. Therefore, longer time series in the high resolution of SSS data would be useful for understanding CDW itself and its effect on physical-biogeochemical interactions in the ECS.

Author Contributions: Conceptualization, D.-W.K. and Y.-H.J.; methodology, D.-W.K.; software, D.-W.K.; validation, D.-W.K.; formal analysis, D.-W.K.; investigation, D.-W.K.; resources, J.-Y.J.; data curation, D.-W.K.; writing—original draft preparation, D.-W.K.; writing—review and editing, Y.-H.J., Y.-J.P. and J.-Y.J.; visualization, D.-W.K.; supervision, Y.-H.J.; project administration, Y.-H.J.; funding acquisition, Y.-H.J. All authors have read and agreed to the published version of the manuscript.

Funding: This research was funded by the project titled Technology development for Practical Applications of Multi-Satellite data to maritime issues funded by the Ministry of Oceans and Fisheries, Korea. The project titled "Construction of Ocean Research Station and their Application Studies" funded by the Ministry of Oceans and Fisheries, Korea had provided with the I-ORS data.

Conflicts of Interest: Authors declare no conflict of interest.

References

1. Chen, C.; Zhu, J.; Beardsley, R.C.; Franks, P.J. Physical-biological sources for dense algal blooms near the Changjiang River. *Geophys. Res. Lett.* **2003**, *30*, 1515–1518. [CrossRef]
2. Lee, N.K.; Suh, Y.S.; Kim, Y.S. Satellite remote sensing to monitor seasonal horizontal distribution of resuspended sediments in the East China Sea. *J. Korean Assoc. Geogr. Inf. Stud.* **2003**, *6*, 151–161.
3. Moon, J.H.; Pang, I.C.; Yoon, J.H. Response of the Changjiang diluted water around Jeju Island to external forcings: A modeling study of 2002 and 2006. *Cont. Shelf Res.* **2009**, *29*, 1549–1564. [CrossRef]
4. Kim, H.C.; Yamaguchi, H.; Yoo, S.; Zhu, J.; Okamura, K.; Kiyomoto, Y.; Tanaka, K.; Kim, S.W.; Park, T.; Ishizaka, J. Distribution of Changjiang diluted water detected by satellite chlorophyll-a and its interannual variation during 1998–2007. *J. Oceanogr.* **2009**, *65*, 129–135. [CrossRef]
5. Moon, J.H.; Hirose, N.; Yoon, J.H.; Pang, I.C. Offshore detachment process of the low-salinity water around Changjiang Bank in the East China Sea. *J. Phys. Oceanogr.* **2010**, *40*, 1035–1053. [CrossRef]

6. Bai, Y.; Pan, D.; Cai, W.J.; He, X.; Wang, D.; Tao, B.; Zhu, Q. Remote sensing of salinity from satellite-derived CDOM in the Changjiang River dominated East China Sea. *J. Geophys. Res. Ocean.* **2013**, *118*, 227–243. [CrossRef]

7. Kim, S.B.; Lee, J.H.; Matthaeis, P.; Yueh, S.; Hong, C.S.; Lee, J.H.; Lagerloef, G. Sea surface salinity variability in the E ast C hina S ea observed by the A quarius instrument. *J. Geophys. Res. Ocean.* **2014**, *119*, 7016–7028. [CrossRef]

8. Lee, D.K.; Kwon, J.I.; Son, S. Horizontal distribution of Changjiang Diluted Water in summer inferred from total suspended sediment in the Yellow Sea and East China Sea. *Acta Oceanol. Sin.* **2015**, *34*, 44–50. [CrossRef]

9. Moon, J.-H.; Kim, T.; Son, Y.B.; Hong, J.-S.; Lee, J.-H.; Chang, P.-H.; Kim, S.-K. Contribution of low-salinity water to sea surface warming in the summer of 2016. *Prog. Oceanogr.* **2019**, *175*, 68–88.

10. Suh, Y.S.; Jang, L.H.; Lee, N.K. Detection of low salinity water in the northern East China Sea during summer using ocean color remote sensing. *Korean J. Remote Sens.* **2004**, *20*, 153–162. (in Korean).

11. Yoon, H.J.; Cho, H.K. A study on the diluted water from the Yangtze River in the East China Sea using satellite data. *J. Korean Assoc. Geogr. Inf. Stud.* **2005**, *8*, 33–43. (in Korean).

12. Zhou, F.; Xuan, J.L.; Ni, X.B.; Huang, D.J. A preliminary study of variations of the Changjiang Diluted Water between August of 1999 and 2006. *Acta Oceanol. Sin.* **2009**, *28*, 1–11.

13. KHOA. High sea temperature phenomenon of August 2016. In *Unusual ocean analysis report*; KHOA (Korea Hydrographic and Oceanographic Agency): Busan, Republic of Korea, 2016; Volume 1. (in Korean)

14. Koblinsky, C.J.; Hildebrand, P.; LeVine, D.; Pellerano, F.; Chao, Y.; Wilson, W.; Yueh, S.; Lagerloef, G. Sea surface salinity from space: Science goals and measurement approach. *Radio Sci.* **2003**, *38*. [CrossRef]

15. Entekhabi, D.; Njoku, E.G.; O'Neill, P.E.; Kellogg, K.H.; Crow, W.T.; Edelstein, W.N.; Kimball, J. The soil moisture active passive (SMAP) mission. *Proc. IEEE* **2010**, *98*, 704–716. [CrossRef]

16. Font, J.; Camps, A.; Borges, A.; Martín-Neira, M.; Boutin, J.; Reul, N.; Kerr, Y.H.; Hahne, A.; Mecklenburg, S. SMOS: The challenging sea surface salinity measurement from space. *Proc. IEEE* **2009**, *98*, 649–665. [CrossRef]

17. Kerr, Y.H.; Waldteufel, P.; Wigneron, J.P.; Delwart, S.; Cabot, F.; Boutin, J.; Escorihuela, M.J.; Font, J.; Reul, N.; Juglea, S.E.; et al. The SMOS mission: New tool for monitoring key elements of the global water cycle. *Proc. IEEE* **2010**, *98*, 666–687. [CrossRef]

18. Geiger, E.F.; Grossi, M.D.; Trembanis, A.C.; Kohut, J.T.; Oliver, M.J. Satellite-derived coastal ocean and estuarine salinity in the Mid-Atlantic. *Cont. Shelf Res.* **2013**, *63*, S235–S242. [CrossRef]

19. Chen, S.; Hu, C. Estimating sea surface salinity in the northern Gulf of Mexico from satellite ocean color measurements. *Remote Sens. Environ.* **2017**, *201*, 115–132. [CrossRef]

20. Mohammed, P.N.; Aksoy, M.; Piepmeier, J.R.; Johnson, J.T.; Bringer, A. SMAP L-band microwave radiometer: RFI mitigation prelaunch analysis and first year on-orbit observations. *IEEE Trans. Geosci. Remote. Sens.* **2016**, *54*, 6035–6047. [CrossRef]

21. Nakada, S.; Kobayashi, S.; Hayashi, M.; Ishizaka, J.; Akiyama, S.; Fuchi, M.; Nakajima, M. High-resolution surface salinity maps in coastal oceans based on geostationary ocean color images: quantitative analysis of river plume dynamics. *J. Oceanogr.* **2018**, *74*, 287–304. [CrossRef]

22. Chen, R.F. In situ fluorescence measurements in coastal waters. *Org. Geochem.* **1999**, *30*, 397–409. [CrossRef]

23. Binding, C.E.; Bowers, D.G. Measuring the salinity of the Clyde Sea from remotely sensed ocean colour. *Estuar. Coast. Shelf Sci.* **2003**, *57*, 605–611. [CrossRef]

24. Del Vecchio, R.; Blough, N.V. Spatial and seasonal distribution of chromophoric dissolved organic matter and dissolved organic carbon in the Middle Atlantic Bight. *Mar. Chem.* **2004**, *89*, 169–187. [CrossRef]

25. Sasaki, H.; Siswanto, E.; Nishiuchi, K.; Tanaka, K.; Hasegawa, T.; Ishizaka, J. Mapping the low salinity Changjiang Diluted Water using satellite-retrieved colored dissolved organic matter (CDOM) in the East China Sea during high river flow season. *Geophys. Res. Lett.* **2008**, *35*. [CrossRef]

26. Ahn, Y.H.; Shanmugam, P.; Moon, J.E.; Ryu, J.H. Satellite remote sensing of a low-salinity water plume in the East China Sea. *Ann. Geophys. Atmos. Hydrospheres Space Sci.* **2008**, *26*, 2019–2035. [CrossRef]

27. Liu, R.; Zhang, J.; Yao, H.; Cui, T.; Wang, N.; Zhang, Y.; An, J. Hourly changes in sea surface salinity in coastal waters recorded by Geostationary Ocean Color Imager. *Estuarine Coast. Shelf Sci.* **2017**, *196*, 227–236. [CrossRef]

28. Sun, D.; Su, X.; Qiu, Z.; Wang, S.; Mao, Z.; He, Y. Remote Sensing Estimation of Sea Surface Salinity from GOCI Measurements in the Southern Yellow Sea. *Remote Sens.* **2019**, *11*, 775. [CrossRef]

29. Hu, C.; Muller-Karger, F.E.; Biggs, D.C.; Carder, K.L.; Nababan, B.; Nadeau, D.; Vanderbloemen, J. Comparison of ship and satellite bio-optical measurements on the continental margin of the NE Gulf of Mexico. *Int. J. Remote Sens.* **2003**, *24*, 2597–2612. [CrossRef]

30. Choi, J.K.; Park, Y.J.; Ahn, J.H.; Lim, H.S.; Eom, J.; Ryu, J.H. GOCI, the world's first geostationary ocean color observation satellite, for the monitoring of temporal variability in coastal water turbidity. *J. Geophys. Res. Ocean.* **2012**, *117*. [CrossRef]

31. Doxaran, D.; Lamquin, N.; Park, Y.J.; Mazeran, C.; Ryu, J.H.; Wang, M.; Poteau, A. Retrieval of the seawater reflectance for suspended solids monitoring in the East China Sea using MODIS, MERIS and GOCI satellite data. *Remote Sens. Environ.* **2014**, *146*, 36–48. [CrossRef]

32. Son, Y.B.; Gardner, W.D.; Richardson, M.J.; Ishizaka, J.; Ryu, J.H.; Kim, S.H.; Lee, S.H. Tracing offshore low-salinity plumes in the Northeastern Gulf of Mexico during the summer season by use of multispectral remote-sensing data. *J. Oceanogr.* **2012**, *68*, 743–760. [CrossRef]

33. O'Reilly, J.E.; Maritorena, S.; Mitchell, B.G.; Siegel, D.A.; Carder, K.L.; Garver, S.A.; Kahru, M.; McClain, C. Ocean color chlorophyll algorithms for SeaWiFS. *J. Geophys. Res. Ocean.* **1998**, *103*, 24937–24953. [CrossRef]

34. Ryu, J.H.; Han, H.J.; Cho, S.; Park, Y.J.; Ahn, Y.H. Overview of geostationary ocean color imager (GOCI) and GOCI data processing system (GDPS). *Ocean Sci. J.* **2012**, *47*, 223–233. [CrossRef]

35. Lie, H.J.; Cho, C.H.; Lee, J.H.; Lee, S. Structure and eastward extension of the Changjiang River plume in the East China Sea. *J. Geophys. Res. Ocean.* **2003**, *108*. [CrossRef]

36. Ha, K.J.; Nam, S.; Jeong, J.Y.; Moon, I.J.; Lee, M.; Yun, J.; Jang, C.J.; Kim, Y.S.; Byun, D.S.; Shim, J.S.; et al. Observations utilizing Korean Ocean Research Stations and their Applications for Process Studies. *Bull. Am. Meteorol. Soc.* **2019**, *100*, 2061–2075. [CrossRef]

37. Wu, Q.; Wang, X.; He, X.; Liang, W. Validation and Application of SMAP SSS Observation in Chinese Coastal Seas. In *Coastal Environment, Disaster, and Infrastructure-A Case Study of China's Coastline*; Liang, X.S., Zhang, Y., Eds.; IntechOpen: London, UK, 2018; pp. 273–284.

38. Lee, M.S.; Park, K.A.; Moon, J.E.; Kim, W.; Park, Y.J. Spatial and temporal characteristics and removal methodology of suspended particulate matter speckles from Geostationary Ocean Color Imager data. *Int. J. Remote Sens.* **2019**, *40*, 3808–3834. [CrossRef]

39. Wei, J.; Lee, Z.; Shang, S. A system to measure the data quality of spectral remote-sensing reflectance of aquatic environments. *J. Geophys. Res. Ocean.* **2016**, *121*, 8189–8207. [CrossRef]

40. Kingma, D.P.; Ba, J. Adam: A method for stochastic optimization. *arXiv* **2014**, arXiv:1412.6980.

41. Moh, T.; Cho, J.H.; Jung, S.K.; Kim, S.H.; Son, Y.B. Monitoring of the Changjiang River Plume in the East China Sea using a Wave Glider. *J. Coast. Res.* **2018**, *85* (sp1), 26–30. [CrossRef]

42. Beardsley, R.C.; Limeburner, R.; Yu, H.; Cannon, G.A. Discharge of the Changjiang (Yangtze river) into the East China sea. *Cont. Shelf Res.* **1985**, *4*, 57–76. [CrossRef]

43. Gitelson, A.A.; Schalles, J.F.; Hladik, C.M. Remote chlorophyll-a retrieval in turbid, productive estuaries: Chesapeake Bay case study. *Remote Sens. Environ.* **2007**, *109*, 464–472. [CrossRef]

44. Shi, W.; Wang, M. Satellite observations of flood-driven Mississippi River plume in the spring of 2008. *Geophys. Res. Lett.* **2009**, *36*. [CrossRef]

45. Shen, F.; Verhoef, W.; Zhou, Y.; Salama, M.S.; Liu, X. Satellite estimates of wide-range suspended sediment concentrations in Changjiang (Yangtze) estuary using MERIS data. *Estuaries Coasts* **2010**, *33*, 1420–1429. [CrossRef]

46. Zhang, M.; Tang, J.; Dong, Q.; Song, Q.; Ding, J. Retrieval of total suspended matter concentration in the Yellow and East China Seas from MODIS imagery. *Remote Sens. Environ.* **2010**, *114*, 392–403. [CrossRef]

47. Son, S.; Wang, M.; Shon, J.K. Satellite observations of optical and biological properties of the Korean dump site of the Yellow Sea. *Remote Sens. Environ.* **2011**, *115*, 562–572. [CrossRef]

48. Son, S.; Wang, M. Water properties in Chesapeake Bay from MODIS-Aqua measurements. *Remote Sens. Environ.* **2012**, *123*, 163–174. [CrossRef]

49. Shi, W.; Wang, M. Ocean reflectance spectra at the red, near-infrared, and shortwave infrared from highly turbid waters: A study in the Bohai Sea, Yellow Sea, and East China Sea. *Limnol. Oceanogr.* **2014**, *59*, 427–444. [CrossRef]

50. Bricaud, A.; Claustre, H.; Ras, J.; Oubelkheir, K. Natural variability of phytoplanktonic absorption in oceanic waters: Influence of the size structure of algal populations. *J. Geophys. Res. Ocean.* **2004**, *109*. [CrossRef]

Article

Assessment and Improvement of Global Gridded Sea Surface Temperature Datasets in the Yellow Sea Using In Situ Ocean Buoy and Research Vessel Observations

Kyungman Kwon [1,2], Byoung-Ju Choi [1,*], Sung-Dae Kim [3], Sang-Ho Lee [4] and Kyung-Ae Park [5]

[1] Department of Oceanography, Chonnam National University, Gwangju 61186, Korea; hydra82@jnu.ac.kr
[2] Research Institute for Basic Science, Chonnam National University, Gwangju 61186, Korea
[3] Korea Institute of Ocean Science and Technology, Busan 49111, Korea; sdkim@kiost.ac.kr
[4] Department of Oceanography, Kunsan National University, Gunsan 54150, Korea; sghlee@kunsan.ac.kr
[5] Department of Earth Science Education/Research Institute of Oceanography, Seoul National University, Seoul 08826, Korea; kapark@snu.ac.kr
* Correspondence: bchoi@jnu.ac.kr; Tel.: +82-62-530-3471

Received: 31 December 2019; Accepted: 22 February 2020; Published: 26 February 2020

Abstract: The sea surface temperature (SST) is essential data for the ocean and atmospheric prediction systems and climate change studies. Five global gridded sea surface temperature products were evaluated with independent in situ SST data of the Yellow Sea (YS) from 2010 to 2013 and the sources of SST error were identified. On average, SST from the gridded optimally interpolated level 4 (L4) datasets had a root mean square difference (RMSD) of less than 1 °C compared to the in situ observation data of the YS. However, the RMSD was relatively high (2.3 °C) in the shallow coastal region in June and July and this RMSD was mostly attributed to the large warm bias (>2 °C). The level 3 (L3) SST data were frequently missing in early summer because of frequent sea fog formation and a strong (>1.2 °C/12 km) spatial temperature gradient across the tidal mixing front in the eastern YS. The missing data were optimally interpolated from the SST observation in offshore warm water and warm biased SST climatology in the region. To fundamentally improve the accuracy of the L4 gridded SST data, it is necessary to increase the number of SST observation data in the tidally well mixed region. As an interim solution to the warm bias in the gridded SST datasets in the eastern YS, the SST climatology for the optimal interpolation can be improved based on long-term in situ observation data. To reduce the warm bias in the gridded SST products, two bias correction methods were suggested and compared. Bias correction methods using a simple analytical function and using climatological observation data reduced the RMSD by 19–29% and 37–49%, respectively, in June.

Keywords: sea surface temperature; global gridded dataset; validation; evaluation; Yellow Sea; bias correction

1. Introduction

Sea surface temperature (SST) is one of the main physical variables that provides information regarding the current state of the ocean [1]. SST data have been used for data assimilation in ocean circulation models and as a bottom boundary condition for atmospheric prediction models [2–4]. They are also essential for climate modeling and ocean–atmosphere interaction studies [5]. Satellite-based SST datasets have been produced by various organizations and are available at large spatial scales and at small temporal intervals [6–9]. Several previous studies have compared these global SST products with in situ observation data. Evaluations of the global SST datasets have mostly been performed at large scales [6,10,11] and a limited number of studies have been conducted on the SST datasets for regional and marginal seas [12].

The SST in the Yellow Sea (YS) is influenced by local factors such as the complex terrain in coastal regions, strong tidal currents, and the East Asian summer monsoon [13]. The East Asian summer monsoon begins to develop over the Indochina Peninsula from early to mid-May and rapidly expands to the Yangtze River Basin, western and southern Japan and the Southwestern Philippine Sea. It extends to North China, Korea and part of Japan by mid-June. Sea fog occurs frequently in the Eastern YS from June to July, and the sea fog forms when the surface air temperature (SAT) is higher than the SST [14–17]. This difference in the SAT and SST is higher in the regions of the Eastern YS where strong tidal mixing occurs [14].

Although satellite observation-based SST data of the YS have been used for operational ocean predictions and climate change studies, the accuracy of the satellite-based SST datasets is not yet known. The efficient analysis of errors in the satellite-based SST data of the YS is difficult because the availability of in situ SST observation data is very limited. In situ SST observation data of the YS are available from a limited number of buoys and vessel conductivity-temperature-depth (CTD) profiles, but no Argo float data exist for the region. Satellite observations of SST are therefore important for ocean prediction systems based on numerical models and data assimilation for the YS [2,18,19].

Hu et al. [20] analyzed satellite-based SST products for the South China Sea and the surrounding areas in 2008 and 2009 and found that the Operational Sea Surface Temperature and Sea Ice Analysis (OSTIA) data were the most accurate. For the South China Sea, the bias in the SST products increased in the shallow coastal region and during poor weather conditions. Xie et al. [8] evaluated optimally interpolated gridded SST products in the East Asian marginal seas and found a warm bias and large root mean square difference (RMSD) in satellite-based SST products in the regions where the bottom depth was shallower than 40 m. These authors developed a bias correction scheme depending on the depth and applied the bias correction method, which significantly reduced the bias; however, the RMSD was not significantly reduced. Kim et al. [21] compared the satellite-based SST with in situ SST observations for the East China Sea and Northwestern Pacific in summer. The RMSDs for microwave band-based Advanced Microwave Scanning Radiometer for the Earth Observing System (AMSR-E) SST and the New Generation Sea Surface Temperature for Open Ocean (NGSST-O) dataset were 0.55 °C and 0.7 °C, respectively. The microwave sensor-based SST data were better than the NGSST-O data in the study regions. However, the RMSD of the NGSST-O decreased when wind speeds were greater than 6 m/s. When the wind speed was lower and solar radiation was higher, the temperature difference between the skin and sub-skin was observed to become as high as 1.5 °C [22–25]. Kwak et al. [26] investigated seasonal changes in the optimal interpolation sea surface temperature (OISST) and in the SSTs measured using CTD profiles over a 30-year period, and found that the RMSD was higher (2 °C) near the coastline in the Eastern YS.

In this paper, we compared the OISST, Merged Satellite and in situ Data Global Daily Sea Surface Temperature (MGDSST), OSTIA, Remote Sensing Systems Microwave and Infrared (MWIR), and Group for High Resolution SST (GHRSST) Multi-Product Ensemble (GMPE) gridded gap-free (Level 4 or L4) SST datasets with the in situ SST data collected by ocean buoys and CTD profiles for the period from 2010 to 2013. We calculated the RMSD, bias and correlation coefficient to provide a quantitative analysis. The objectives of this study were to determine spatio-temporal characteristics of errors (RMSD and bias) in the L4 gridded SST products of the YS and to identify physical processes that affect the SST observation and the potential sources of error in the L4 gridded SST products. Bias correction methods were developed to reduce the RMSD of the gridded SST data for the vertically well-mixed coastal region during summer. The gridded SST products and in situ data used in this study are introduced in Section 2. In Section 3, inter-comparison results between the five gridded SST datasets and the in situ SST dataset are described, and the bias correction methods are proposed. In Section 4, several error sources for the L4 gridded SST products in the coastal mixed regions of the Southeastern YS are presented. The conclusions are given in Section 5.

2. Materials and Methods

2.1. Satellite SST Products

The main characteristics of the five L4 gridded SST datasets are summarized in Tables 1 and 2. The datasets have different horizontal grid spacing and represent different SST types. The OISST data were generated by the National Climatic Data Center (NCDC) of National Oceanic and Atmospheric Administration (NOAA) with a spatial grid spacing of 0.25° and a daily temporal resolution [27]. The OISST product uses operational Advanced Very High Resolution Radiometer (AVHRR) satellite data. Both Infrared (IR)-Only and Infrared–Microwave (IR–MW) products are available for the OISST datasets. The IR-Only OISST dataset was used in this study. The MGDSST data were provided by the Japan Meteorological Agency (JMA) using an optimal interpolation (OI) technique. The MGDSST dataset also has a spatial grid resolution of 0.25° and a daily temporal resolution. The product uses operational AVHRR and AMSR-E satellite data [28]. The OSTIA dataset was produced by the UK Met Office and uses satellite observation data together with in situ observations to determine the SST [29]. The SST analysis was performed using a variant of the OI technique described by Reynolds et al. [27]. The OSTIA SST has a daily temporal scale and a resolution of 0.05°. The MWIR SST dataset uses the OI technique on SST data from the operational Tropical Rainfall Measuring Mission Microwave Imager (TMI), AMSR-E and WindSAT microwave sensors as well as the Moderate Resolution Imaging Spectroradiometer (MODIS) Terra and MODIS Aqua IR sensors [30,31]. The MWIR SST dataset has a spatial grid resolution of 9 km and a daily temporal resolution. The GMPE system was developed and is operated by the UK Met Office. The dataset consists of a median and standard deviation of a daily ensemble at a resolution of 0.25°. Twelve SST datasets contributed to the ensemble median [6]. Both daytime and nighttime SST measurements were used for the L4 gridded SST datasets, and the diurnal warming was removed from the daytime SST using different methods [27–30].

Table 1. Characteristics of the global gridded sea surface temperature (SST) products.

Product	Grid Spacing & SST Type	Institution and Country	Reference
OISST	0.25° 0.5-m bulk	NCDC/NOAA, USA	Reynolds et al. [27]
MGDSST	0.25° Foundation	Japan Meteorological Agency, Japan	Kurihara et al. [28]
OSTIA	0.05° Foundation	Met Office, UK	Donlon et al. [29]
MWIR	9 km Foundation	Remote Sensing Systems, USA	Gentemann et al. [30]
GMPE	0.25° Ensemble	Met Office, UK	Martin et al. [6]

Table 2. Instruments and sensors used to produce the global gridded SST products.

Product	Argo Floats	Buoys GTS	Ships GTS	AVHRR NOAA	AVHRR MetOP	MODIS Aqua,Terra	SEVIRI MSG	TMI TRMM	WindSat
OISST				•	•				
MGDSST	•	•	•	•	•				
OSTIA		•	•	•	•		•	•	
MWIR						•		•	•

Solid dot symbol represents use of data from the observation platform for each L4 SST product. GTS, Global Telecommunications System; MetOP, Meteorological Operational Satellite Program of Europe; SEVIRI, Meteosat Second Generation (MSG) Spinning Enhanced Visible and Infrared Radiometer; TMI, Tropical Rainfall Mapping Mission (TRMM) Microwave Instrument.

The AVHRR Pathfinder Version 5.3 (PFV53) Level 3 (L3C) SST dataset is a collection of global, twice-daily (Day and Night) 4 km SST data produced by the NOAA National Centers for Environmental

Information (NCEI). The PFV53 dataset was computed with data from the AVHRR instruments on board NOAA's polar orbiting satellite series using an entirely modernized system based on SeaDAS. The PFV53 data were collected through the operational periods of the NOAA-7 through NOAA-19 Polar Operational Environmental Satellites, and are available from 1981 to the present. The L3C SST dataset for the period from 2010 to 2013 was examined in this study.

The Multifunctional Transport Satellites (MTSAT) is a Japanese geostationary dual-function satellite program, procured by the Japan Civil Aviation Bureau and the JMA. The satellite data from the MTSAT has short-wave infrared (SWIR, 3.8 μm) and long-wave infrared (LWIR, 10.8 μm) channels, which were used to monitor the area of fog. The MTSAT has five channels that consist of four IR window channels and one visible window channel, and makes observations using a 4 km IR resolution, which is sufficient for monitoring sea fog. The MTSAT data consist of snapshots taken at 30 min intervals. Images from the loaded 3.8 μm sensor and the IR differential images (3.8–10.8 μm) from the MTSAT can detect nighttime fog and show its distribution over a wide area. The sea fog images from the MTSAT for the years of 2010–2013 were used to detect sea fog in this study.

2.2. In Situ SST Data in the Yellow Sea

The in situ temperature observation data were obtained from ship CTD profiles and ocean buoys in the YS to evaluate the five L4 gridded SST datasets. The CTD observation data were provided by the National Institute of Fisheries Science (NIFS) of Korea; the data were obtained along 25 lines and 207 stations in the Eastern YS (Figure 1). The surface temperature was measured six times per year. Another set of CTD observation data was provided by the Korea Marine Environment Management Corporation (KOEM). These CTD data were obtained from 417 stations off the Korean coastlines, measured four times per year. The surface observation depth of the CTD profiles was 1.5–2.0 m and these measurements were used as the CTD SSTs in this study.

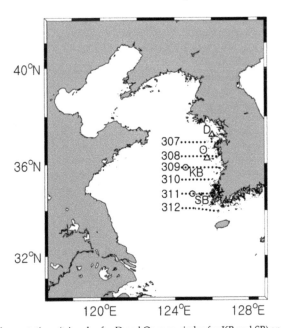

Figure 1. Ocean buoy stations (triangles for D and O; open circles for KB and SB) and the bi-monthly routine hydrographic observation stations (dots) of the National Institute of Fisheries Sciences, Korea. The number indicates bi-monthly routine oceanographic observation lines.

Surface buoy D is located near Deokjeok Island and the surface buoy O is located near Oeyeon Island. The SST data measured by buoys D and O were provided by the Korea Meteorological Administration (KMA). The buoys measured the SST every hour and the daily mean SSTs were calculated using the hourly data. Buoy KB was located in the middle of the YS. The SST data at KB was measured every 10 min and daily mean SSTs were calculated from the 10 min data. Surface buoy SB was located near the southern boundary of the YS and hourly mean data from this buoy were used to produce the daily mean data (Figure 1). The buoys measured the temperatures at a depth of 0.5–1.0 m. A total of 1249 and 2031 SST observations were collected from the CTD profiles and ocean buoys, respectively. The SST data from the ocean buoys (KB and SB) and the bi-monthly routine hydrographic observation stations were independent data and they were not used to produce the global gridded SST datasets, whereas the OSTIA and MGDSST datasets were produced using SST data from buoys D and O. The validations were performed by in situ datasets, and confidence intervals for statistical results were evaluated by the bootstrapping procedure [32].

2.3. Bias Correction Method of the L4 SST Products

To reduce the high RMSD in the coastal region of the Eastern YS, two spatial-temporal bias correction methods were designed in this study. In the first method, the bias correction function (ΔT) consisted of an exponential function depending on distance from the land and a cosine function depending on time as follows:

$$\Delta T(r,t) = A_0 e^{-(r/45)^2} \cos[2\pi/240(t-180)], \quad 120 < t < 240, \quad 0 < r < 50 \tag{1}$$

where t represents the number of days from January 1 of each year and r is the distance from the land in km. The SST bias correction function was designed to fit the five L4 SST datasets with the in situ surface temperature data. A_0 is the maximum bias near the coast in June and was calculated to be 3.64, 3.21, 2.65, 5.33, and 4.10 °C for the OISST, MGDSST, OSTIA, MWIR and GMPE SST, respectively.

In the second bias correction method, the bi-monthly horizontal distributions of the in situ CTD observation data from 2010 to 2013 were used to construct daily bias correction functions from June to July. First, the monthly climatologies were constructed using the CTD observation and the original L4 gridded SST data. Daily climatologies were then calculated from the monthly climatologies by linear interpolation in time. The daily bias correction function (ΔT) was defined as the difference between the two daily climatology datasets on day t:

$$\Delta T(\text{lon}, \text{lat}, t) = \overline{SST_{Level4}(\text{lon}, \text{lat}, t)} - \overline{SST_{CTD}(\text{lon}, \text{lat}, t)}, \quad 120 < t < 240 \tag{2}$$

where t represents the number of days from January 1 of each year. The overbar represents the average of the four years at (lon, lat) on day t and (lon, lat) represents the horizontal location. The daily bias correction function at time t was calculated based on the quantitative comparison of the daily OISST, MGDSST, OSTIA, MWIR and GMPE SST climatologies with the daily in situ CTD SST climatology.

2.4. OSTIA SST Interpolation Scheme

The following is a short description of the OSTIA SST analysis procedure [29]. The background field $\mathbf{x}^b_{i,k}$ at grid point i and time k is defined as:

$$x^b_{i,k} = \lambda_{i,k}(x^a_{i,k-1} - x^c_{i,k-1}) + x^c_{i,k} \tag{3}$$

where $\lambda_{i,k}$ is a scalar less than 1, $x^a_{i,k-1}$ is the previous analysis, and $x^c_{i,k}$ is a reference climatology valid for the same time of year at time k. Relaxation time scale at grid point i and time k is 30 days for the determination of $\lambda_{i,k}$. The background field was calculated using Equation (3) and the bias corrected measurements were then used to produce an analysis using a multi-scale OI type scheme. The OI equation is given by:

$$x^a_k = x^b_k + BH^T_k\left[H_kBH^T_k + R_k\right]^{-1}(y_k - H_kx^b_k) \tag{4}$$

where x_k is a vector containing all values $x_{i,k}$, $i = 1, \ldots, N$ on the analysis grids at time k, y_k is a vector containing the observations, B is the background error covariance matrix, R_k is the observation error covariance matrix, and H_k is the observation operator that interpolates from the model grid to the observation locations. This equation was solved using the analysis correction method [33,34]. The background error covariance matrix B was calculated using model outputs from a three-year numerical integration using the Forecasting Ocean Assimilation Model (FOAM) of the UK Met Office [35,36]. Covariances were split into two components, and two error correlation length scales of 10 km and 100 km were used for mesoscale and large scale errors, respectively. The data in the observation error covariance matrix R were assumed to be uncorrelated with each other and were obtained from the information supplied by the GHRSST single sensor error statistic values. A simple quadratic bottom friction (C_d) value of 0.00125 was used to crudely parametrize tidal mixing in the FOAM [37].

3. Results

3.1. Comparison with In Situ SST Observations

To evaluate the accuracy of L4 gridded SST datasets (OISST, MGDSST, OSTIA, MWIR, and GMPE) from five operational systems, the RMSD and bias of the gridded SST datasets were calculated relative to the in situ SST obtained from the ocean buoys and routine hydrographic observation stations in the Eastern YS. The correlation coefficients between the buoy and L4 gridded SST data were 0.97–0.99 (Figure 2). The RMSD of the gridded SST data was the lowest (0.74 °C) for the GMPE and the highest (1.04 °C) for the OISST. The blue dots lying above the center line of the OISST and GMPE datasets in August imply that the L4 gridded SSTs were higher on some days than the in situ SSTs. The L4 gridded SSTs in June (red dots) were aligned with the in situ ocean buoy SSTs from the deep stratified region.

Compared with the CTD SST data, some of the red and blue dots were above the center line (Figure 3), which indicated that the SST from the satellite-based observations were warmer than that from the in situ CTD data in June and August. The correlation coefficients were 0.93–0.98. The RMSD of the satellite observations was the lowest (0.96 °C) for the OSTIA SST and the highest (1.39 °C) for the MWIR SST. These comparisons suggest that the SST from the OSTIA, GMPE, and MGDSST datasets had a RMSD of less than 1.00 °C compared with the in situ CTD observations in the Eastern YS (Table 3).

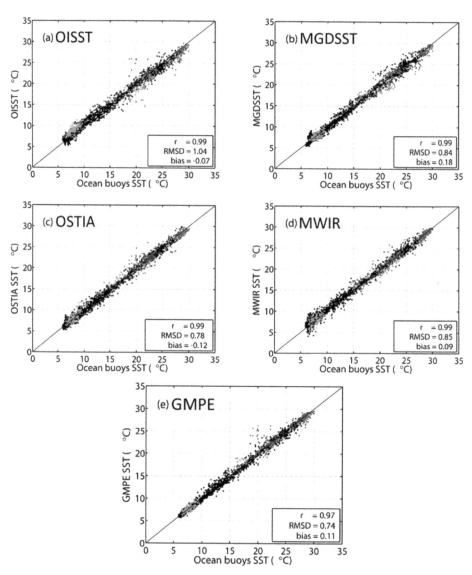

Figure 2. Comparison between the SST (°C) data measured by ocean buoys (KB and SB) and the global gridded SST products (OISST, MGDSST, OSTIA, MWIR, and GMPE). Red, blue, green, and black dots correspond to the SSTs observed in June, August, December, and other months, respectively. r is the correlation coefficient.

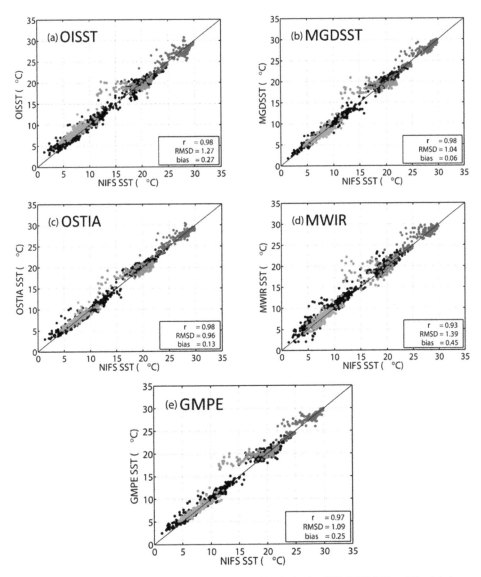

Figure 3. Comparison between the SST data measured by the in situ CTD (NIFS SST) and the global gridded datasets (**a**) OISST, (**b**) MGDSST, (**c**) OSTIA, (**d**) MWIR, and (**e**) GMPE. Red, blue, green, and black dots correspond to the SSTs observed in June, August, December, and other months, respectively. r is the correlation coefficient.

Table 3. Root mean square difference (RMSD) and bias of SST data obtained from in situ (the ocean buoys and conductivity-temperature-depth (CTD) measurements and remote sensing (satellite observation). The smallest RMSD and bias relative to KB, SB, and NIFS observations are given in bold face. The number of matchup data is given in parentheses.

Location	OISST RMSD	OISST Bias	MGDSST RMSD	MGDSST Bias	OSTIA RMSD	OSTIA Bias	MWIR RMSD	MWIR Bias	GMPE RMSD	GMPE Bias
KB (1399)	1.00 ±0.04	**−0.08** ±**0.05**	0.79 ±0.04	−0.20 ±0.04	0.77 ±0.03	−0.11 ±0.04	0.83 ±0.03	0.17 ±0.04	**0.68** ±**0.03**	−0.12 ±0.04
SB (632)	1.13 ±0.06	**−0.04** ±**0.09**	0.95 ±0.05	0.20 ±0.07	**0.82** ±**0.04**	−0.14 ±0.06	0.87 ±0.04	0.17 ±0.07	0.87 ±0.05	0.11 ±0.07
D (1240)	1.97 ±0.07	1.37 ±0.08	1.46 ±0.05	1.02 ±0.06	0.88 ±0.04	0.41 ±0.04	2.22 ±0.09	1.41 ±0.10	1.54 ±0.06	1.07 ±0.06
O (1283)	1.04 ±0.04	0.36 ±0.05	0.78 ±0.03	0.39 ±0.06	0.45 ±0.02	0.04 ±0.02	0.98 ±0.04	0.50 ±0.05	0.76 ±0.03	0.38 ±0.04
NIFS (1246)	1.26 ±0.05	0.27 ±0.07	1.02 ±0.04	**0.06** ±**0.06**	**0.96** ±**0.04**	0.13 ±0.05	1.39 ±0.06	0.45 ±0.07	1.04 ±0.05	0.25 ±0.06
Mean	1.17	0.38	0.95	0.29	0.88	0.12	1.12	0.54	0.91	0.34

3.2. Spatial and Temporal Variations of the SST RMSD

3.2.1. Spatial Distribution of the SST RMSD

The spatial distribution of the RMSD between the SST from the global gridded products and those from the in situ research vessel observations was calculated at each observation station over the four years (Figure 4). About 24 values of SST data were compared at each station (black dot in Figure 4). The RMSD between the satellite-based SSTs and the CTD SSTs was found to be comparatively higher in the shallow and vertically well-mixed region or in the shallow coastal region within 50 km from the land and lower in the offshore region. The RMSD of the MWIR SST (Figure 4d) was the highest and that of the OSTIA SST was the lowest (Figure 4c). The annual mean RMSDs for the shallow coastal region within 50 km from the land were 1.50, 1.17, 1.14, 1.72, and 1.27 °C for the OISST, MGDSST, OSTIA, MWR, and GMPE datasets, respectively (Figure 4 and Table 4). The RMSDs were larger in June–August than in other months (Table 4). Spatial distribution of the four-years mean CTD SST in June showed relatively cooler coastal water (T < 18 °C) near the coast (Figure 4f). The cool water bounded by approximately 18 °C isotherm is located in the shallow coastal region within 50 km from the land due to strong tidal stirring, which forms a tidal mixing front along the 30–50 m isobaths. This region coincided with the high RMSD area of the L4 gridded SST datasets. The annual mean RMSDs for the offshore region were 0.88, 0.73, 0.64, 0.74, and 0.59 °C for the OISST, MGDSST, OSTIA, MWIR, and GMPE, respectively (Figure 4 and Table 4). The RMSD of the GMPE SST was the lowest and that of the OISST was the highest.

Table 4. Seasonal variations in the RMSD between the L4 gridded SST products (mostly based on satellite observation) and the in situ measurements (ocean buoys and research vessels) in the shallow and vertically well-mixed region. The lowest RMSD in each column is given in bold face.

Mixedregion	March–May Buoy	March–May CTD	June–August Buoy	June–August CTD	September–November Buoy	September–November CTD	December–February Buoy	December–February CTD
OISST	1.50 ±0.07	1.50 ±0.16	2.09 ±0.11	2.43 ±0.30	0.79 ±0.04	0.86 ±0.12	0.95 ±0.04	1.20 ±0.14
MGDSST	0.70 ±0.04	**0.85** ±**0.12**	1.55 ±0.08	2.19 ±0.26	0.90 ±0.05	0.79 ±0.11	0.92 ±0.05	**0.86** ±**0.10**
OSTIA	**0.59** ±**0.03**	0.94 ±0.14	**0.93** ±**0.06**	**1.82** ±**0.21**	**0.49** ±**0.03**	**0.72** ±**0.09**	**0.45** ±**0.02**	1.06 ±0.12
MWIR	1.07 ±0.05	0.86 ±0.12	2.26 ±0.14	2.85 ±0.35	1.02 ±0.06	1.49 ±0.23	1.43 ±0.07	1.68 ±0.19
GMPE	0.79 ±0.04	0.88 ±0.13	1.78 ±0.10	2.40 ±0.30	0.64 ±0.04	0.85 ±0.13	0.77 ±0.04	0.95 ±0.12
Mean	0.93	1.01	1.72	2.34	0.77	0.94	0.90	1.15

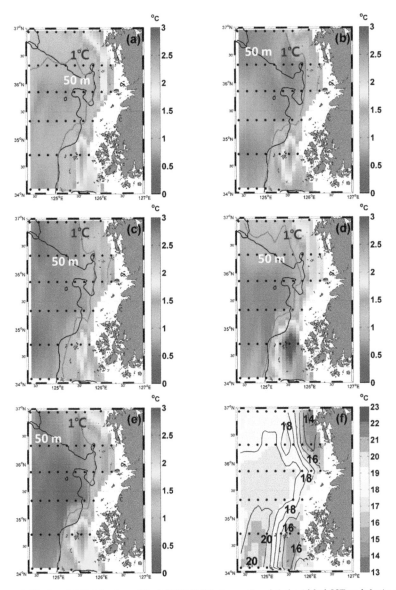

Figure 4. Horizontal distribution of the RMSD (°C) between the global gridded SST and the in situ CTD SST: (**a**) OISST (**b**) MGDSST, (**c**) OSTIA, (**d**) MWIR, and (**e**) GMPE. The black contour represents the 50-m isobath. The red contour is plotted along the 1 °C RMSD. (**f**) Four-year mean distribution of the CTD SST in June. The contour interval for the surface temperature in (**f**) is 1 °C.

3.2.2. Seasonal Variation of the RMSD and Bias

There were relatively large differences between the L4 gridded SSTs and the in situ SSTs in June and August (Figure 3). To examine seasonal the variations in the RMSD, monthly mean RMSD of the L4 gridded SST data were calculated at the ocean buoy stations (KB, SB, D, and O), the NIFS observation stations in the deep stratified region (NIFS-S) where the bottom depth was deeper than 50 m, and the NIFS observation stations in the shallow vertically mixed region (NIFS-M) where the bottom depth was shallower than 50 m, for four years (Figure 5). The ocean buoys at KB and SB were located in the relatively deep stratified region while the ocean buoy at D was located in the relatively shallow and vertically well-mixed region. The ocean buoy at O was located near the boundary between the stratified and the mixed regions (Figure 1). In general, the RMSDs of the OISST, MGDSST, OSTIA, and GMPE SST datasets were higher from June to August in the Eastern YS. However, the RMSD of the MWIR dataset did not have a clear seasonal variation in the deep stratified region. The RMSD was higher in the shallow and vertically well-mixed region (D and NIFS-M) from June to July (Figure 5e,f). In June and July, a cold bias in the global gridded SST data appeared in the deep offshore region such as at the KB, SB, and NIFS-S stations, while a warm bias (1–4 °C) appeared in the shallow coastal region, such as at the D and NIFS-M stations (Figure 6). The relatively large RMSD in the shallow and vertically well-mixed region was related to the warm bias in June–July. In the shallow mixed region, the annual mean biases of the OISST, MGDSST, OSTIA, MWIR, and GMPE SST data compared with the CTD SST data in June were 1.96, 1.73, 1.17, 2.28, and 2.27 °C, respectively (Figure 6f).

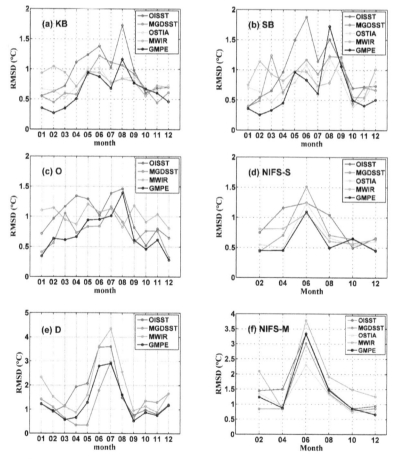

Figure 5. Seasonal variations in the monthly mean RMSD (°C) between the in situ SST and the global gridded SST data. (**a**) KB, (**b**) SB, (**c**) O, and (**e**) D represent the in situ observations collected at the ocean surface buoy stations. (**d**) NIFS-S and (**f**) NIFS-M represent the in situ CTD observations from the deep stratified region (h > 50 m) and the shallow vertically mixed region (h < 50 m) stations, respectively, along the routine bimonthly observation lines of the research vessels. Note that the vertical ranges in (**e**) and (**f**) are different from those in the other panels.

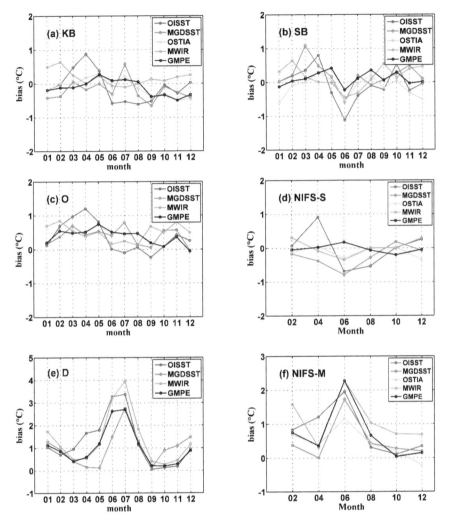

Figure 6. Seasonal variation of monthly mean bias (°C) of the gridded SST data relative to the in situ observation. (**a**) KB, (**b**) SB, (**c**) O, and (**e**) D represent the in situ observations at the ocean surface buoys. (**d**) NIFS-S and (**f**) NIFS-M represent the in situ observations at the deep stratified region (>50 m) and the shallow vertically mixed region (<50 m) stations, respectively, along the routine bimonthly observation lines of research vessels. Note that the vertical ranges in (**e**) and (**f**) are different from that in the other panels.

To examine the seasonal characteristics of each SST dataset, the RMSDs between the L4 gridded SST datasets and the in situ data were divided for MM (March–May), JA (June–August), SN (September–November), and DF (December–February), and were examined in the vertically well-mixed shallow region and the stratified region (Tables 4 and 5). In the shallow and vertically well-mixed region, the RMSD between the MGDSST and the in situ CTD SST was the lowest in MM and DF while the RMSD between the OSTIA SST and the in situ CTD SST was the lowest in JA and SN (Table 4). This indicates that the MGDSST and OSTIA SST datasets are the best for the shallow mixed region of the Eastern YS throughout the year.

Table 5. Seasonal variations in the RMSD between the L4 gridded SST products (mostly based on satellite observation) and the in situ measurements (ocean buoys and research vessels) in the deep stratified region. The lowest RMSD in each column is given in bold face.

Stratifiedregion	March–May		June–August		September–November		December–February	
	Buoy	**CTD**	**Buoy**	**CTD**	**Buoy**	**CTD**	**Buoy**	**CTD**
OISST	1.04	1.16	1.45	1.14	0.80	0.50	0.60	0.71
	±0.05	±0.11	±0.09	±0.09	±0.05	±0.06	±0.03	±0.06
MGDSST	0.82	0.70	1.13	1.11	0.74	0.64	0.53	0.45
	±0.04	±0.07	±0.06	±0.10	±0.05	±0.06	±0.03	±0.04
OSTIA	0.70	0.50	0.94	0.95	**0.66**	**0.54**	0.70	0.58
	±0.04	±0.05	±0.05	±0.08	**±0.04**	**±0.06**	±0.04	±0.05
MWIR	0.89	0.81	**0.85**	0.84	0.74	0.57	0.93	0.73
	±0.04	±0.07	**±0.05**	±0.07	±0.05	±0.07	±0.05	±0.06
GMPE	**0.59**	**0.45**	0.98	**0.80**	0.67	0.65	**0.37**	**0.45**
	±0.04	**±0.05**	±0.07	**±0.09**	±0.04	±0.08	**±0.02**	**±0.04**
Mean	0.81	0.72	1.07	0.97	0.72	0.58	0.63	0.58

In the stratified offshore region, the RMSD between the GMPE SST and the in situ CTD SST was the lowest in MM, JA, and DF, and that between the OSTIA SST and the in situ CTD SST was the lowest in SN (Table 5). When the OISST (IR band product) and the MWIR (microwave band product) were compared with the in situ data in June–August, the RMSDs of the OISST and MWIR were 1.14–1.45 °C and 0.84–0.85 °C, respectively, in the deep stratified region (Table 5). This implies that the microwave band observations are more effective in the stratified region during the summer than the IR band observation (Figure 5a–d).

3.3. Bias Correction of the L4 SST Products in the YS

The RMSDs of the OSTIA SST data were 1.14 and 0.64 °C in the shallow mixed and deep stratified regions, respectively. The relatively high RMSD in the shallow and vertically well-mixed region in June and July was largely contributed by the warm bias (Figures 5e,f and 6e,f). To improve the accuracy of the L4 gridded SST data in June and July, two bias correction methods were developed as described in Section 2.3. After applying the first bias correction method (Equation (1)) based on the analytical functions of the OISST dataset, the bias corrected OISST (OISSTabc) data were compared with the SST from ocean buoy D and NIFS-M (Figure 7). The OSTIA dataset was also corrected using the bias correction function and the bias corrected OSTIA (OSTIAabc) data were compared with NIFS-M SST data. The OSTIA SST dataset was produced using SST data from ocean buoy D, therefore, it was not compared with the SST data from this buoy. The RMSDs of the OISST and OISSTabc datasets were compared with the in situ SST data from buoy D and showed that the RMSD was reduced by 54% in May–August. The RMSD of the OISST (OSTIA) dataset was decreased by 28% (19%) with the bias correction at NIFS-M stations in June (Figure 7). Although not shown in the figure, the RMSD of the MGDSST, MWIR and GMPE datasets decreased by 23%, 26% and 29%, respectively, at NIFS-M stations in June. After the second bias correction method (Equation (2)) was applied to the L4 gridded SST datasets, the bias corrected SST (OISSTrbc and OSTIArbc) data were compared with the in situ SST from the NIFS-M and ocean buoy D (Figure 8). Comparing the RMSDs of the original OISST and bias corrected OISSTrbc data with the in situ SST data from buoy D, the RMSD was reduced by 58% in May–August. The RMSD of the OISST (OSTIA) dataset decreased by 41% (37%) with the bias correction at NIFS-M stations in June. Although not shown in the figure, the RMSD of the MGDSST, MWIR and GMPE datasets decreased by 43%, 47% and 49%, respectively, at NIFS-M stations in June.

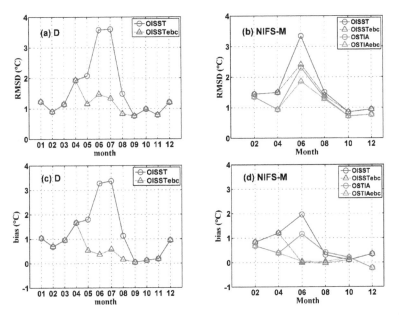

Figure 7. Seasonal variations in the monthly mean (**a,b**) SST RMSD and (**c,d**) bias (°C) between the in situ data (D and NIFS-M) and the gridded datasets (OISST and OSTIA) before and after applying the first bias correction method using the analytical (exponential and cosine) functions. The circles represent the original SST data and triangle represents the bias corrected data.

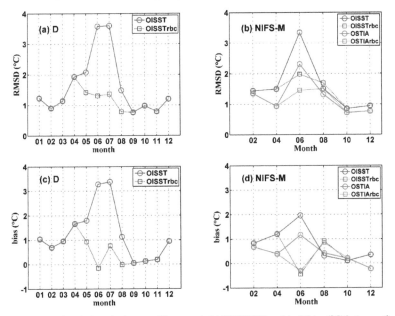

Figure 8. Seasonal variations in the monthly mean (**a,b**) SST RMSD and (**c,d**) bias (°C) between the in situ data (D and NIFS-M) and the gridded datasets (OISST and OSTIA) before and after applying the second bias correction using the in situ CTD observation data. Circle symbols represent the RMSD and bias of the original SST data. Square symbols represent the RMSD of the data corrected using the second bias correction method.

When the second bias correction method was applied to the L4 gridded OSTIA SST dataset in June, the RMSD reduced in both the shallow coastal region and the deep stratified region (Figure 9). This suggests that there were large warm biases in the background field of the L4 gridded SST datasets in June (Figures 4 and 9) and these could be corrected using the long-term climatology of in situ CTD SST observation data for the Eastern YS. The bias correction reduced not only bias but also the RMSD for the Eastern YS (Figures 7–9).

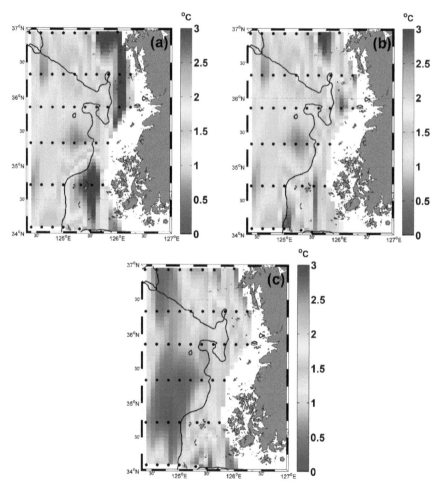

Figure 9. Spatial distribution of the RMSDs (°C) between (**a**) the original OSTIA SST, (**b**) the OSTIAabc SST, and (**c**) the OSTIArbc SST compared with the in situ CTD SST for June from 2010 to 2013. The black contour represents the 50-m isobaths in the Eastern YS.

4. Discussion

The RMSD between the L4 gridded SST and in situ SST had a seasonal variation and it was relatively large in June and July. The large RMSD in June and July (Figure 5) was contributed by the warm bias in the shallow and vertically well-mixed region (Figure 6). Since the bias contributes to a large portion of the RMSD in the YS during summer, it is important to find the causes that induce the SST bias in the L4 gridded SST datasets. There were several error (the RMSD and bias) sources of the L4 gridded SST products in the coastal mixed regions.

To identify oceanographic conditions that affect the relatively large RMSD and bias in the Eastern YS, (i) spatial distribution of the warm bias in the L4 gridded SST datasets, (ii) frequent sea fog formation, and (iii) the number of available L3 SST data were examined in this section. (iv) How to reduce the bias during the optimal interpolation procedure of the L4 gridded SST is sought for the Eastern YS.

4.1. Spatial Pattern of Warm Bias in the Shallow Region

To find the characteristics of the warm bias in the L4 gridded SST data, spatial distributions of the SST from the OSTIA, OISST, and MWIR dataset were compared with that of the SST from CTD profiles in the Eastern YS for June from 2010 to 2013 (Figure 10). The SST distributions from the OISST and MWIR dataset were distinctly different and had a warm bias in the tidally well-mixed and cool water region. However, the OSTIA SST had a smaller bias than the other SSTs because it uses the in situ data from the ocean buoys during the OI process (Figure 10b). The original SST rather than the bias was plotted in Figure 10 to show both spatial pattern and horizontal gradient of the SST from the L4 gridded datasets. The horizontal gradient was higher in the in situ SST than the L4 gridded SSTs. The spatial pattern of bias were similar to those of RMSD in Figure 4.

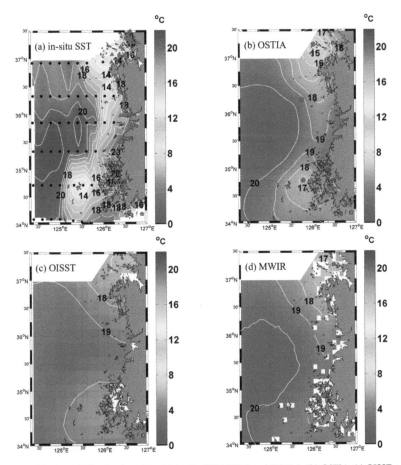

Figure 10. Horizontal distribution of the (**a**) in situ CTD (NIFS and KOEM), (**b**) OSTIA, (**c**) OISST, and (**d**) MWIR SST for June from 2010 to 2013. The blue points in (**b**) are the locations of the KMA buoys, which observe SST.

The combined effect of tidal mixing and stratification produces the tidal mixing front along the 30–50 m depth isobaths and relatively cool water occupies the shallow coastal region in summer. The OISST data are based on satellite IR band observations, which are prone to cloud and fog shielding, and underestimate tidal cooling over the shallow mixed region [38]. Using microwave sensors, the AMSR-E can measure SST through clouds and capture the tidal cooling effect well. However, the microwave sensors have a large observation error near land and islands. The MWIR SST also had a warm bias in the shallow mixed region (Figure 10d). This was because of the presence of land inference errors due to the characteristic natures of the microwave sensors [31]. Xie et al. [8] accessed satellite-based SST data in the shelf and coastal seas around China and suggested a depth-dependent bias correction method. Using the method, the bias in the SST data in the region shallower than 40 m depth was significantly corrected but the reduction in the RMSD was relatively small (6–14%). In their bias correction method the seasonal variation of the SST bias was not considered.

4.2. Sea Fog Formation in the Eastern YS

Sea fog was frequently observed as the cold patches along the Kuril Islands from July to August due to a tidal cooling effect, and the AVHRR SST was lower than AMSR-E SST [38]. The SST based on the satellite IR sensors in the Kashevarov Bank has cloud and fog shielding, which results in a warm bias larger than 5 °C and an underestimation of tidal cooling. Satellites with a microwave sensor can measure the SST through the clouds and also in the regions under the influence of strong tidal mixing. The satellites with IR sensors could not observe SST in the tidally well-mixed and cool water region because of lower clouds and sea fog in the Okhotsk Sea [38].

Inter-comparison of the in situ (ship and buoy) SST data with satellite-based SST products for the shelf and coastal seas around China and in the Northwest Pacific over a one-year period indicate that the RMSD increases sharply in the coastal region [3]. A comparison of the OISST data with the CTD SST observation data over 30 years for the Eastern YS revealed that the RMSD was 2–3 °C in the coastal area of the Eastern YS [26]. However, a reason for the large RMSD near the coastal shallow region has not yet been explained. The sea fog frequently forms in June and July and hinders measurement of SST by satellites IR-based sensors in the Eastern YS [14,39]. Sea fog forms frequently in June and July (Figure 11). To investigate why fog forms frequently in June and July in the Southeastern YS, the monthly mean wind speed, relative humidity, and wind vector measured by the ocean buoys D and O were compared for the four years (Figure 11). The monthly mean relative humidity in June and July were 88% and 91%, respectively. The wind speeds in June and July were 3.16 and 4.96 m/s, respectively. The sea fog forms frequently because of the temperature difference between the warm and humid air, and the relatively cold sea surface with some aid from the southerly wind in June and July. The ocean buoy D was located where the temperature difference between the SAT and SST was 2.1 °C. Since ocean buoy O was located at the boundary between the stratified and mixed regions, the temperature difference was 0.8 °C. In contrast, the ocean buoys KB and SB were located in the relatively deep stratified region and the temperature differences at KB and SB were −0.4 and −0.3 °C, respectively (Figure 11).

Although the relative humidity, wind speed, and wind vector in August were similar to those in July, the number of foggy days (triangles and circles in Figure 11) drastically decreased in August. Zhang et al. [40] studied the reasons for the occurrence and disappearance of sea fog in the YS and East China Sea. The YS and East China Sea are situated between the subtropical low and mid-latitude high and therefore experience an easterly wind shift in the seasonal march from July to August. As the intensification of atmospheric convection over the subtropical Northwestern Pacific region excites a barotropic wave train from July to August, the meridional dipole pattern of a geopotential decrease over the subtropics and an increase in the mid-latitudes causes the prevailing winds to shift from southerly to easterly over the YS. This wind shift terminates the southerly wind and the northward advection of warm and moist air that sustains the YS fog from April to July. The atmospheric environment is therefore foggy in June and July, which obstructs the satellite-based observation of SST for the Eastern YS.

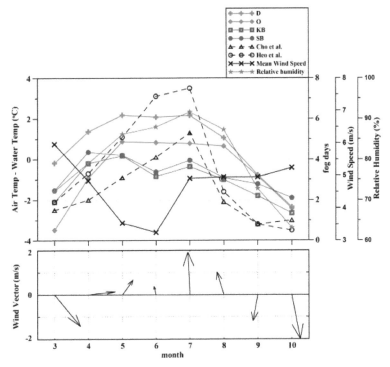

Figure 11. Monthly evolution of differences (°C) between the surface air temperature and the SST measured by the ocean buoys D (cross), O (diamond), KB (square), and SB (closed dot). Temperature differences were calculated from the hourly data. The monthly variation of sea foggy days from March to October from Cho et al. [14] and Heo et al. [39]. The monthly variation of relative humidity (green star) and wind speed (black cross) are in the upper panel. Variations of monthly mean wind vector are in the lower panel.

4.3. Availabily of a L3 SST in the YS

The AVHRR PFV53 L3 SST data are one of the most widely used data and they have five internal quality flags, such as best, acceptable, low, bad, and worst. The L4 gridded SST data were generated by the optimal interpolation of available L3 SST data with the best and acceptable quality [29]. For the Eastern YS, the number of AVHRR PFV53 L3 SST data with each quality flag was counted for June and July from 2010 to 2013 (Figure 12). In general, in June and July, the number of SST observations with the best and acceptable flags was relatively smaller in the shallow and vertically well-mixed regions. The SST data with the best and acceptable quality flags represented 8.2% of the total observations in the tidally well-mixed region (Figure 12a,b) and 70.9% of the data had the worst quality flags.

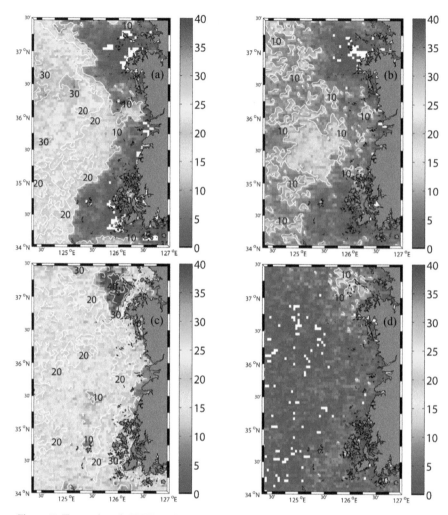

Figure 12. The number of AVHRR PFV53 L3 satellite SST observation data with (**a**) best, (**b**) acceptable, (**c**) low, and (**d**) bad quality flags in June and July from 2010 to 2013.

To evaluate the satellite-based L3 SST product in the shallow and vertically well-mixed regions, the RMSD and bias of the L3 SST data were calculated at ocean buoys D and O in June and July from 2010 to 2013 (Figure 13). There were 17, 12, 48, 22, and 346 observational data available with the best, acceptable, low, bad, and worst quality flags, respectively. None of the satellite-based L3 SST data at buoy D were assigned as best quality data in June and July for the four years (Figure 13a). The RMSDs between the L3 SST data with the best, acceptable, low, bad, and worst quality flags and in situ SST data were found to be 0.60, 1.06, 2.37, 4.05, and 15.26 °C, respectively. The biases were found to be −0.38, 0.07, −0.22, −0.09, and −11.90 °C, respectively. Only the SST data classified as the best and acceptable quality were found to constitute relatively reliable data with an RMSD lower than 0.83 °C and the corresponding bias was smaller than −0.16 °C.

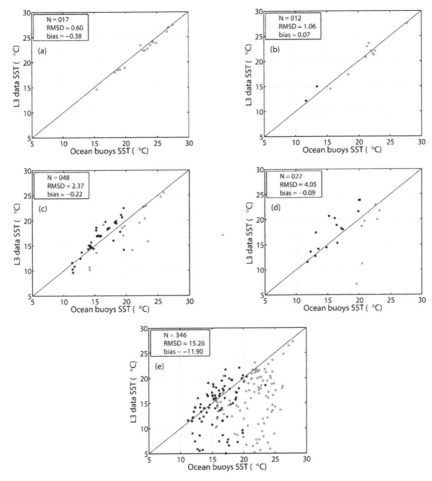

Figure 13. Comparison of the satellite-based L3 SST data with (**a**) best, (**b**) acceptable, (**c**) low, (**d**) bad, and (**e**) worst quality flags and the SST data from ocean buoys D and O in June and July from 2010 to 2013. N is the number of matchup data points. Red and blue dots represent the SST data from ocean buoys O and D, respectively.

Some of the satellite-based L3 SST data plotted near the ideal regression line were classified as the worst quality data (Figure 13e). It is necessary to know how the quality flags of the AVHRR pathfinder L3 SST data are assigned to recover the data with incorrectly assigned flags. The AVHRR pathfinder L3 SST data processing performs a series of tests to determine whether a pixel contains SST data of suspicious quality including cloud contamination and uniformity tests [41]. The maximum and minimum brightness temperature values for channels 4 and 5 were calculated for 3 × 3 boxes centered around the pixel being classified. The difference between the maximum and minimum brightness temperatures for both channels must be less than 1.2 °C [31]. This uniformity test or SST gradient test seeks to identify contamination by small clouds and is based on the assumption that SSTs are relatively uniform at small scales (e.g., 3 × 3 pixels). However, pixels in the areas of sharp frontal features on scales smaller than 12 km might be erroneously identified as suspicious data by this test. The temperature gradient in the Eastern YS during the summer was calculated using the in situ SST data in June from 2010 to 2013 (Figure 14). The horizontal temperature gradient was higher than 1.2 °C across 12 km along the tidal mixing front region in the Eastern YS (Figures 10a and 14). This strong SST gradient could lead the uniformity test in the Pathfinder algorithm to classify good quality SST data in the frontal region as worst quality data.

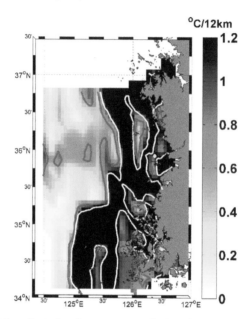

Figure 14. Horizontal gradient of surface temperature for June from 2010 to 2013. The unit of the temperature gradient is °C/12 km. Blue and white contours represent 0.7 and 1.2 °C/12 km, respectively.

To investigate the relationship between sea fog formation and the L3 SST data quality flags, the L3 SST data flags were compared with the detected sea fog events from the MTSAT images at the locations of buoys D and O in June and July from 2010 to 2013. There were 445 in situ SST observations from the buoys and 488 values from the matching satellite-based L3 SST data. The number of buoy SST data was smaller than that of the L3 SST data because there were some days when the buoys could not measure SST. When the L3 SST data had the best and acceptable quality flags, no sea fog was found in the MTSAT images (Table 6). When sea fog formed in the lower column of the atmosphere, 37% of the L3 SST data had low or bad quality flags and 63% had the worst quality flags. This implied that sea fog formation degrades the quality of the satellite-based L3 SST data. Most (64%) of the low and bad quality data were related to sea fog and 36% of the low and bad quality data were not related to sea fog. However, most (78%) of the worst quality L3 SST data were not related to sea fog formation and only 22% of the L3 SST data with the worst quality flag were related to sea fog formation. This indicated that there were other causes, such as the strong horizontal SST gradient, which fails the uniformity test in the Pathfinder algorithm, which incorrectly assigned the worst quality flag to the good or acceptable quality L3 SST data (Figure 13e).

Table 6. Comparison of the L3 SST data flags (best, acceptable, low, bad, and worst) and sea fog appearances detected by the MTSAT at the locations of ocean buoys D and O for June and July from 2010 to 2013.

Quality Flag	Best	Acceptable	Low	Bad	Worst	Total Number
Sea fog formation	0	0	31	14	77	122
No sea fog	17	12	17	8	269	323
Total number	17	12	48	22	346	445

In the coastal region of the Southeastern YS, it was difficult to observe SSTs owing to poor atmospheric conditions in June and July (Figure 11 and Table 6). Even if the SSTs are observed by satellite sensors in this coastal region, good or acceptable SST data may be classified as the worst quality data because of the strong horizontal temperature gradient higher than 1.2 °C/12 km across the tidal mixing front (Figure 14). While the L4 gridded SST data were produced using the OI method, there were insufficient good and acceptable L3 SST data in this coastal region. Then, the missing data points in the L4 gridded SST dataset were filled with a combination of SST climatology at the point and offshore SST data in warm water using the OI algorithm [29].

4.4. Background and Covariances Used for the Optimal Interpolation

Since the L3 SST data were sparse in the Eastern YS during summer as shown in the previous section, it was necessary to understand the process of optimal interpolation for the missing data for the L4 gridded SST dataset. As the OSTIA SST dataset was found to be the best among the five gridded SST datasets when compared to the in situ data, in this section, we used this dataset as an example of the L4 gridded SST data and examined its production procedure.

The L4 gridded SST is less reliable where the observation data are sparse because the missing SST at a grid point is interpolated from the data at the surrounding grid points. Especially in the coastal region of the YS, the tidal front develops along the coast and induces a strong horizontal SST gradient, causing more observational data errors (Figures 10 and 14). In June and July, most (93%) of the L3 SST data in the coastal region of the Southeastern YS were assigned as low, bad or worst quality data (Table 6), which resulted in the interpolation of SST data in the coastal region during the production of the L4 gridded SST datasets. The missing SST data in the coastal region were not interpolated from the nearby grid points but they were interpolated using the offshore observation data far away (10–70 km) in warm water and the climatology based on Equationa (3) and (4).

The L4 gridded SST datasets had the warm bias in the Eastern YS and the suggested two bias correction methods might be useful. However, it is not a fundamental solution. The background ($x_{i,k}^b$) or climatology ($x_{i,k}^c$) in Equation (3) needs to be improved using the long-term in situ observation data in this region. The background error covariance matrix **B** in Equation (4) has limited skill in the coastal regions, such as the Eastern YS, because the FOAM model does not include the tides, which are important forcing to produce the well-mixed cold water region in the Eastern YS. Near the tidal front in the Eastern YS, covariance **B** can be another error source during the interpolation process for the L4 gridded SST datasets.

Currently, the regularly obtained in situ SST data near the coast of YS is sparse. In the Southeastern YS where a strong SST gradient develops and satellite observations are technically difficult, the errors in the L4 gridded SST observation can be fundamentally reduced by installing more ocean buoys to measure SST.

5. Conclusions

The SST is an essential data for data assimilation of the ocean and atmospheric prediction systems. To evaluate SST datasets that cover the YS and to identify the error sources of the SST data, five L4 gridded SST datasets were selected and the spatio-temporal characteristics of the L4 gridded SST datasets were analyzed from January 2010 to December 2013. The SST from the OSTIA, GMPE and MGDSST datasets had a RMSD lower than 1 °C. The spatial distribution of RMSDs between the CTD SST and the five L4 gridded SST datasets revealed that the RMSD was higher in the vertically mixed coastal region within 50 km of the coastline or islands and was relatively lower in the offshore stratified region. In the mixed region, the RMSD was lowest for the OSTIA SST dataset and highest for the MWIR SST dataset. In the stratified region, the RMSDs of all L4 gridded SST datasets was less than 1 °C.

The RMSD of the L4 gridded SST data compared to in situ SST data was higher in June and July because of the warm bias in the shallow mixed region. To identify the cause of warm bias in the Southeastern YS, the satellite-based L3 SST data were examined. The L3 SST data were found to be rare, i.e., the number of high quality data were less than 30 in June and July for the four years at a grid point in the shallow mixed region. The L3 SST data were frequently missing because of sea fog formation in the coastal region and the large SST gradient in the region. The missing SST data were interpolated from the SST observation data in offshore warm water and warm biased SST climatology in the shallow coastal region. The optimal interpolation procedure for the L4 gridded SST was the main cause of the warm bias. To fundamentally improve the accuracy of the L4 gridded SST data, it is necessary to increase the number of SST observation data in the tidally well-mixed region. One way to improve the L4 gridded SST data in the Eastern YS is to loosen the horizontal gradient test threshold (1.2 °C/12 km) during the L3 data processing for June and July. Another way to achieve this would be to add more in situ observation data from ocean surface observation buoys in the tidal front zone and the shallow mixed region within 50 km from the land and islands. As a temporary solution to the warm bias in the gridded SST datasets in the Eastern YS, it is possible to improve the SST climatology for the optimal interpolation based on long-term field observation data.

Although the OSTIA SST dataset was the best in the YS, the RMSDs in the shallow mixed and stratified regions were 1.24 and 0.68 °C, respectively. To cut back the RMSD in the shallow mixed region, two bias correction methods were suggested. The bias correction reduced the RMSDs of the OISST, MGDSST, OSTIA, MWIR, and GMPE datasets by 41%, 43%, 37%, 47%, and 49%, respectively, in June.

Author Contributions: Conceptualization, B.-J.C. and K.K.; methodology, K.K.; software, K.K.; validation, K.K., S.-D.K. and B.-J.C.; formal analysis, K.K. and B.-J.C.; investigation, K.-A.P.; resources, B.-J.C.; data curation, S.-D.K.; writing—original draft preparation, K.K. and B.-J.C.; writing—review and editing, B.-J.C. and K.-A.P.; visualization, K.K.; supervision, S.-H.L.; project administration, S.-H.L.; funding acquisition, B.-J.C. and S.-H.L. All authors have read and agreed to the published version of the manuscript.

Funding: This research was supported by the Basic Science Research Program through the National Research Foundation of Korea (NRF) funded by the Ministry of Education (NRF-2016R1A6A1A03012647). This research was part of the project titled 'Improvements of ocean prediction accuracy using numerical modeling and artificial intelligence technology', funded by the Ministry of Oceans and Fisheries, Korea.

Acknowledgments: The authors would like to thank the NIFS of Korea for providing in situ CTD observation data. We thank the anonymous reviewers for their valuable comments and suggestions.

Conflicts of Interest: The authors declare no conflicts of interest.

References

1. O'Carroll, A.G.; Armstrong, E.M.; Beggs, H.M.; Bouali, M.; Casey, K.S.; Corlett, G.K.; Dash, P.; Donlon, C.J.; Gentemann, C.L.; Hoeyer, J.; et al. Observational needs of sea surface temperature. *Front. Mar. Sci.* **2019**, *6*, 420. [CrossRef]

2. Kwon, K.M.; Choi, B.-J.; Lee, S.-H.; Kim, Y.H.; Seo, G.-H.; Cho, Y.-K. Effect of model error representation in the Yellow and East China Sea modeling system based on the ensemble Kalman filter. *Ocean Dyn.* **2016**, *66*, 263–283. [CrossRef]

3. Kwon, K.; Choi, B.-J.; Lee, S.-H. Assimilation of different SST datasets to a coastal ocean modeling system in the Yellow and East China Sea. *J. Coast. Res.* **2018**, *85*, 1041–1045. [CrossRef]

4. Browne, P.; Rosnay, P.D.; Zuo, H.; Bennett, A.; Dawson, A. Weakly coupled ocean–atmosphere data assimilation in the ECMWF NWP system. *Remote Sens.* **2019**, *11*, 234. [CrossRef]

5. Xie, S.-P. Ocean–atmosphere interaction and tropical climate. In *Tropical Meteorology*; Wang, Y., Ed.; The Encyclopedia of Life Support Systems (EOLSS): Paris, France, 2009; pp. 189–201.

6. Martin, M.; Dash, P.; Ignatov, A.; Banzon, V.; Beggs, H.; Brasnett, B.; Cayula, J.-F.; Cummings, J.; Donlon, C.; Gentemann, C.; et al. Group for high resolution sea surface temperature (GHRSST) analysis fields inter-comparisons. Part 1: A GHRSST multi-product ensemble (GMPE). *Deep Sea Res. Part II Top. Stud. Oceanogr.* **2012**, *77*, 21–30.

7. Dash, P.; Ignatov, A.; Martin, M.; Donlon, C.; Brasnett, B.; Reynolds, R.W.; Banzon, V.; Beggs, H.; Cayula, J.F.; Chao, Y.; et al. Group for high resolution sea surface temperature (GHRSST) analysis fields inter-comparisons. Part 2: Near real time web-based Level 4 SST quality monitor (L4-SQUAM). *Deep Sea Res. Part II Top. Stud. Oceanogr.* **2012**, *77*, 31–43.

8. Xie, J.; Zhu, J.; Yan, L. Assessment and inter-comparison of five high-resolution sea surface temperature products in the shelf and coastal seas around China. *Cont. Shelf Res.* **2008**, *28*, 1286–1293. [CrossRef]

9. Castro, S.L.; Wick, G.A.; Steele, M. Validation of satellite sea surface temperature analyses in the Beaufort Sea using UpTempO buoys. *Remote Sens. Environ.* **2016**, *187*, 458–475. [CrossRef]

10. Reynolds, R.W.; Rayner, N.A.; Smith, T.M.; Stokes, D.C.; Wang, W. An improved in situ and satellite SST analysis for climate. *J. Clim.* **2002**, *15*, 1609–1625. [CrossRef]

11. Reynolds, R.W.; Zhang, H.-M.; Smith, T.M.; Gentemann, C.L.; Wentz, F. Impacts of in situ and additional satellite data on the accuracy of a sea-surface temperature analysis for climate. *Int. J. Clim.* **2005**, *25*, 857–864. [CrossRef]

12. Pisano, A.; Nardelli, B.B.; Tronconi, C.; Santoleri, R. The new Mediterranean optimally interpolated pathfinder AVHRR SST dataset (1982–2012). *Remote Sens. Environ.* **2016**, *176*, 107–116. [CrossRef]

13. Ding, Y.; Chan, J.C.L. The East Asian summer monsoon: An overview. *Meteorol. Atmos. Phys.* **2005**, *89*, 117–142.

14. Cho, Y.-K.; Kim, M.-O.; Kim, B.-C. Sea fog around the Korean peninsula. *J. Appl. Meteor.* **2000**, *39*, 2473–2479.

15. Fu, G.; Guo, J.; Xie, S.-P.; Duan, Y.; Zhang, M. Analysis and high-resolution modeling of a dense sea fog event over the Yellow Sea. *Atmos. Res.* **2006**, *81*, 293–303. [CrossRef]

16. Gao, S.H.; Lin, H.; Shen, B.; Fu, G. A heavy sea fog event over the Yellow Sea in March 2005: Analysis and numerical modeling. *Adv. Atmos. Sci.* **2007**, *24*, 65–81.

17. Zhang, H.M.; Reynolds, R.W.; Lumpkin, R.; Molinari, R.; Arzayus, K.; Johnson, M.; Smith, T.M. An integrated global observing system for sea surface temperature using satellites and in situ data: Research to operations. *Bull. Am. Meteorol. Soc.* **2009**, *90*, 31–38. [CrossRef]

18. Seo, G.-H.; Cho, Y.-K.; Choi, B.-J. Variations of heat transport in the northwestern Pacific marginal seas inferred from high-resolution reanalysis. *Prog. Oceanogr.* **2014**, *121*, 98–108. [CrossRef]

19. Seo, G.-H.; Choi, B.-J.; Cho, Y.-K.; Kim, Y.H.; Kim, S. Evaluation of a regional ocean reanalysis system for the east Asian marginal seas based on the ensemble Kalman filter. *Ocean Sci. J.* **2015**, *50*, 29–48. [CrossRef]

20. Hu, X.; Zhang, C.; Shang, S. Validation and inter-comparison of multi-satellite merged sea surface temperature products in the South China Sea and its adjacent waters. *J. Remote Sens.* **2015**, *19*, 328–338. (In Chinese)

21. Kim, E.J.; Kang, S.K.; Jang, S.T.; Lee, J.H.; Kim, Y.H.; Kang, H.W.; Kwon, Y.Y.; Seung, Y.H. Satellite-derived SST validation based on in-situ data during summer in the East China Sea and western North Pacific. *Ocean Sci. J.* **2010**, *45*, 159–170. [CrossRef]

22. Donlon, C.J.; Nightingale, T.J.; Sheasby, T.; Turner, J.; Robinson, I.S.; Emery, W.J. Implications of the oceanic thermal skin temperature deviation at high wind speeds. *Geophys. Res. Lett.* **1999**, *26*, 2505–2508. [CrossRef]

23. Donlon, C.J.; Minnett, P.J.; Gentemann, C.; Nightingale, T.J.; Barton, I.J.; Ward, B.; Murray, M.J. Toward improved validation of satellite sea surface skin temperature measurements for climate research. *J. Clim.* **2002**, *15*, 353–369. [CrossRef]

24. Barton, I.J. Interpretation of satellite-derived sea surface temperatures. *Adv. Space Res.* **2001**, *28*, 165–170. [CrossRef]

25. Murray, M.J.; Allen, M.R.; Merchant, C.J.; Harris, A.R.; Donlon, C.J. Direct observations of skin-depth SST variability. *Geophys. Res. Lett.* **2000**, *27*, 1171–1174. [CrossRef]

26. Kwak, M.-T.; Seo, G.-H.; Cho, Y.-K.; Kim, B.-G.; You, S.H.; Seo, J.-W. Long-term Comparison of satellite and in-situ sea surface temperatures around the Korean Peninsula. *Ocean Sci. J.* **2015**, *50*, 109–117.

27. Reynolds, R.W.; Smith, T.M.; Liu, C.; Chelton, D.B.; Casey, K.C.; Schlax, M.G. Daily high-resolution blended analyses for sea surface temperature. *J. Clim.* **2007**, *23*, 5473–5496. [CrossRef]

28. Kurihara, Y.; Sakurai, T.; Kuragano, T. Global daily sea surface temperature analysis using data from satellite microwave radiometer, satellite infrared radiometer and in-situ observations. *Weath. Bull.* **2006**, *73*, 1–18. (In Japanese)

29. Donlon, C.J.; Martin, M.; Stark, J.D.; Robert-Jones, J.; Fiedler, E.; Wimmer, W. The operational sea surface temperature and sea ice analysis (OSTIA) system. *Remote Sens. Environ.* **2012**, *116*, 140–158.

30. Gentemann, C.L.; Wentz, F.J.; DeMaria, M. Near real time global optimum interpolated microwave SSTs: Applications to hurricane intensity forecasting. In Proceedings of the 27th Conference on Hurricanes and Tropical Meteorology, Monterey, CA, USA, 23–28 April 2006.

31. Gentemann, C.L.; Wentz, F.J.; Brewer, M.; Hilburn, K.; Smith, D. Passive Microwave Remote Sensing of the Ocean: An Overview. In *Oceanography from Space*; Barale, V., Gower, J.F.R., Alberotanza, L., Eds.; Springer: The Hague, The Netherlands, 2010; pp. 13–33.

32. Efron, B.; Tibshirani, R.J. An Introduction to the Bootstrap. In *Monographs on Statistics and Applied Probability*; Chapman & Hall/CRC: New York, NY, USA, 1993; pp. 1–436.

33. Lorenc, A.C.; Bell, R.S.; Macpherson, B. The Meteorological Office analysis correction data assimilation scheme. *Quart. J. R. Meteor. Soc.* **1991**, *117*, 59–89. [CrossRef]

34. Martin, M.J.; Hines, A.; Bell, M.J. Data assimilation in the FOAM operational short-range ocean forecasting system: A description of the scheme and its impact. *Quart. J. R. Meteor. Soc.* **2007**, *133*, 981–995. [CrossRef]

35. Bell, M.J.; Forbes, R.M.; Hines, A. Assessment of the FOAM global data assimilation system for real-time operational ocean forecasting. *J. Mar. Syst.* **2000**, *25*, 1–22. [CrossRef]

36. Bell, M.; Barciela, R.; Hines, A.; Martin, M.; McCulloch, M.; Storkey, D. The forecasting ocean assimilation model (FOAM) system. In *Building the European Capacity in Operational Oceanography*; Dahlin, H., Flemming, N.C., Nittis, K., Petersson, S.E., Eds.; Elsevier: Athens, Greece, 2003; pp. 197–202.

37. Bell, M.J.; Barciela, R.; Hines, A.; Martin, M.; Sellar, A.; Storkey, D. The Forecasting Ocean Assimilation Model (Foam) System. In *Ocean Weather Forecasting*; Chassignet, E.P., Verron, J., Eds.; Springer: Dordrecht, The Netherlands, 2006; pp. 397–411.

38. Tokinaga, H.; Xie, S.-P. Ocean tidal cooling effect on summer sea fog over the Okhotsk Sea. *J. Geophys. Res.* **2009**, *114*, D14102. [CrossRef]

39. Heo, K.; Ha, K. A coupled model study on the formation and dissipation of sea fogs. *Mon. Weather Rev.* **2010**, *138*, 1186–1205. [CrossRef]

40. Zhang, S.-P.; Xie, S.-P.; Liu, Q.-Y.; Yang, Y.-Q.; Wang, X.-G.; Ren, Z.-P. Seasonal variations of Yellow Sea fog: Observations and mechanisms. *J. Clim.* **2009**, *22*, 6758–6772.

41. Kilpatrick, K.A.; Podesta, G.P.; Evans, R. Overview of the NOAA/NASA advanced very high resolution radiometer Pathfinder algorithm for sea surface temperature and associated matchup database. *J. Geophys. Res.* **2001**, *106*, 9179–9198. [CrossRef]

Technical Note

Estimation of the Particulate Organic Carbon to Chlorophyll-*a* Ratio Using MODIS-Aqua in the East/Japan Sea, South Korea

Dabin Lee [1], SeungHyun Son [2], HuiTae Joo [3], Kwanwoo Kim [1], Myung Joon Kim [1], Hyo Keun Jang [1], Mi Sun Yun [4], Chang-Keun Kang [5] and Sang Heon Lee [1,*]

[1] Department of Oceanography, Pusan National University, Geumjeong-gu, Busan 46241, Korea; ldb1370@pusan.ac.kr (D.L.); goanwoo7@pusan.ac.kr (K.K.); mjune@pusan.ac.kr (M.J.K.); janghk@pusan.ac.kr (H.K.J.)
[2] CIRA, Colorado State University, Fort Collins, CO 80523, USA; ssnocean@gmail.com
[3] Oceanic Climate and Ecology Research Division, National Institute of Fisheries Science, Busan 46083, Korea; huitae@pusan.ac.kr
[4] College of Marine and Environmental Sciences, Tianjin University of Science and Technology, Tianjin 300457, China; misunyun@pusan.ac.kr
[5] School of Earth Sciences & Environmental Engineering, Gwangju Institute of Science and Technology, Gwangju 61005, Korea; ckkang@gist.ac.kr
* Correspondence: sanglee@pusan.ac.kr; Tel.: +82-51-510-2256

Received: 30 December 2019; Accepted: 3 March 2020; Published: 5 March 2020

Abstract: In recent years, the change of marine environment due to climate change and declining primary productivity have been big concerns in the East/Japan Sea, Korea. However, the main causes for the recent changes are still not revealed clearly. The particulate organic carbon (POC) to chlorophyll-*a* (chl-*a*) ratio (POC:chl-*a*) could be a useful indicator for ecological and physiological conditions of phytoplankton communities and thus help us to understand the recent reduction of primary productivity in the East/Japan Sea. To derive the POC in the East/Japan Sea from a satellite dataset, the new regional POC algorithm was empirically derived with in-situ measured POC concentrations. A strong positive linear relationship ($R^2 = 0.6579$) was observed between the estimated and in-situ measured POC concentrations. Our new POC algorithm proved a better performance in the East/Japan Sea compared to the previous one for the global ocean. Based on the new algorithm, long-term POC:chl-*a* ratios were obtained in the entire East/Japan Sea from 2003 to 2018. The POC:chl-*a* showed a strong seasonal variability in the East/Japan Sea. The spring and fall blooms of phytoplankton mainly driven by the growth of large diatoms seem to be a major factor for the seasonal variability in the POC:chl-*a*. Our new regional POC algorithm modified for the East/Japan Sea could potentially contribute to long-term monitoring for the climate-associated ecosystem changes in the East/Japan Sea. Although the new regional POC algorithm shows a good correspondence with in-situ observed POC concentrations, the algorithm should be further improved with continuous field surveys.

Keywords: particulate organic carbon; chlorophyll-*a*; phytoplankton; ocean color; East/Japan Sea

---

1. Introduction

The East/Japan Sea is a semi-marginal sea located in the northwestern Pacific, and it is bordered by Korea, Japan, and Russia. Many previous studies reported that the East/Japan Sea is a productive region, especially the Ulleung Basin located in the southwestern East/Japan Sea [1–7]. Recently, not only the changes in environmental conditions but also alterations in biological characteristics were reported in many previous studies [4–6,8]. Especially, Joo et al. [5] addressed a significant declining trend of the annual primary production in the Ulleung Basin which is the biological hotspot in the East/Japan Sea. However, although many other studies have observed the changes in the East/Japan Sea, the main causes driving the recent changes remain unclear.

The particulate organic carbon (POC) to chlorophyll-a (chl-*a*) ratio (POC:chl-*a*) has been used as an important indicator for ecological and physiological state of phytoplankton. Generally, the concentrations of photosynthetic pigments in phytoplankton cells depend on the environmental factors and the physiological conditions of phytoplankton [9–14]. For instance, the POC:chl-*a* can be increased under high light intensity and low nitrogen supply conditions [1,5,6]. Moreover, the POC:chl-*a* varies according to the size structure of the phytoplankton community [15–17]. Therefore, understanding of spatial and temporal variations in the POC:chl-*a* can help us to determine the ecological and physiological conditions of a phytoplankton community. However, the existing POC-deriving algorithm using satellite ocean color data [18] is not validated in the East/Japan Sea. Although the algorithm is fully validated in global ocean, it is hard to directly apply to this small regional sea. Thus, the algorithm needs to be calibrated and the accuracy should be evaluated before applying to our study area.

Therefore, this study aims to (1) develop a new regional POC algorithm using an ocean color satellite and (2) investigate spatiotemporal variability of the POC:chl-*a* in the East/Japan Sea.

2. Materials and Methods

2.1. Study Area and Sampling

Total chl-*a* and POC concentrations were measured at 41 stations in the East/Japan Sea (16 stations in 2012, 11 stations in 2013, six stations in 2014, and eight stations in 2015, respectively; Figure 1). Northern and southern regions of the East/Japan Sea were defined as shown in Figure 1 to investigate spatial variations of POC:chl-*a*.

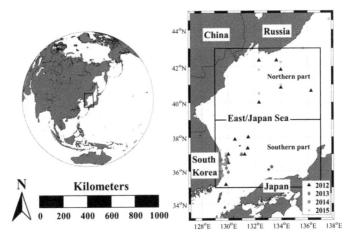

Figure 1. In-situ measurement stations in the East/Japan Sea from 2012 to 2015. Black triangles, blue diamonds, red circles, and orange squares indicate the stations measured in 2012, 2013, 2014, and 2015, respectively. Black lines indicate domains for northern and southern regions of the East/Japan Sea.

Sampling and analysis of total chl-*a* and POC concentrations were conducted based on Lee et al. [19]. Water samples for the total chl-*a* and POC concentrations of phytoplankton were collected from the surface layer using a rosette sampler with Niskin bottles. The collected water samples were immediately filtered on Whatman® glass microfiber filters (precombusted; Grade GF/F, diameter = 24 mm) using a vacuum pressure lower than 5 in. Hg. The filtered samples were frozen immediately and preserved until analysis at the laboratory. The chl-*a* concentrations were measured with a precalibrated fluorometer (10-AU, Turner Designs) after extraction in 90% acetone in a freezer for 24 h based on Parsons et al. [20]. The filters for POC concentrations were frozen immediately and preserved for mass spectrometric analysis at the Alaska Stable Isotope Facility of the University of Alaska Fairbanks, USA.

2.2. Satellite Datasets

We obtained the chl-*a* and remote sensing reflectance (Rrs) data from the MODIS (Moderate Resolution Imaging Spectroradiometer) onboard the satellite Aqua provided by the OBPG (Ocean Biology Processing Group at NASA Goddard Space Flight Center; https://oceandata.sci.gsfc.nasa.gov/MODIS-Aqua/). We used the Level-3 daily composite datasets covering the East/Japan Sea from July 2002 to December 2018 at 4-km of spatial resolution [21,22].

The POC derivation algorithm for the East/Japan Sea was empirically derived with our in-situ observation data. Based on the Stramski et al. [18], a power–law relationship between a blue-to-green band ratio of Rrs and POC were used to estimate POC concentrations. The equation for the algorithm is expressed below:

$$POC = a \times \left(\frac{Rrs(443)}{Rrs \text{ between 547 and 565 nm}} \right)^b \tag{1}$$

where *a* and *b* are constants. Constants were determined empirically by regression analysis with our in-situ dataset. The input wavelength for the green band can be replaced with available band between 547 and 565 nm. In this study, Rrs(443) and Rrs(547) which are available bands of MODIS were used as input wavelengths for blue-to-green band ratio for the POC algorithm.

POC:chl-*a* ratios were calculated by dividing our estimated POC concentrations by remotely sensed chl-*a* concentrations. Monthly composited data for POC and POC:chl-*a* were obtained by averaging daily data for each month.

3. Results

3.1. POC Algorithm Derivation

In-situ measured POC concentrations ranged from 84.07 to 713.69 mg m^{-3}, and the average was 262.93 ± 205.82 mg m^{-3}. The blue-to-green band Rrs ratio were extracted from MODIS-Aqua monthly composite datasets. To determine two constants in the POC algorithm, *a* and *b*, the curve fitting using nonlinear regression between Rrs ratio and in-situ POC concentration was conducted (Figure 2a).

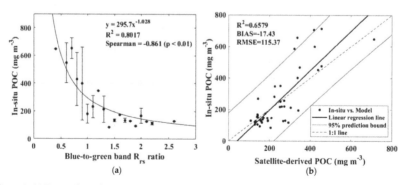

Figure 2. (a) Power–law relationship between blue-to-green band ratio and in-situ measured particulate organic carbon (POC) concentrations and (b) linear relationship between POC derived by the algorithm and in-situ measured POC concentrations.

From this result, the constants *a* and *b* were determined as 295.7 and −1.028, respectively. Based on these two constants, the POC algorithm was derived as follows:

$$POC = 295.7 \times \left(\frac{Rrs(443)}{Rrs(547)}\right)^{-1.028} \tag{2}$$

The determination coefficient (R^2) and Spearman's correlation coefficient for the relationship between Rrs ratio and in-situ POC concentrations were 0.8017 and −0.861, respectively (Figure 2a). The POC concentrations derived from the regional model also showed a strong linear relationship with in-situ measured POC concentrations ($R^2 = 0.6579$), and most of the satellite-derived POC were plotted within 95% prediction bounds (Figure 2b).

The new POC algorithm showed lower RMSE and bias in comparison to the existing algorithm [18] in the East/Japan Sea (Figure 3). The RMSE and bias of the new algorithm were 115.37 and −17.43, respectively, and those of Stramski et al. [18] were 161.47 and −97.29, respectively.

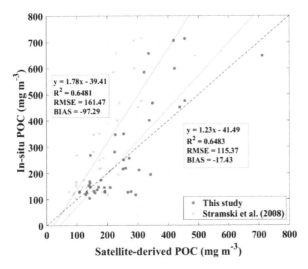

Figure 3. Validation results for the new POC algorithm (blue circles) and Stramski et al. [18] (orange crosses). Black dashed line represents 1:1 line. Blue and orange solid lines represent the linear regression lines between in-situ POC and the modeled POC derived from the new algorithm and Stramski et al. [18], respectively.

The POC concentrations were derived for the East/Japan Sea from our new regional algorithm (Figure 4). The climatological monthly mean POC (January 2003–December 2018) showed a seasonal variation of the POC concentration in the East/Japan Sea (Figure 4). Generally, the POC concentrations were relatively higher during spring season, and the lowest POC concentrations were observed during summer season.

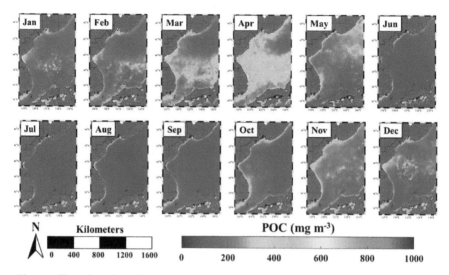

Figure 4. Climatological monthly mean distribution of the POC in the East/Japan Sea (2003 January–2018 December).

3.2. POC:chl-a

The climatological monthly distribution of the POC:chl-*a* (January 2003–December 2018) in the East/Japan Sea showed strong seasonal variations (Figure 5). In contrast to the seasonal pattern of POC, relatively lower POC:chl-*a* ratios were observed during spring and autumn compared to those during winter and summer.

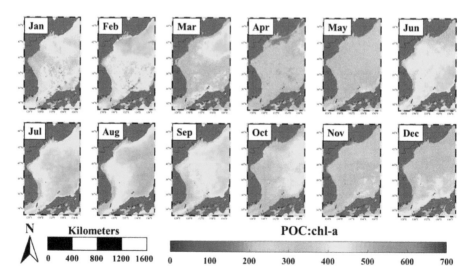

Figure 5. Climatological monthly mean distribution of the POC:chl-*a* in the East/Japan Sea (January 2003–December 2018).

The ranges of the mean POC:chl-*a* in the northern and southern East/Japan Sea were 169.6–528.4 and 172.4–549.3, respectively (Figure 6). Domains for the two regions are shown in Figure 1. The average of POC:chl-*a* in the northern and southern regions of the East/Japan Sea were 377.6 ± 80.4 and 388.1 ± 69.7, respectively. No statistically significant difference in POC:chl-*a* was observed between the northern and southern regions (*t*-test, $p > 0.05$).

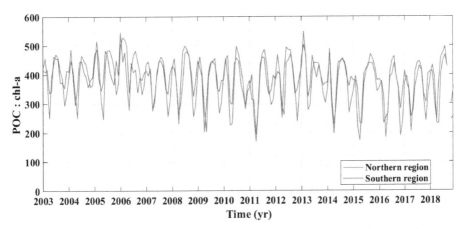

Figure 6. Time-series of monthly mean POC:chl-*a* in the northern (blue) and southern (orange) East/Japan Sea (January 2003–December 2018).

The POC:chl-*a* showed a strong seasonal variation (Figure 7). Climatological monthly mean POC:chl-*a* showed the lowest values during April (270.3 ± 74.7 and 261.9 ± 60.9 for the northern and southern East/Japan Sea, respectively), and the highest values were observed during August (461.0 ± 18.1) and July (451.5 ± 23.6) for the northern and southern East/Japan Sea, respectively (Figure 7). The POC:chl-*a* during spring and autumn were significantly lower than summer and winter (*t*-test, $p < 0.01$).

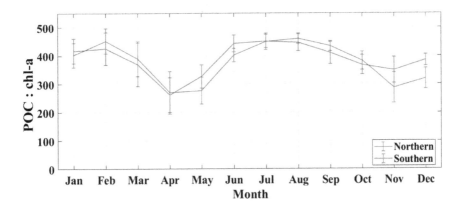

Figure 7. Time-series of climatological monthly mean POC:chl-*a* in the northern (blue) and southern (orange) East/Japan Sea (January 2003–December 2018). Vertical lines indicate standard deviations of the climatological monthly mean POC:chl-*a*.

4. Discussion

4.1. New Regional POC Algorithm

Since the POC algorithm reported by Stramski et al. [18] was developed for the eastern South Pacific and eastern Atlantic Oceans, we validated and modified the algorithm to derive a suitable POC model for the East/Japan Sea. The previously reported POC algorithm by Stramski et al. [18] tends to underestimate the POC concentrations in the East/Japan Sea (Figure 3). However, our modified regional POC algorithm showed significantly improved accuracy in the East/Japan Sea, and the POC concentrations estimated from the regional algorithm derived in this study showed a strong linear relationship with in-situ measured POC concentrations and its linear regression line located near the 1:1 line (Figure 3).

However, only 41 data points were used to derive the POC model in this study, and in fact, the number of data points could not be considered sufficient. Nevertheless, the modified POC model showed a good correspondence with field measured POC concentrations. If continuous field observations and the calibration and validation of the algorithm are conducted, the POC algorithm with a better performance would be derived.

4.2. Spatial and Temporal Variability of the POC:chl-a

The seasonal variation of the POC:chl-*a* in the East/Japan Sea would be closely related with the physiological conditions of phytoplankton communities. Generally, phytoplankton tend to accumulate more carbon in their cells under the high light intensity and nutrient depleted conditions [9,13,14, 23,24]. Phytoplankton increase their chl-*a* contents to maximize light absorption under the low light condition [11,25,26]. Moreover, many previous studies reported that the carbon to chl-*a* ratio of phytoplankton decreases with increased growth rate [14,27–30]. The size structure of the phytoplankton community also can affect the carbon to chl-*a* ratio [15–17]. Small cell-sized phytoplankton such as flagellates usually show higher carbon to chl-*a* ratio than large cell-sized phytoplankton such as diatoms [15–17]. However, the daily carbon uptake rate by phytoplankton tends to be lowered when the productivity contribution of picoplankton to the total primary production is high [31,32]. It suggests that the investigation of POC:chl-*a* in the East/Japan Sea can provide some potential clues on the recent changes of primary productivity in the East/Japan Sea.

In general, there are two phytoplankton blooms per year in the East/Japan Sea; spring bloom and fall bloom [4,5,33–35]. Signals of the blooms were also observed in the climatological monthly mean distribution of the POC (Figure 3). The strongest bloom occurs during spring season while the weaker blooms appears in fall. Both spring and fall blooms are mainly caused by the massive growth of diatoms [3,4,33,35,36]. During spring and fall blooms, the size distribution of phytoplankton cells would shift from smaller to larger size due to the rapid growth of diatoms. Consequently, lower POC:chl-*a* during spring and fall might be caused by the bloom mainly driven by the growth of diatoms. However, these suggestions are only our hypothesis based on several previous studies. To understand the temporal variation of the POC:chl-*a* in the East/Japan Sea, further research with field observations is needed.

On the other hand, the timing of spring bloom showed a difference in the northern and southern regions of the East/Japan Sea (Figures 3 and 4). POC concentrations were highest in April in the southern region and highest in May in the northern region. Additionally, the distribution of POC:chl-*a* during April and May appeared to be the opposite of POC concentrations. Other previous studies have also observed a similar spatial distribution of chl-*a* in the East/Japan Sea with satellite datasets [37,38]. Those spatiotemporal distributions of POC concentrations and POC:chl-*a* suggest that the spring bloom occurs earlier in the southern regions of the East/Japan Sea.

5. Summary and Conclusions

In this study, the regional POC algorithm for the East/Japan Sea was derived empirically using in-situ measured POC and MODIS-Aqua satellite datasets. In-situ measured POC concentrations at the 41 stations in the East/Japan Sea were used to calibrate and validate our new regional POC algorithm. The power–law relationship between POC concentration and blue-to-green band of remote sensing reflectance, Rrs(443)/Rrs(547), was used to derive the algorithm based on Stramski et al. [18] (Figure 2a).

Remote Sens. **2020**, *12*, 840

The algorithm showed a good correspondence with in-situ measured POC concentrations in the East/Japan Sea (Figure 2b). Based on the result, the monthly mean POC concentration and POC:chl-*a* were derived in the East/Japan Sea.

Both POC concentration and POC:chl-*a* showed strong seasonality in the East/Japan Sea. POC concentrations were relatively higher and POC:chl-*a* ratios were lower in spring and fall seasons which are typical blooming periods in the East/Japan Sea. The seasonal variation of POC:chl-*a* in the East/Japan Sea might be closely related with the community structure of phytoplankton. Many previous studies previously reported that the POC:chl-*a* of large-sized phytoplankton tends to be relatively lower than that of small cell-sized phytoplankton [15–17]. Spring and fall blooms in the East/Japan Sea are generally triggered by the intense growth of diatoms [3,4,33,35,36]. The size structure of phytoplankton communities can shift from small to large due to the massive growth of diatoms during the bloom periods, resulting in relatively lower POC:chl-*a*.

Although the POC algorithm derived in this study showed good agreement with field observation results, the number of in-situ data is not sufficient. The regional POC algorithm suggested in this study should be considered as a pilot version, and continuous field observations must be conducted to improve the regional POC algorithm.

Author Contributions: Conceptualization, D.L. and S.H.L.; Data curation, D.L. and H.J.; Funding acquisition, C.-K.K. and S.H.L.; Methodology, D.L.; Investigation, D.L., H.J., K.K., M.J.K., H.K.J., and M.S.Y.; Project administration, C.-K.K. and S.H.L.; Supervision, S.S. and S.H.L.; Validation, D.L.; Visualization, D.L.; Writing—original draft, D.L.; Writing—review and editing, S.S. and S.H.L. All authors have read and agreed to the published version of the manuscript.

Funding: This research was supported by "Long-term change of structure and function in marine ecosystems of Korea", "Improvements of ocean prediction accuracy using numerical modeling and artificial intelligence technology" and "Technology development for Practical Applications of Multi-Satellite data to maritime issues" funded by the Ministry of Ocean and Fisheries, Korea.

Acknowledgments: The authors would like to thank the anonymous reviewers and the handling editors who dedicated their time to providing the authors with constructive and valuable recommendations.

Conflicts of Interest: The authors declare no conflict of interest.

References

1. Yamada, K.; Ishizaka, J.; Nagata, H. Spatial and temporal variability of satellite primary production in the Japan Sea from 1998 to 2002. *J. Oceanogr.* **2005**, *61*, 857–869. [CrossRef]

2. Yoo, S.; Park, J. Why is the southwest the most productive region of the East Sea/Sea of Japan? *J. Mar. Syst.* **2009**, *78*, 301–315. [CrossRef]

3. Kwak, J.H.; Lee, S.H.; Park, H.J.; Choy, E.J.; Jeong, H.D.; Kim, K.R.; Kang, C.K. Monthly measured primary and new productivities in the Ulleung Basin as a biological "hot spot" in the East/Japan Sea. *Biogeosciences* **2013**, *10*, 4405–4417. [CrossRef]

4. Lee, S.H.; Son, S.; Dahms, H.U.; Park, J.W.; Lim, J.H.; Noh, J.H.; Kwon, J.I.; Joo, H.T.; Jeong, J.Y.; Kang, C.K. Decadal changes of phytoplankton chlorophyll-a in the East Sea/Sea of Japan. *Oceanology* **2014**, *54*, 771–779. [CrossRef]

5. Joo, H.; Park, J.W.; Son, S.; Noh, J.H.; Jeong, J.Y.; Kwak, J.H.; Saux-Picart, S.; Choi, J.H.; Kang, C.K.; Lee, S.H. Long-term annual primary production in the Ulleung Basin as a biological hot spot in the East/Japan Sea. *J. Geophys. Res. Oceans* **2014**, *119*, 3002–3011. [CrossRef]

6. Kim, K.; Kim, K.R.; Min, D.H.; Volkov, Y.; Yoon, J.H.; Takematsu, M. Warming and structural changes in the east (Japan) sea: A clue to future changes in global oceans? *Geophys. Res. Lett.* **2001**, *28*, 3293–3296. [CrossRef]

7. Jo, N.; Kang, J.J.; Park, W.G.; Lee, B.R.; Yun, M.S.; Lee, J.H.; Kim, S.M.; Lee, D.; Joo, H.T.; Lee, J.H.; et al. Seasonal variation in the biochemical compositions of phytoplankton and zooplankton communities in the southwestern East/Japan Sea. *Deep. Res. Part II Top. Stud. Oceanogr.* **2017**, *143*, 82–90. [CrossRef]

8. Kang, D.J.; Park, S.; Kim, Y.G.; Kim, K.; Kim, K.R. A moving-boundary box model (MBBM) for oceans in change: An application to the East/Japan Sea. *Geophys. Res. Lett.* **2003**, *30*. [CrossRef]

9. Geider, R.J. Light and temperature dependence of the carbon to chlorophyll a ratio in microalgae and cyanobacteria: Implications for physiology and growth of phytoplankton. *New Phytol.* **1987**, *106*, 1–34. [CrossRef]

10. Riemann, B.; Simonsen, P.; Stensgaard, L. The carbon and chlorophyll content of phytoplankton from various nutrient regimes. *J. Plankton Res.* **1989**, *11*, 1037–1045. [CrossRef]

11. Jakobsen, H.H.; Markager, S. Carbon-to-chlorophyll ratio for phytoplankton in temperate coastal waters: Seasonal patterns and relationship to nutrients. *Limnol. Oceanogr.* **2016**, *61*, 1853–1868. [CrossRef]

12. Lee, D.; Jeong, J.-Y.; Jang, H.K.; Min, J.-O.; Kim, M.J.; Youn, S.H.; Lee, T.; Lee, S.H. Comparison of Particulate Organic Carbon to Chlorophyll-a Ratio Based on the Ocean Color Satellite Data at the Ieodo and Socheongcho Ocean Research Stations. *J. Coast. Res.* **2019**, *90*, 267. [CrossRef]

13. Thompson, P.A.; Guo, M.-x.; Harrison, P.J. Effects of variation in temperature. I. on the biochemical composition of eight species of marine phytoplankton. *J. Phycol.* **1992**, *28*, 481–488. [CrossRef]

14. Wang, X.J.; Behrenfeld, M.; Le Borgne, R.; Murtugudde, R.; Boss, E. Regulation of phytoplankton carbon to chlorophyll ratio by light, nutrients and temperature in the equatorial pacific ocean: A basin-scale model. *Biogeosciences* **2009**, *6*, 391–404. [CrossRef]

15. Stramski, D. Refractive index of planktonic cells as a measure of cellular carbon and chlorophyll a content. *Deep. Res. Part I Oceanogr. Res. Pap.* **1999**, *46*, 335–351. [CrossRef]

16. Sathyendranath, S.; Stuart, V.; Nair, A.; Oka, K.; Nakane, T.; Bouman, H.; Forget, M.H.; Maass, H.; Platt, T. Carbon-to-chlorophyll ratio and growth rate of phytoplankton in the sea. *Mar. Ecol. Prog. Ser.* **2009**, *383*, 73–84. [CrossRef]

17. Roy, S.; Sathyendranath, S.; Platt, T. Size-partitioned phytoplankton carbon and carbon-to-chlorophyll ratio from ocean colour by an absorption-based bio-optical algorithm. *Remote Sens. Environ.* **2017**, *194*, 177–189. [CrossRef]

18. Stramski, D.; Reynolds, R.A.; Babin, M.; Kaczmarek, S.; Lewis, M.R.; Röttgers, R.; Sciandra, A.; Stramska, M.; Twardowski, M.S.; Franz, B.A.; et al. Relationships between the surface concentration of particulate organic carbon and optical properties in the eastern South Pacific and eastern Atlantic Oceans. *Biogeosciences* **2008**, *5*, 171–201. [CrossRef]

19. Lee, S.H.; Joo, H.T.; Lee, J.H.; Lee, J.H.; Kang, J.J.; Lee, H.W.; Lee, D.; Kang, C.K. Seasonal carbon uptake rates of phytoplankton in the northern East/Japan Sea. *Deep. Res. Part II Top. Stud. Oceanogr.* **2017**, *143*, 45–53. [CrossRef]

20. Parsons, T.R.; Maita, C.M.; Lalli, C.M. *A Manual of Chemical & Biological Methods for Seawater Analysis*; Pergamon Press: New York, NY, USA, 1984.

21. NASA Goddard Space Flight Center, Ocean Ecology Laboratory, Ocean Biology Processing Group. *Moderate-Resolution Imaging Spectroradiometer (MODIS) Aqua Chlorophyll Data*; NASA OB.DAAC: Greenbelt, MD, USA, 2018.

22. NASA Goddard Space Flight Center, Ocean Ecology Laboratory, Ocean Biology Processing Group. *Moderate-Resolution Imaging Spectroradiometer (MODIS) Aqua Remote-Sensing Reflectance Data*; NASA OB.DAAC: Greenbelt, MD, USA, 2018.

23. Morgan, K.C.; Kalff, J. Effect of light and temperature interactions on growth of cryptomonas erosa (cryptophyceae) 1. *J. Phycol.* **1979**, *15*, 127–134. [CrossRef]

24. Terry, K.L.; Hirata, J.; Laws, E.A. Light-limited growth of two strains of the marine diatom Phaeodactylum tricornutum Bohlin: Chemical composition, carbon partitioning and the diel periodicity of physiological processes. *J. Exp. Mar. Biol. Ecol.* **1983**, *68*, 209–227. [CrossRef]

25. Beardall, J.; Morris, I. The concept of light intensity adaptation in marine phytoplankton: Some experiments with Phaeodactylum tricornutum. *Mar. Biol.* **1976**, *37*, 377–387. [CrossRef]

26. Markager, S. Light absorption and quantum yield for growth in five species of marine macroalga. *J. Phycol.* **1993**, *29*, 54–63. [CrossRef]

27. Geider, R.J.; MacIntyre, H.L.; Kana, T.M. A dynamic regulatory model of phytoplanktonic acclimation to light, nutrients, and temperature. *Limnol. Oceanogr.* **1998**, *43*, 679–694. [CrossRef]

28. Le Bouteiller, A.; Leynaert, A.; Landry, M.R.; Le Borgne, R.; Neveux, J.; Rodier, M.; Blanchot, J.; Brown, S.L. Primary production, new production, and growth rate in the equatorial Pacific: Changes from mesotrophic to oligotrophic regime. *J. Geophys. Res. C Oceans* **2003**, *108*, 8141. [CrossRef]

29. Behrenfeld, M.J.; Boss, E.; Siegel, D.A.; Shea, D.M. Carbon-based ocean productivity and phytoplankton physiology from space. *Glob. Biogeochem. Cycles* **2005**, *19*, 1–14. [CrossRef]

30. Armstrong, R.A. Optimality-based modeling of nitrogen allocation and photoacclimation in photosynthesis. *Deep. Res. Part II Top. Stud. Oceanogr.* **2006**, *53*, 513–531. [CrossRef]

31. Agawin, N.S.R.; Duarte, C.M.; Agustí, S. Nutrient and temperature control of the contribution of picoplankton to phytoplankton biomass and production. *Limnol. Oceanogr.* **2000**, *45*, 591–600. [CrossRef]

32. Lee, S.H.; Kim, B.K.; Lim, Y.J.; Joo, H.T.; Kang, J.J.; Lee, D.; Park, J.; Ha, S.Y.; Lee, S.H. Small phytoplankton contribution to the standing stocks and the total primary production in the Amundsen Sea. *Biogeosciences* **2017**, *14*, 3705–3713. [CrossRef]

33. Lee, J.-Y.; Kang, D.-J.; Kim, I.-N.; Rho, T.; Lee, T.; Kang, C.-K.; Kim, K.-R. Spatial and temporal variability in the pelagic ecosystem of the East Sea (Sea of Japan): A review. *J. Mar. Syst.* **2009**, *78*, 288–300. [CrossRef]

34. Joo, H.T.; Son, S.H.; Park, J.W.; Kang, J.J.; Jeong, J.Y.; Lee, C.; Kang, C.K.; Lee, S.H. Long-term pattern of primary productivity in the East/Japan sea based on ocean color data derived from MODIS-Aqua. *Remote Sens.* **2016**, *8*, 25. [CrossRef]

35. Chang, K.I.; Zhang, C.I.; Park, C.; Kang, D.J.; Ju, S.J.; Lee, S.H.; Wimbush, M. *Oceanography of the East Sea (Japan Sea)*; Chang, K.-I., Zhang, C.-I., Park, C., Kang, D.-J., Ju, S.-J., Lee, S.-H., Wimbush, M., Eds.; Springer International Publishing: Cham, Switzerland, 2016; ISBN 978-3-319-22719-1.

36. Zuenko, Y.; Selina, M.; Stonik, I. On Conditions of Phytoplankton Blooms in the Coastal Waters of the North-Western East/Japan Sea. *Ocean Sci. J.* **2006**, *41*, 31–41. [CrossRef]

37. Yamada, K.; Ishizaka, J.; Yoo, S.; Kim, H.C.; Chiba, S. Seasonal and interannual variability of sea surface chlorophyll a concentration in the Japan/East Sea (JES). *Prog. Oceanogr.* **2004**, *61*, 193–211. [CrossRef]

38. Jo, C.O.; Park, S.; Kim, Y.H.; Park, K.A.; Park, J.J.; Park, M.K.; Li, S.; Kim, J.Y.; Park, J.E.; Kim, J.Y.; et al. Spatial distribution of seasonality of SeaWiFS chlorophyll-a concentrations in the East/Japan Sea. *J. Mar. Syst.* **2014**, *139*, 288–298. [CrossRef]

MDPI

St. Alban-Anlage 66

4052 Basel

Switzerland

Tel. +41 61 683 77 34

Fax +41 61 302 89 18

www.mdpi.com

Remote Sensing Editorial Office

E-mail: remotesensing@mdpi.com

www.mdpi.com/journal/remotesensing

CPSIA information can be obtained
at www.ICGtesting.com
Printed in the USA
LVHW011800310523
748331LV00059B/50